Geosphere-Biosphere Interactions and Climate

Geosphere-Biosphere Interactions and Climate brings together many of the world's leading environmental scientists to discuss the interaction between the geosphere-biosphere and climate. The volume arises from a working group of the prestigious Pontifical Academy of Sciences and the International Geosphere-Biosphere Programme. The chapters give a state-of-the-art summary of our current understanding of the present climate and environment by exploring Earth's past, analyzing human influence on the climate, describing climate and its relation to the Earth's surface, ocean, and atmosphere, and making future predictions of climate variability. This volume will be invaluable for researchers and graduate students in climate studies who wish to gain a multidisciplinary perspective of our current understanding of the interaction between the geosphere-biosphere and climate.

Lennart O. Bengtsson is a professor at the Max-Planck-Institut für Meteorologie, Hamburg.

Claus U. Hammer is a professor in the Department of Geophysics, Niels Bohr Institute, University of Copenhagen.

Geosphere-Biosphere Interactions and Climate

Edited by

LENNART O. BENGTSSON
Max-Planck-Institut für Meteorologie

CLAUS U. HAMMER
University of Copenhagen

CAMBRIDGE
UNIVERSITY PRESS

PUBLISHED BY THE PRESS SYNDICATE OF THE UNIVERSITY OF CAMBRIDGE
The Pitt Building, Trumpington Street, Cambridge, United Kingdom

CAMBRIDGE UNIVERSITY PRESS
The Edinburgh Building, Cambridge CB2 2RU, UK
40 West 20th Street, New York, NY 10011-4211, USA
10 Stamford Road, Oakleigh, VIC 3166, Australia
Ruiz de Alarcón 13, 28014 Madrid, Spain
Dock House, The Waterfront, Cape Town 8001, South Africa

http://www.cambridge.org

First published 2001

Printed in the United States of America

Typeface Ehrhardt 11/13.5 pt.   *System* LaTeX 2_ε [TB]

A catalog record for this book is available from the British Library.

Library of Congress Cataloging in Publication data
Geosphere-biosphere interactions and climate / edited by Lennart O. Bengtsson, Claus U. Hammer.
     p.   cm.
   Includes index.
   ISBN 0-521-78238-4
   1. Climatology – Congresses.   2. Nature – Effect of human beings on – Congresses.
   3. Biogeochemical cycles – Congresses.   I. Bengtsson, Lennart.   II. Hammer, Claus U.
   QC980 .G46   2001
   577.2′2 – dc21   00-064184

   ISBN 0 521 78238 4 hardback

Contents

Dedication to Hans Oeschger

Hans Oeschger was a professor emeritus of physics at the University of Bern, Switzerland, when he planned the Vatican Conference on Geosphere-Biosphere Interactions and Climate. Following his retirement in 1992 his health had its ups and downs, but his concern for the global climate and environment did not prevent him from pursuing this important theme. He did not spare himself, and many of his international colleagues were concerned about his health, when they noted his many travels and engagements as an emeritus.

His last achievement – this conference at the Vatican in Casa Pius IV, a beautiful and inspiring location – missed him as a major participant. His illness had grown worse, and he was hospitalized during the conference. A few weeks later we sadly learned that Hans Oeschger had died.

Born on April 2, 1927, in Ottenbach, Switzerland, and a doctor of science from the University of Bern in 1955, he became associated with the University of Bern until his retirement as a professor emeritus in 1992.

During his years at the university he and his colleagues developed techniques for measuring radiocarbon on very small samples of carbon dioxide. Hans Oeschger's group became world famous for its studies on gas samples, and in the late sixties he entered a fruitful collaboration with Professor Willi Dansgaard, University of Copenhagen, and Chester C. Langway, Jr. (later professor at State University of New York at Buffalo). The setting was ready for the next two decades of ice core research.

In the Greenland Ice Sheet Project (GISP), now called GISP 1, a collaboration between the United States, Denmark, and Switzerland, the second deep ice core drilling in Greenland at Dye 3, 1979–1981, successfully ended 10 years of climate and environmental work on the Greenland Ice Sheet. It is interesting to note that this development took place over the same decade as the scientists realized that atmospheric changes due to anthropogenic activities were of great importance.

When Hans became an emeritus in 1992 it actually was in the same year as the Greenland Ice Core Project (GRIP) drilling reached the bedrock 3028 meters below the ice surface of the Summit region in Greenland. The results of the GRIP and GISP 2 drillings are now widely known, and together with Professor Willi Dansgaard (who also retired in 1992) Hans could read in the international journals that the rapid changes of climate observed in the ice cores now were called the "Dansgaard-Oeschger" events.

Hans obtained several honors for his important work on climate and gases, such as The Harald C. Urey Medal from the European Association of Geochemistry, the Seligman Crystal from the International Glaciological Society, and, together with Willi Dansgaard and Claude Lorius, the U.S. Tyler Prize for Environmental Achievement in 1996.

The participants of the conference hereby dedicate this volume to the honor of the late Professor Hans Oeschger.

List of Contributors

Lennart O. Bengtsson
Max-Planck-Institute for Meteorology
Bundesstr. 55
20146 Hamburg
Germany

Claus U. Hammer
University of Copenhagen
Niels Bohr Institute
Juliance Maries Vej 30
DK-2100 Copenhagen N
Denmark

Paul J. Crutzen
Max-Planck-Institute for Chemistry
P.O. Box 3060
55020 Mainz
Germany

Meinrat O. Andreae
Max-Planck-Institute for Chemistry
Biogeochemistry Department
P.O. Box 3060
55020 Mainz
Germany

Inez Fung
Center for Atmospheric Sciences
University of California
Berkeley, CA 94720-4767
U.S.A.

Martin Heimann
Max-Planck-Institute for Biogeochemistry
P.O. Box 100164
07701 Jena
Germany

Wallace S. Broecker
Lamont-Doherty Earth Observatory of
    Columbia University
Palisades, NY 10964
U.S.A.

Stephen H. Schneider
Department of Biological Sciences
Stanford University
Stanford, CA 94305-5020
U.S.A.

David Schimel
Max-Planck-Institute for Biogeochemistry
P.O. Box 100164
07701 Jena
Germany

André Berger
Institut d'Astronomie et de Géophysique
    G. Lemaître
Université catholique de Louvain
2 Chemin du Cyclotron
1348 Louvain-la-Neuve
Belgium

Thomas F. Stocker
Climate and Environmental Physics
   Institute
University of Bern
Sidlerstr. 5
3012 Bern
Switzerland

J. E. Kutzbach
Center for Climatic Research
University of Wisconsin-Madison
1225 West Dayton St.
Madison, WI 53706-1695
U.S.A.

Iain Colin Prentice
Max-Planck-Institute for Biogeochemistry
P.O. Box 100164
07701 Jena
Germany

W. R. Peltier
Department of Physics
University of Toronto
Toronto, Ontario, M5S 1A7
Canada

James F. Kasting
Department of Geosciences
443 Deike
Pennsylvania State University
University Park, PA 16802
U.S.A.

J. C. Duplessy
Laboratoire des Sciences du Climat
   et de l'Environnement
Laboratoire mixte CEA-CNRS
Parc du CNRS
91198 Gif sur Yvette cedex
France

Jean Jouzel
Laboratoire des Sciences du Climat
   et de l'Environnement
UMR CEA-CNRS 1572
CEA Saclay
91198 Gif sur Yvette cedex
France

Will Steffen
IGBP Secretariat
The Royal Swedish Academy
   of Sciences
Box 50005
104 05 Stockholm
Sweden

Hartmut Grassl
Max-Planck-Institute for
   Meteorology
Bundesstr. 55
20146 Hamburg
Germany

Preface

A Study Conference on "Geosphere-Biosphere Interactions and Climate" was held at the Pontifical Academy of Sciences, 9–13 November 1998. The purpose of the Study Conference was to examine the role of the biogeochemical cycles and climate, to identify the key scientific issues in this field of research, and to outline a long-term scientific strategy.

Biogeochemical cycles play a key role in the way the Earth's climate both influences and is influenced by interactions with the geosphere and the biosphere. It is probably no overstatement to say that the elucidation of the biogeochemical cycles in Nature constitutes one of the greatest scientific challenges of our time.

The greenhouse gases – carbon dioxide, methane, nitrous oxide, and so on – together with aerosol particles, albeit in tiny concentrations, play a crucial role in determining the Earth's climate. Their concentrations are determined by physical, chemical, and biological processes in the terrestrial and oceanic biospheres and in the ocean as well as by chemical interactions within the atmosphere itself. Thus, to discover how atmospheric trace gas composition is regulated requires an understanding of the complex interactive system that sustains life on Earth, in the face of variations imposed both externally (e.g., orbital variations and changes in the solar irradiation) and by human activities, such as fossil fuel burning, deforestation, and agricultural practices.

Research during recent decades has shown that the climate of the Earth, interacting with chemical processes occurring within the atmosphere and the biosphere, constitutes a complex interwoven and integrated system. A change within some part of such a system may affect the system as a whole in such a way that the initial disturbance is amplified through the action of various feedback processes. For example, changes in the concentration of greenhouse gases in the atmosphere together with various geophysical factors have influenced the transition from glacial to interglacial periods, in turn leading to large-scale changes in vegetation and associated effects on biological processes. Today, these natural processes are further influenced by anthropogenic effects, which have now reached such magnitude that some of the natural biogeochemical cycles are playing a subordinate role.

The greenhouse gas concentration in the atmosphere has now reached a level that probably is the highest during the past one million years, and it is increasing several hundred times faster than it did due to natural processes in the past.

Among the topics that were discussed at the Study Conference, the first was the anthropogenic issue. What has become apparent during the past two to three decades is the substantial anthropogenic effect on the chemical composition of the atmosphere. Only during the past 30 years has the climate forcing from the greenhouse gases increased totally by some 25%; only half is coming from carbon dioxide, and the rest from methane, nitrous oxide, and the chlorofluorocarbons (CFCs). Under certain low-temperature conditions the CFCs have the additional property of destroying stratospheric ozone, which is dramatically demonstrated every spring over Antarctica. Whether this may take place more generally also over the Northern Hemisphere is a matter of great concern, in particular because there are distinct indications that the temperature in the stratosphere is decreasing, based on satellite data for the past 20 years, by some 0.5 °C/decade. Radiosonde observations from the 1950s onward suggest that the fall in stratospheric temperature has gone on for an even longer time, but observations may be open to criticism. If we look upon some of the other more chemically active gases with much shorter residence time in the atmosphere, such as NO_x and SO_2, we note that the anthropogenic emission of these gases is already higher than the natural emission, with increasing tendency in particular in the developing countries. In addition to influencing the radiative balance of the atmosphere and thus the climate, these gases are also influencing the biosphere and hence are affecting the Earth's system as a whole.

The climate system is highly complex and regulated by series of interwoven feedback processes. It is incompletely known because of the lack of relevant data, and many important processes are not yet well understood. This means that anthropogenic influences on climate can never be exactly predicted, and humankind must be prepared for unexpected events. Professor Paul Crutzen's account of the ozone hole in Chapter 1 of this book is an example of such an event.

A second important topic that was discussed at the Study Conference was the modeling of the Earth's system. Modeling of the climate system, including the climate of the past, has made great strides in recent years, and major efforts are going on both to improve individual components of the system – atmosphere, ocean, land surfaces, and so on – and to couple the various parts. Particularly challenging is the handling of abrupt events such as those showing up as rapid transitions between different phases of thermo-haline circulations.

A third broad topic addressed during the Study Conference was to explore information from past climates and thereby obtain a more comprehensive understanding of the Earth's system.

Reconstructions based on botanical and zoological data suggest that there have been periods when high-latitude regions were significantly warmer than they are today, with correspondingly much smaller changes in the Tropics. Certain data, although difficult to interpret, suggest that these warmer climates were associated with very high concentrations of carbon dioxide, perhaps 10 times as high as the present value.

Fluctuations in the greenhouse gas concentrations during the past 200,000 years or so have been determined from isotope analyses of ice cores obtained from Greenland and Antarctica. The concentration of carbon dioxide has evidently varied from about 180 ppm during the most intensive glaciations to about 300 ppm during interglacial periods. The variation of methane has been similar. No one today seriously argues – as Arrhenius originally suggested – that temporal variations in greenhouse gases are the cause of recent glaciation cycles, which are now regarded as arising in response to external forcing caused by orbital variations. Instead the greenhouse gases act to amplify this forcing. But the response is by no means simple, and it has not yet been fully explained, for carbon dioxide and methane interact quite differently with the oceans and the biosphere.

Ice core measurements from Greenland and sediment studies from the North Atlantic Ocean also indicate that abrupt climate changes have occurred on time scales ranging from decades to centuries, possibly associated with changes in the thermo–haline circulation of the North Atlantic. The extent to which variations in the large-scale atmospheric circulation in its interaction with land surface processes could have influenced such events is a matter for further investigation.

The final topic of the Study Conference was to discuss an overall research strategy and ways in which the scientific community should best meet the challenge of climate change.

The objectives of modern research on biogeochemical cycles and climate can be expressed in terms of a series of scientific issues, such as the following.

How will the terrestrial CO_2 cycle respond to changes in temperature and precipitation in different parts of the world? Where will there be positive and negative feedbacks? What effect will this have on the nitrogen cycle? Will the emission of N_2O and NO increase or decrease? What are the possible anthropogenic effects?

How will the ocean CO_2 cycle respond to climate change? What are the critical processes in this respect?

How can we estimate the effect of minor climatic variations due to volcanic eruptions or to El Niño events on the CO_2 and CH_4 variations?

How will the biosphere, particularly in the Tropics, change due to anthropogenic and other effects, and what consequences will this have on tropospheric ozone, the natural "chemical detergent" OH in the atmosphere, and on the development and distribution of cloud condensation nuclei and hence on cloud distribution and climate?

How can we explain the annual fluctuations in the atmospheric trace gas concentrations? What is the relative role of the oceans and the atmosphere in this respect? What conclusions can we draw from the natural variation, prior to any anthropogenic emissions, as revealed in paleorecords?

Changes in the atmospheric chemistry, according to all available information, have added an appreciable positive feedback on the orbitally driven solar insolation forcing. When it was cold, all greenhouse gases had low concentrations, and also the atmospheric loading of aerosol (soil dust) was much higher, increasing the planetary albedo. What were the mechanisms behind the changes in the atmospheric chemistry and aerosol loading, and what was the strength of this feedback?

Over very long periods of time, some 1 billion to 100 million years ago, the radiation of the sun has increased by some 25%. In the same period, a notable cooling of the atmosphere and of the surface of the Earth has taken place. Is this condition due to interactions between tectonic forces and climate whereby the concentration of carbon dioxide exercises a special control function? What is the role of the biosphere in this time perspective?

There is clear evidence from paleorecords (e.g., during Miocene, Eocene, and Cretaceous) that the climate changes differ from place to place in middle and high latitudes. There are differences between subpolar and semitropical climate changes that cannot be accounted for in terms of changes in astronomical forcing caused by orbital variations. What role does the biosphere have in controlling the climate of the Earth on these time scales, and what in particular are the upper and lower bounds on the exchange of carbon between the atmosphere, the oceans, and the biosphere? What is the possible maximum and minimum concentration of carbon dioxide in the atmosphere? Can we a priori exclude the prospect of "runaway" climate changes? Indeed there are many questions. They are difficult to answer, but they serve to outline the interdisciplinary character of our task.

Lennart Bengtsson and Claus Hammer

Introduction

Human influence on the global environment has increased considerably this century. In the early 1900s there were severe local environmental problems, mainly in densely populated areas, but the influence on the global environment was hardly noticeable. The situation at the end of the century has changed drastically. Anthropogenic influences on the overall composition of the atmosphere are already partly dominating natural processes. The stratospheric ozone destruction by CFCs and other artificial chemical constituents and the climate warming caused by increased greenhouse gas emission are problems that are now facing humankind. It is to be expected that the environmental problems will be more severe in the future because of the increased population and the likelihood that advanced industrialization will encompass the whole world.

Environmental problems are by their nature complex, because they often include series of intricate feedback processes. They cannot therefore be predicted exactly, and major surprises in the way the environment will respond to anthropogenic and other influences must be expected. An area that requires much more scientific attention is the interaction between the biosphere, the geosphere, and the climate. That such major interactions occur is well documented from palaeorecords, which show that the past climate of the Earth was very different from what it is today.

To address these issues a Study Conference on "Geosphere-Biosphere Interactions and Climate" was organized by the Pontifical Academy of Sciences 9–13 November 1998 at the headquarters of the Academy in Vatican City. A group of some 25 scientists, experts on different aspects of these problem areas, was invited to determine the present state of our scientific knowledge and to outline a long-term strategy for future research and possible lines of action that society can take based on what we now know. The specific aim of the conference that served to guide the individual contributions is described in the Preface.

This book contains the revised version of 18 contributions from the Study Conference on "Geosphere-Biosphere Interactions and Climate." They have been arranged by a broad subdivision of the topic into five major parts, examining first the problem with the increasing anthropogenic influences on the climate and the global environment and second the specific human perspective of climate change. This is followed by reviews of the key methodologies that are central in understanding the interactions between the biosphere, geosphere, and climate.

The third part deals with the mathematical modeling of the Earth's system. This approach is currently the main tool to predict weather and climate, including simulations of future climate change. This methodology is now being extended to address a wider range of problems, including the dynamical changes of the biosphere and the geosphere. This incorporates modeling studies of the climates that existed far back in the early history of our planet.

The fourth part of this volume deals with the reconstruction of past climates with the help of different paleo data. There have been spectacular contributions in recent years, particularly from the analyses of ice cores in Greenland and the Antarctic, which have significantly changed our concept of past climate. This includes indications of rapid climate changes over periods of a few decades for which, so far, there is no comprehensive explanation. An understanding of such events requires a more in-depth understanding of the feedback processes between climate, biosphere, and geosphere.

In the fifth part, summaries are given of the strategies for organizing the future science program and the particular roles of WCRP (World Climate Research Programme) and IGBP (International Geophysical and Biological Programme).

We would like to express our deep appreciation to the Pontifical Academy of Science for organizing the Conference on "Geosphere-Biosphere Interactions and Climate" and for the Academy's support toward this publication.

We would also like to acknowledge the valuable assistance of Dr. Annette Kirk in the editorial work.

Finally, we express the hope that publication of these proceedings will advance the knowledge and further the understanding of the complex issues addressed by the contributors.

The Editors

CONCEPTUAL MODEL of Earth System processes operating on timescales of decades to centuries

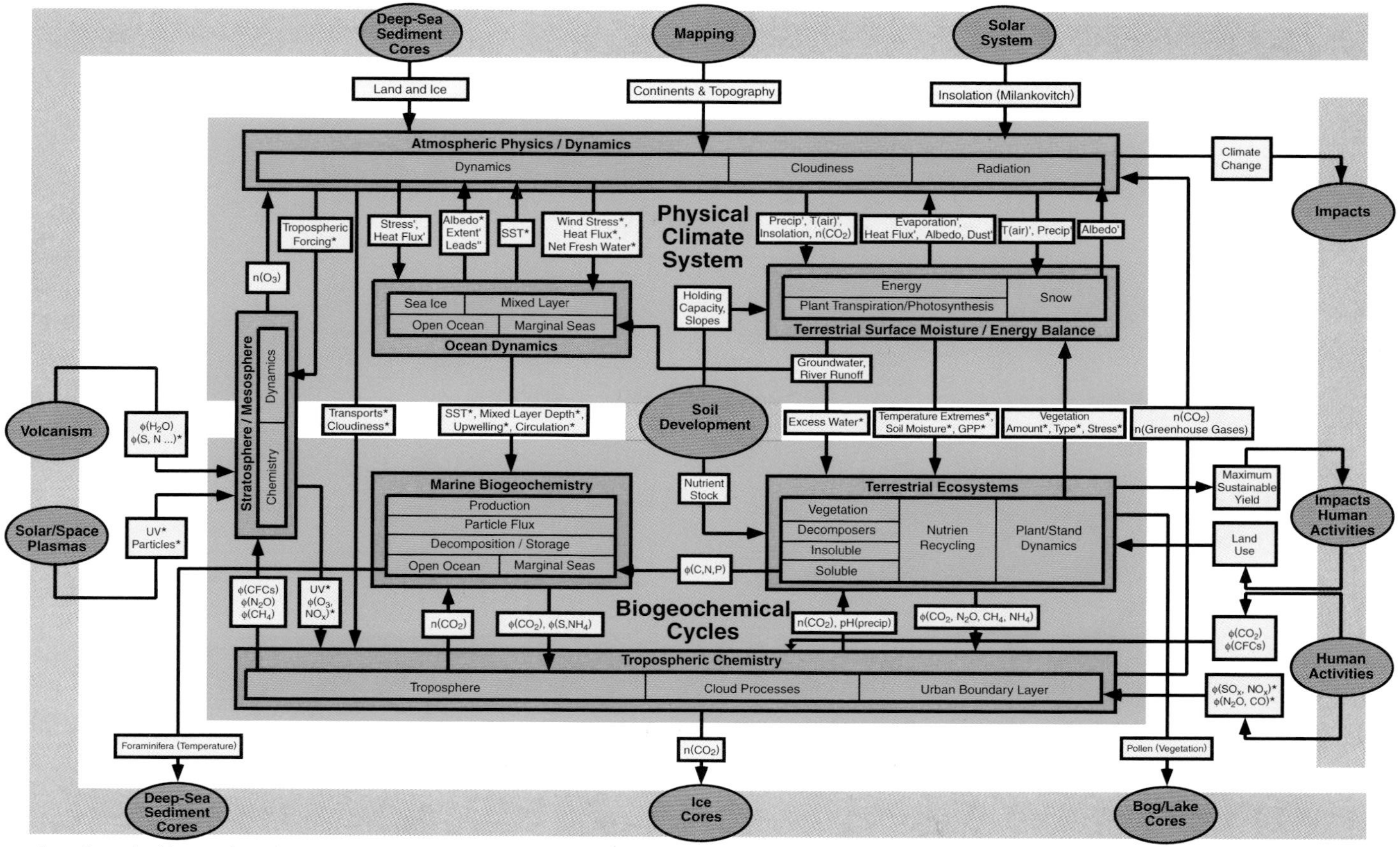

' = on timescale of hours to days   * = on timescale of months to seasons   φ= flux   n = concentration

Figure 2.1

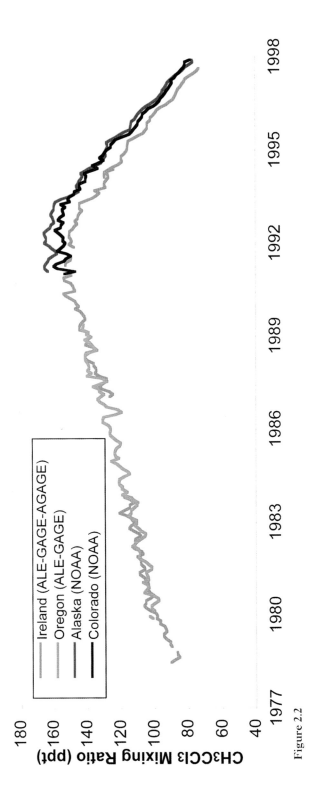

Figure 2.2

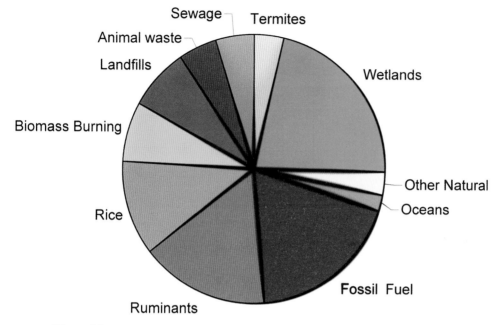

Figure 2.3a

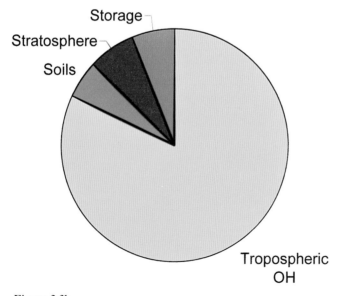

Figure 2.3b

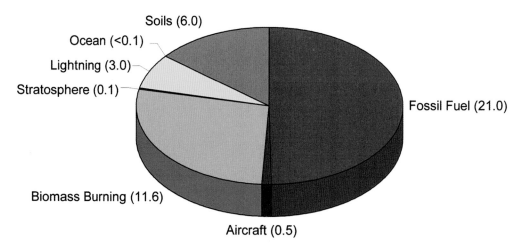

NO$_x$ Emissions in Tg N(NO)/yr

Soils (6.0)

Ocean (<0.1)

Lightning (3.0)

Stratosphere (0.1)

Fossil Fuel (21.0)

Biomass Burning (11.6)

Aircraft (0.5)

Figure 2.4

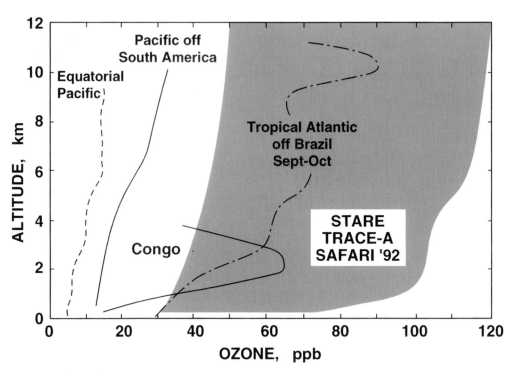

Figure 2.5

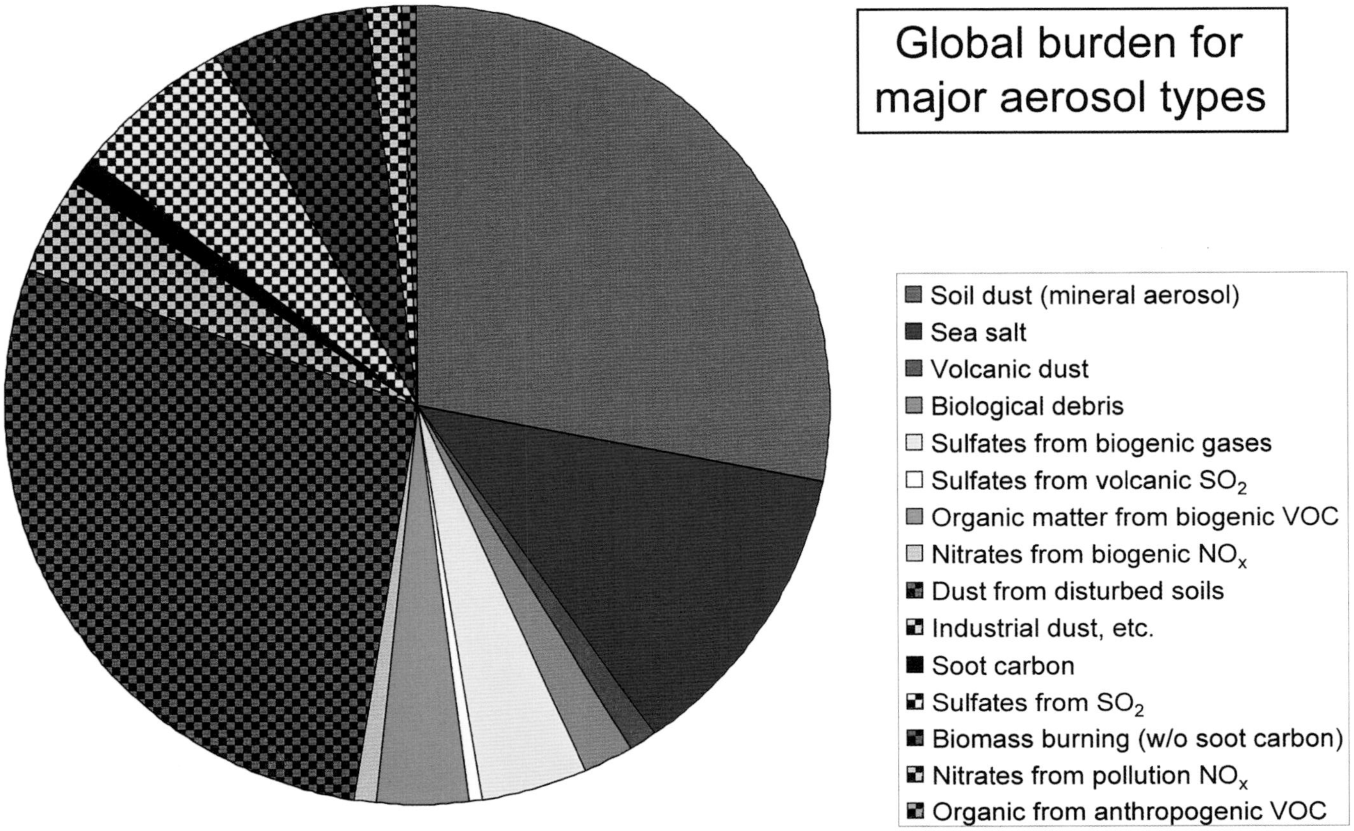

Global burden for major aerosol types

Soil dust (mineral aerosol)
Sea salt
Volcanic dust
Biological debris
Sulfates from biogenic gases
Sulfates from volcanic SO_2
Organic matter from biogenic VOC
Nitrates from biogenic NO_x
Dust from disturbed soils
Industrial dust, etc.
Soot carbon
Sulfates from SO_2
Biomass burning (w/o soot carbon)
Nitrates from pollution NO_x
Organic from anthropogenic VOC

Figure 2.6

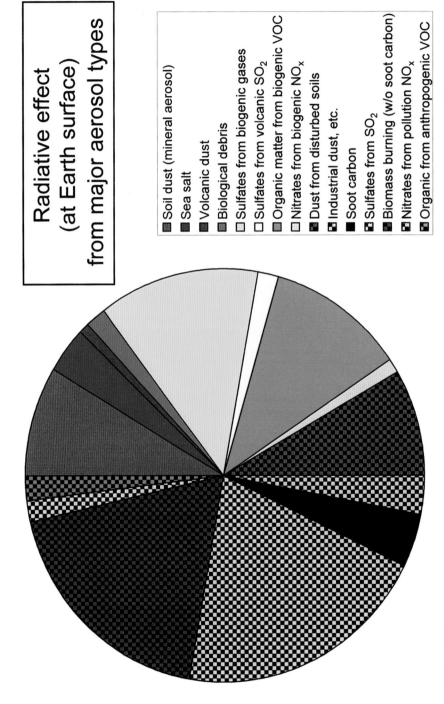

Radiative effect
(at Earth surface)
from major aerosol types

■ Soil dust (mineral aerosol)
■ Sea salt
■ Volcanic dust
■ Biological debris
□ Sulfates from biogenic gases
□ Sulfates from volcanic SO_2
■ Organic matter from biogenic VOC
□ Nitrates from biogenic NO_x
■ Dust from disturbed soils
■ Industrial dust, etc.
■ Soot carbon
■ Sulfates from SO_2
■ Biomass burning (w/o soot carbon)
■ Nitrates from pollution NO_x
■ Organic from anthropogenic VOC

Figure 2.7

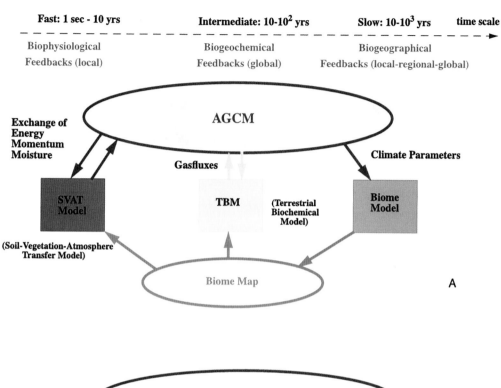

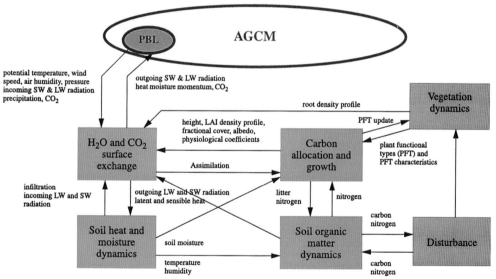

Figure 4.1

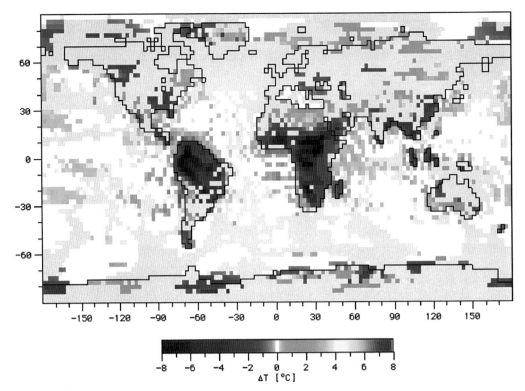

Figure 4.2a

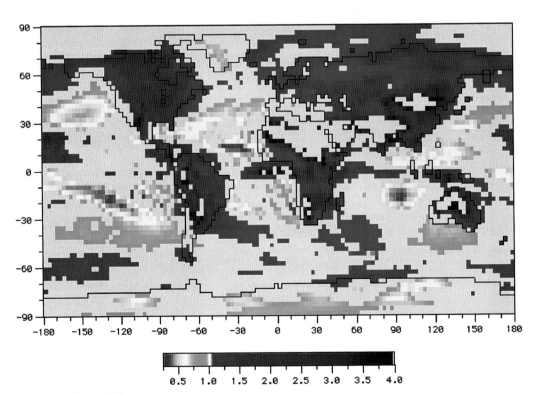

Figure 4.2b

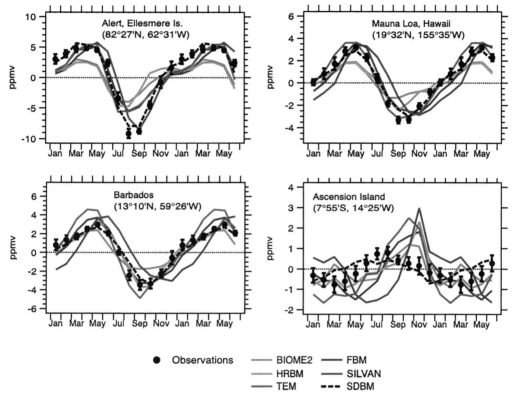

Figure 4.3

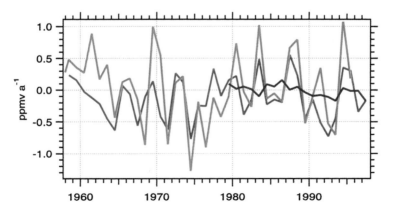

Figure 4.4

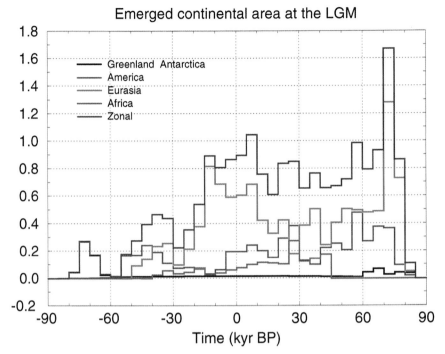

Figure 8.4

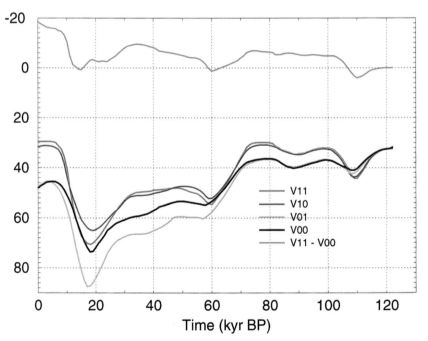

Figure 8.5

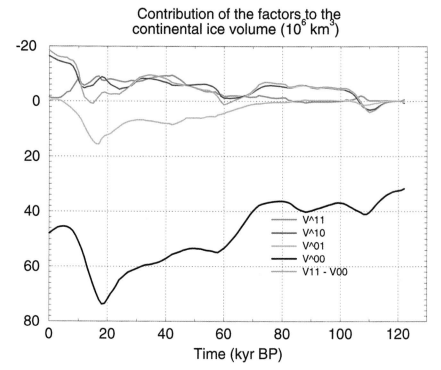

Figure 8.6

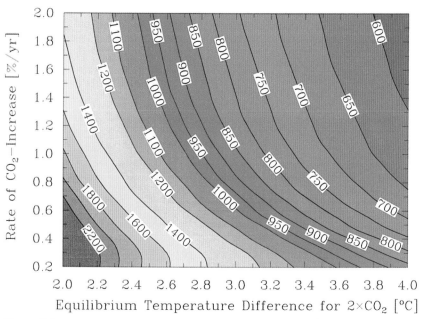

Figure 9.6

1 The Antarctic Ozone Hole, a Human-Caused Chemical Instability in the Stratosphere

What Should We Learn from It?

PAUL J. CRUTZEN

ABSTRACT

Atmospheric ozone plays a critical role in limiting the penetration of biologically harmful, solar ultraviolet radiation to the Earth surface. Furthermore, the absorption of ultraviolet radiation from the Sun and infrared radiation emitted from the Earth's warm surface influence temperatures in the lower stratosphere, creating dynamically stable conditions with strongly reduced vertical exchange. Through industrial emissions, ozone-depleting catalysts have increasingly been produced in the stratosphere, leading to reductions in ozone. The situation is especially grave during springtime over Antarctica, where, since the 1980s, each year almost all ozone in the 14–22 km height region is chemically destroyed. This so-called "ozone hole" was not predicted by any model and came as a total surprise to all scientists. The ozone hole developed at a least likely location. Through the emissions of chlorofluorocarbons, humankind has created a chemical instability, leading to rapid loss of ozone. A question is whether there may be other instabilities that might be triggered in the environment by human activities.

1.1 Introduction

The study of the chemistry of the atmosphere is both of immediate scientific interest and of high social relevance. We first note that the gases that are most significant for atmospheric chemistry and for the Earth's climate are not its main components – nitrogen (N_2), oxygen (O_2), and argon (Ar), which together with variable amounts of water vapor make up greater than 99.9% of the molecules in the Earth's atmosphere – but rather are many gases that are found only in very low concentrations. The main gases cannot be influenced significantly by human activities. The minor gases can. Several of them play important roles in climate and atmospheric chemistry. Carbon dioxide (CO_2), which currently has a concentration of approximately 360 among 1 million air molecules, is of crucial importance in that, together with water vapor and sunlight, it builds the organic molecules of living matter. Carbon dioxide is also of great significance for the Earth's climate, an important theme of this conference. However, despite these important aspects, CO_2 plays no significant direct role in atmospheric chemistry.

The chemically reactive gases have even much lower abundances in the atmosphere than CO_2. Several of them also act as greenhouse gases. One among these is ozone (O_3),

1

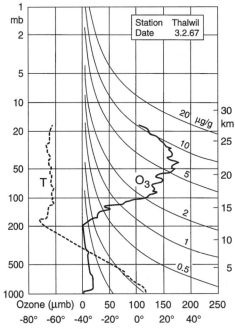

Figure 1.1. Measured ozone and temperature profiles over Thalwil, Switzerland, 1967.

which, next to water vapor, is the most important gas for the photochemistry of the atmosphere. Without ozone, the chemistry and chemical composition of the atmosphere would be totally different.

Looking at the temperature and ozone profiles of the atmosphere, as shown in Figure 1.1, we recognize that in the troposphere – that part of the atmosphere where temperatures decrease with height – ozone concentrations are quite low. Higher up, in the stratosphere, ozone concentrations rise steeply with altitude until 25–30 km, and temperatures no longer decrease. Because of this even temperature distribution with height, vertical mixing in the stratosphere is much suppressed. This is also why the stratosphere is characterized by quiet "weather." There are also few clouds in the stratosphere; however, the exceptions are important.

The stratospheric temperature structure is directly connected to the ozone distribution. The ozone that accumulates in the stable layer of the stratosphere absorbs upwelling "warm" infrared radiation from the Earth's surface as well as ultraviolet (UV) radiation from the Sun. The latter is the same process that largely protects life on Earth from this potentially harmful radiation. This absorption of radiation from above and below provides an important energy source for the stratosphere and explains why temperatures do not decrease with height. Ozone concentrations and temperatures in the stratosphere are very closely coupled. Stable meteorological conditions keep most ozone in the stratosphere, limiting the flow of this poisonous gas to the Earth's surface and largely confining the hydrological cycle to the troposphere. The strong coupling of ozone and temperature in the stratosphere is in my opinion a very important property of the atmosphere that is insufficiently recognized by the climate research community.

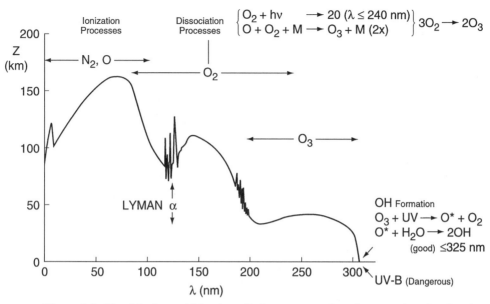

Figure 1.2. The altitude to which solar radiation penetrates into the atmosphere is a function of wavelength. Radiation shorter than 195 nm is absorbed in the mesosphere above 50 km. The longer-wavelength ultraviolet is mainly absorbed in the stratosphere by ozone.

In the troposphere there is very little ozone. Until about 20 years ago it was thought that the troposphere contained only ozone that had been transported down from the stratosphere. At that time tropospheric ozone was considered to be interesting only in the study of atmospheric transport, and its enormous importance for the chemistry of the troposphere was not recognized. Tropospheric ozone makes up only about 10% of all ozone in the atmosphere, with an average volume mixing ratio of about 40 nmol/mol (nanomole per mole, n = nano = 10^{-9}). However, as we discuss in this chapter, in the absence of tropospheric ozone the chemical composition of the atmosphere would be totally different.

If we look at the altitude to which solar radiation penetrates into the atmosphere, we see that the very short wavelengths – shorter than 200 nm – are to a large degree removed by 50 km (see Figure 1.2). This happens primarily through the absorption of the radiation by atomic and molecular forms of oxygen (O) and nitrogen (N). But these main gases do not absorb beyond about 240 nm. Fortunately, ozone does so very strongly in the 200–300 nm wavelength range. Were it not for atmospheric ozone, this radiation would penetrate to the Earth's surface. For the Earth's current biosphere, this would have had catastrophic consequences. Only during the past one-third of the Earth's age has the atmosphere contained comparable amounts of ozone (and oxygen) as at present. The Earth has thus been without the protective shield of oxygen and ozone during most of its existence. This must have forced primitive life to develop only in dark hideaways shielded from sun's damaging ultraviolet rays. The average concentration of ozone in the atmosphere amounts to only about 0.3 per million air molecules, but it nevertheless suffices to absorb the main part of the dangerous UV radiation.

Beyond 300 nm, the absorption ability of ozone becomes weaker so that UV radiation at longer wavelengths can penetrate to the Earth's surface. It is the radiation up to 320 nm, also called UV-B radiation (*B* stands for "biologically active") that still poses a problem for life on Earth. Light-skinned people are all familiar with the fact that when they stay too long unprotected in the sun, they get sunburned, and from frequent exposures skin cancer may develop. Plants can also be affected by this radiation. On the other hand, we know from research conducted during the past 25 years that this same radiation is also very important for keeping our atmospheric environment clean. The reason is the following: Up to wavelengths of about 335 nm, UV radiation is capable of splitting an ozone molecule into an oxygen molecule and an excited oxygen atom (O^*). The latter has enough energy to react with atmospheric water vapor to produce hydroxyl radicals, with the chemical formula OH (Levy, 1971).

$$R1a \quad O_3 + h\nu \quad \rightarrow O^* + O_2 (<335\,nm)$$
$$R2 \quad O^* + H_2O \rightarrow 2\,OH$$

The $h\nu$ in reaction R1a, and elsewhere in this chapter, symbolizes a photon with frequency ν and energy h, where h is Planck's constant.

The OH radical can be called the "detergent" of the atmosphere, because it is the main species, that reacts with almost all gases, thus removing them from the atmosphere. Without OH radicals, the chemical composition of the atmosphere would be totally different.

There are three factors that are important for the formation of the OH radical: ozone, water vapor, and UV-B radiation. The average concentration of OH amounts to only about 4 out of 10^{14} air molecules (Prinn et al., 1995); negligibly few, one might say, but without this highly reactive radical the chemical composition of the atmosphere would be totally different. Molecular oxygen, which makes up almost 21% of the atmosphere, is not capable of oxidizing any of the atmospheric gases; their oxidation requires initial attack by OH radicals. Ozone in the troposphere is thus not at all the inert gas it was taken for until about 25 years ago, but rather it plays a key role in atmospheric chemistry. Although the role of ozone and hydroxyl in tropospheric chemistry is a fascinating subject, in this chapter we concentrate on stratospheric ozone.

1.2   The Ozone Hole

The main part of this chapter concentrates on the topic of the stratospheric ozone and the dramatic development of the so-called "ozone hole." To explain what has happened, I must start with a short overview of ozone layer chemistry.

Stratospheric ozone is formed through the photolysis of oxygen by solar ultraviolet radiation of wavelengths less than 240 nm, a process that humans cannot influence. The photolysis of O_2 produces two oxygen atoms, each of which combines with oxygen molecules to form ozone (Chapman, 1930).

$$R3 \quad O_2 + h\nu \quad \rightarrow 2\,O(\leq 240\,nm)$$
$$R4 \quad 2 \times (O + O_2 + M \rightarrow O_3 + M); M = N_2 \text{ or } O_2$$
$$\sum : 3\,O_2 \quad \rightarrow 2\,O_3$$

Unless there are chemical reactions going in the opposite direction, most O_2 in the atmosphere would be transformed into ozone in about 10,000 years. For many years it was thought that the return reactions from O_3 to O_2 would involve only oxygen allotropes (Chapman, 1930):

R1b $O_3 + h\nu \rightarrow O + O_2 (\lambda < 1140\,nm)$

R5 $O + O_3 \rightarrow 2\,O_2$

$\sum: 2\,O_3 \rightarrow 3\,O_2$

These reactions too cannot be influenced by humankind. Things changed, however, when it was realized that catalytic reactions could be more important than the previously mentioned Chapman reactions in converting O_3 back to O_2. First, it was hypothesized by Crutzen (1970) that NO and NO_2 could catalyze the destruction of ozone:

R6 $NO + O_3 \rightarrow NO_2 + O_2$

R1 $O_3 + h\nu \rightarrow O + O_2$

R7 $O + NO_2 \rightarrow NO + O_2$

$\sum: 2\,O_3 \rightarrow 3\,O_2$

In the following year Johnston (1971) and Crutzen (1971) independently proposed that the nitric oxide emitted from the large fleets of supersonic transport aircraft, the SSTs – which were planned to be built in the United States, France, Britain, and the Soviet Union – could result in substantial ozone depletion. Only a few SSTs were ever built. However, a few years later, Molina and Rowland (1974) hypothesized that Cl and ClO, released to the atmosphere from the photochemical decay of the chlorofluorocarbon gases ($CFCl_3$ and CF_2Cl_2), could deplete ozone by a similar chain of catalytic reactions as shown earlier with NO and NO_2:

R8 $Cl + O_3 \rightarrow ClO + O_2$

R1 $O_3 + h\nu \rightarrow O + O_2 (<1140\,nm)$

R9 $ClO + O \rightarrow Cl + O_2$

$\sum: 2\,O_3 \rightarrow 3\,O_2$

In particular, since the Second World War, the stratospheric abundance of chlorine-containing gases has increased strongly; consequently, the stratosphere now contains approximately six times more chlorine than the amount provided by methylchloride (CH_3Cl), which is emitted from the oceans. Until 1985, it was thought that ozone destruction via the ClO_x catalytic cycle would take place primarily over the altitude range 30–45 km, whereas at lower elevations, where most ozone is located, much less ozone would be destroyed. However, observations reported in 1985 by researchers of the British Antarctic Survey (Farman et al., 1985) showed that the most dramatic ozone decreases were occurring during September–October principally in the lower layers of the stratosphere over Antarctica, a finding that was totally unexpected. Previously it was believed, and this is true in most situations, that below 30 km reactions between the NO_x and ClO_x catalysts, producing hydrochloric acid (HCl) and chlorine nitrate

($ClONO_2$) via

$$R10 \quad ClO + NO_2 + M \rightarrow ClONO_2 + M$$

and

$$R11 \quad ClO + NO \rightarrow Cl + NO_2$$
$$R12 \quad Cl + CH_4 \rightarrow HCl + CH_3$$

would strongly reduce the concentrations of the "ozone killers" NO_x and ClO_x, thus protecting ozone from otherwise much stronger destruction. Through these reactions, the majority of stratospheric inorganic chlorine is mostly tied up as HCl and $ClONO_2$, which do not react with O_3. That these favorable circumstances do not always exist became clear after Farman et al. (1985) discovered that average springtime (September–October) stratospheric ozone amounts above their research station Halley Bay on the Antarctic continent had been strongly decreasing year by year since the middle of the 1970s. Similar low O_3 values had also been reported by Chubachi (1984) of the Japan Polar Research Institute. From balloon soundings it became clear that rapid and complete ozone loss was taking place within a month in the same height range, 14–22 km, where maximum ozone concentrations are usually found (Figure 1.3). The observations were a total surprise to the stratospheric ozone research community. Until 1985 it had been common wisdom that the ozone in this altitude region was chemically inert. Analyses of satellite observations showed that large ozone decrease occurred over much of Antarctica during the months of September and October. The big question was, how was this possible? After only a few years of intensive research the principal causes became clear. At very cold temperatures, less than around $-80\,^{\circ}C$, which occur

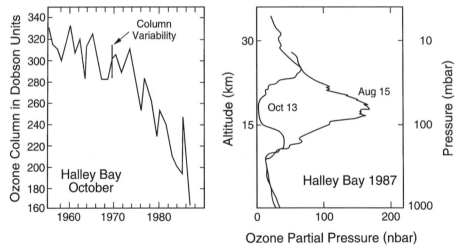

Figure 1.3. The ozone hole. The figure on the left shows the decrease in the total ozone column over the Antarctic (100 Dobson units corresponds to a layer of ozone 1 mm thick as standard temperature and pressure at the earth surface). The right-hand diagram shows the altitude dependence of ozone loss between August and October 1987. Measurements by J. Farman and coworkers of the British Antarctic Survey (1985) and by D. Hofmann and coworkers (1989) of the National Oceanic and Atmospheric Administration (NOAA) in Boulder, Colorado.

at high latitudes in the winter especially over Antarctica, condensation or sublimation of nitric acid (HNO_3) and water vapor takes place on particles (Crutzen and Arnold, 1986; Toon et al., 1986; Dye et al., 1992) that are always present in the stratosphere and that normally consist of water and sulfuric acid (H_2SO_4). This process effectively removes HNO_3 from the gas phase, and with it also the nitrogen oxides (NO and NO_2). What then happens is that HCl and $ClONO_2$, the two most abundant inorganic chlorine species – which do not react with ozone or with each other in the gas phase – react in, or on the surface of, the particles to form Cl_2 and HNO_3 (Solomon et al., 1986; Molina et al., 1987; Tolbert et al., 1987):

$$R13 \quad ClONO_2 + HCl \rightarrow Cl_2 + HNO_3 \text{ (surfaces)}$$

Then, in late winter or early spring, when sunlight returns after the long polar night, the Cl_2 molecules are quickly split, producing Cl atoms. These start a very efficient catalytic chain of reactions (Molina and Molina, 1987), which results in the rapid transformation of two ozone molecules into three oxygen molecules.

$$
\begin{aligned}
&R8 \quad 2 \times (Cl + O_3 \quad \rightarrow ClO + O_2) \\
&R14 \quad ClO + ClO + M \rightarrow Cl_2O_2 + M \\
&R15 \quad Cl_2O_2 + h\nu \quad \rightarrow 2\,Cl + O_2(\lambda < 350\,\text{nm}) \\
&\sum : 2\,O_3 \quad\quad \rightarrow 3\,O_2
\end{aligned}
$$

The second reaction implies that the ozone destruction rate depends on the square of the ClO radical concentration. If we also consider that the formation of chemically active chlorine (Cl and ClO) by reaction R13 involves a reaction between two chlorine-containing species – $ClONO_2$ and HCl – we note that the rate of ozone decomposition could be proportional to between the second and fourth power of the stratospheric chlorine content. With this increasing by 4% per year, as it has been doing until the beginning of the 1990s, the ozone destruction rate could increase by between 8% and 17% per year. The current stratospheric chlorine abundance is about six times greater than that of the natural background that is provided by CH_3Cl, implying at least 36 times faster anthropogenic than natural ozone destruction by the Cl and ClO radicals. In situ observations on a stratospheric research aircraft validated the above explanation for the origin of the ozone hole (Anderson et al., 1989). Precisely in the polar areas, where the stratosphere gets very cold in winter and remains cold during early spring, measurements show high concentrations of ClO radicals and simultaneously rapid ozone destruction (Figure 1.4). It is also important to note here that because of the strong ozone loss, heating of the stratosphere in the ozone-poor air does not take place, leading to lower temperatures and thus enhancing ice or supercooled liquid particle formation, chlorine activation, and ozone depletion, producing a series of positive feedbacks.

In the meantime, it was found that also in the Northern Hemisphere during late winter and early spring, ozone is being increasingly destroyed, although to a lesser extent than over Antarctica because stratospheric temperatures are generally about $10°C$ higher than over Antarctica, thus causing less-efficient particle formation and chlorine activation. During the 1980s, ozone depletion was most evident between January and

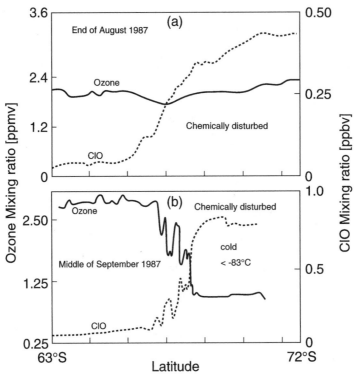

Figure 1.4. High concentrations of ClO radicals and the simultaneous rapid ozone destruction occur in winter when the temperature becomes very cold. Measurements by J. G. Anderson and coworkers (1989) of Harvard University.

April, when the destruction trend approached 1% per year. Ozone destruction took place in other seasons as well but at the lower rate of around 0.4% per year (Stolarski et al., 1991). The declining trend during winter and spring has increased during the present decade over the middle- to high-latitude zone of the Northern Hemisphere. In particular, during the winters 1995/1996 and 1996/1997 similar ozone depletions to those observed over the Antarctic about 15 years earlier were seen (e.g., Müller et al., 1997). It is therefore especially gratifying to note that since early 1986 international agreements have been in effect that forbid the production of CFCs and several other industrial Cl- and Br-containing gases in the developed world, with a decade's respite for the developing world. Hopefully this means that the damage to the ozone layer may not grow much worse in the future. However, even in the best of circumstances, full recovery of the ozone layer will be a slow process. It will take up to half a century before the ozone hole will disappear. The slowness of the repair process is due to the long average atmospheric decay times of the CFC gases, on the order of 50 years for CFC-11 and 110 years for CFC-12.

Nevertheless, as a result of these measures the worst effects on the biosphere have probably been prevented. Thus, estimates by Slaper et al. (1996) would have predicted a fourfold increase in the incidence of skin cancer during the next century if no regulatory measures had been taken. The effects on other parts of the biosphere are harder to quantify, but they could have been important. It would be very surprising if light-skinned people would be the only species to be so strongly damaged by increased UV-B radiation.

Although the ozone layer may thus be recovering, a recent study by Shindell et al. (1998) indicates that as a result of the cooling of the stratosphere due to increasing CO_2 levels, it may be possible that for another two decades stratospheric ozone in both hemispheres may further decrease, an effect that could be especially pronounced over the Arctic, maybe leading there to ozone columns similarly low as over Antarctica. In fact, Waibel et al. (1999) point to the possibility that, because of cooling of the stratosphere by increasing CO_2, even by the third quarter of 2000 severe ozone depletion may still occur.

1.3   Conclusions

Humankind is having a considerable influence on the condition in the atmosphere, even in areas that are very far removed from the pollution sources. Most surprisingly and unexpectedly, over Antarctica during September–October, enormous damage is done to the ozone layer due to a remarkable combination of feedbacks: radiative cooling, giving very low winter and springtime temperatures, and the presence of chlorine gases in the stratosphere at concentrations about six times greater than that of the natural background provided by CH_3Cl. The cold temperatures promote the formation of solid or supercooled liquid polar stratospheric cloud particles consisting of a mixture of H_2SO_4, HNO_3, and H_2O, on whose surfaces, or within which, reactions take place that convert HCl and $ClONO_2$ (which do not react with ozone) to highly reactive radicals Cl and ClO. The latter rapidly remove ozone from the lower stratosphere by catalytic reactions. Nobody predicted this course of events. In fact, until the discovery of the ozone hole, it was generally believed that ozone at high latitudes could not be significantly affected at all by chemical processes and was only subjected to transport. How wrong we all were. Exactly in the part of the stratosphere the farthest away from the industrialized world, and exactly in that altitude region at which until about 1980 maximum concentrations of ozone had always been found, mainly during the month of September, all ozone is going to be destroyed for many more years to come, despite the international agreements that are now in place. This is due to a number of positive feedbacks. This ozone loss should be a warning. It can be difficult or impossible to predict precisely where the weak points in the environment are located. It is therefore important to watch even for seemingly unlikely chains of positive feedbacks leading to major environmental impacts. Examples of potential instabilities were discussed at this workshop, including abrupt climate changes and a weakening of the Atlantic deep-water formation.

1.4   Epilog: And Things Could Have Been Much Worse

Gradually, if the studies by Shindell et al. (1998) and Waibel et al. (1999) turn out not to be valid, over a period of a century or so, stratospheric ozone should largely recover to its natural state. However, it was a close call. Had Farman and his colleagues not persevered in making their measurements in the harsh Antarctic environment for all those years since the International Geophysical Year 1958/1959, the discovery of the ozone hole might have been substantially delayed, and there might have been far less urgency to reach international agreement on the phasing out of CFC production. There might thus have been a substantial risk that an ozone hole could also have developed in the higher latitudes of the Northern Hemisphere.

Furthermore, whereas the establishment of an instability in the O_3/ClO_x system requires chlorine activation by heterogeneous reactions on solid or in supercooled liquid particles, this is not required for inorganic bromine, which, because of gas-phase photochemical reactions, is normally largely present in its activated forms Br and BrO. This makes bromine almost 100 times more dangerous for ozone than chlorine on an atom-to-atom basis. This brings up the nightmarish thought that if the chemical industry had developed organobromine compounds instead of the CFCs – or alternatively, if chlorine chemistry had behaved more like that of bromine – then without any preparedness, we would have been faced with a catastrophic ozone hole everywhere and in all seasons during the 1970s, probably before atmospheric chemists had developed the necessary knowledge to identify the problem and the appropriate techniques for the necessary critical measurements. Noting that nobody had worried about the atmospheric consequences of the release of Cl or Br before 1974, I can only conclude that we have been extremely lucky. This shows that we should always be on our guard for the potential consequences of the release of new products into the environment. Continued surveillance of the composition of the stratosphere, therefore, remains a matter of high priority for many years to come.

REFERENCES

Anderson, J. G., Brune, W. H., and Proffitt, M. H. (1989). Ozone destruction by chlorine radicals within the Antarctic vortex: The spatial and temporal evolution of ClO-O$_3$ anticorrelation based on in situ ER-2 data. J. Geophys. Res., 94, 11465–11479.

Chapman, S. (1930). A theory of upper atmospheric ozone. Mem. R. Soc., 3, 103–125.

Chubachi, S. (1984). Mem. Natl. Inst. Polar Res., special issue, National Institute of Polar Research, Tokyo.

Crutzen, P. J. (1970). The influence of nitrogen oxides on the atmospheric ozone content. Q. J. R. Meteorol. Soc., 96, 320–325.

Crutzen, P. J. (1971). Ozone production rates in an oxygen-hydrogen-nitrogen oxide atmosphere. J. Geophys. Res., 76, 7311–7327.

Crutzen, P. J., and Arnold, F. (1986). Nitric acid cloud formation in the cold Antarctic stratosphere: a major cause for the springtime 'ozone hole.' Nature, 324, 651–655.

Dye, J. E., Baumgardner, D., Gandrud, B. W., Kawa, S. R., Kelly, K. K., Loewenstein, M., Ferry, G. V., Chan, K. R., and Gary, B. L. (1992). Particle size distributions in Arctic Polar stratospheric clouds, growth and freezing of sulfuric acid droplets, and implications for cloud formation. J. Geophys. Res., 97, 8015–8034.

Farman, J. C., Gardiner, B. G., and Shanklin, J. D. (1985). Large losses of total ozone in Antarctica reveal seasonal ClO$_x$/NO$_x$ interaction. Nature, 315, 207–210.

Hofmann, D. J., Harder, J. W., Rosen, J. M., Hereford, J. V., and Carpenter, J. R. (1989). Ozone profile measurements at McMurdo station, Antarctica, during the spring of 1987. J. Geophys. Res., 94, 16527–16536.

Johnston, H. S. (1971). Reduction of stratospheric ozone by nitrogen oxide catalysts from supersonic transport exhaust. Science, **173**, 517–522.

Levy, H. (1971). Normal atmosphere: Large radical and formaldehyde concentrations predicted. Science, **173**, 141–143.

Molina, M. J., and Rowland, F. S. (1974). Stratospheric sink of chlorofluoromethanes: Chlorine atom-catalyzed destruction of ozone. Nature, **249**, 810–812.

Molina, L. T., and Molina, M. J. (1987). Production of Cl_2O_2 from the self-reaction of the ClO radical. J. Phys. Chem., **91**, 433–436.

Molina, M. J., Tso, T.-L., Molina, L. T., and Wang, F. C.-Y. (1987). Antarctic stratospheric chemistry of chlorine nitrate, hydrogen chloride, and ice: Release of active chlorine. Science, **238**, 1253–1257.

Müller, R., Crutzen, P. J., Grooß, J.-U., Brühl, C., Russell, J. M. III, Gernandt, H., McKenna, D. S., and Tuck, A. F. (1997). Nature, **389**, 709–712.

Prinn, R. G., Weiss, R. F., Miller, B. R., Huang, J., Alyea, F. N., Cunnold, D. M., Fraser, P. J., Hartley, D. E., and Simmonds, P. G. (1995). Atmospheric trends and lifetime of CH_3CCl_3 and global OH concentrations. Science, **269**, 187–192.

Shindell, D. T., Rind, D., and Lonergan, P. (1998). Increased polar stratospheric ozone losses and delayed eventual recovery owing to increased greenhouse-gas concentrations. Nature, **392**, 589–592.

Slaper, H., Velders, G. J. M., Daniel, J. S., de Gruij, F. R., and van der Leun, J. C. (1996). Estimates of ozone depletion and skin cancer incidence to examine the Vienna Convention achievements. Nature, **384**, 256–258.

Solomon, S., Garcia, R. R., Rowland, F. S., and Wuebbles, D. J. (1986). On the depletion of Antarctic ozone. Nature, **321**, 755–758.

Stolarski, R. S., Bloomfield, P., McPeters, R. D., and Herman, J. R. (1991). Total ozone trends deduced from Nimbus 7 TOMS data. Geophys. Res. Lett., **18**, 1015–1018.

Tolbert, M. A., Rossi, M. J., Malhotra, R., and Golden, D. M. (1987). Reaction of chlorine nitrate with hydrogen chloride and water at Antarctic stratospheric temperatures. Science, **238**, 1258–1260.

Toon, O. B., Hamill, P., Turco, R. P., and Pinto, J. (1986). Condensation of HNO_3 and HCl in the winter Polar stratospheres. Geophys. Res. Lett., **13**, 1284–1287.

Waibel, A. E., Peter, Th., Carslaw, K. S., Oelhaf, H., Wetzel, G., Crutzen, P. J., Pöschl, U., Tsias, A., Reimer, E., and Fischer, H. (1999). Arctic ozone loss due to denitrification. Science, **283**, 2064–2069.

PART ONE

THE ANTHROPOGENIC PROBLEM

2 Feedbacks and Interactions between Global Change, Atmospheric Chemistry, and the Biosphere

M. O. ANDREAE

ABSTRACT

Human activities are changing the composition of the atmosphere not only directly through the emission of trace gases and aerosols, but also indirectly through perturbations in the physical, chemical, and ecological characteristics of the Earth System. These perturbations in turn influence the rates of production and loss of atmospheric constituents.

The impact of direct anthropogenic emissions on the atmosphere is often relatively easy to assess, especially if they are tied to major industrial activities, where accurate and detailed records are kept for economic reasons. Classical examples are the release of chlorofluorocarbons and the emission of CO_2 from fossil fuel combustion. There are also cases, however, in which it is much more difficult to obtain accurate emission estimates. An example is biomass burning, for which no economic incentive for record keeping exists, and which takes on many forms, each with a different emission profile.

A more complex case exists in which human activities release a precursor compound, which is transformed in the atmosphere to a climatically active substance. This can be illustrated using the example of SO_2, from which sulfate aerosol can be formed. The actual amount of radiatively active sulfate aerosol produced, however, is determined by a complex interplay of atmospheric transport processes, chemical processes in the gas phase, and interactions with other aerosol species.

Some of the most important anthropogenic modifications of the atmosphere, however, are the indirect results of human-caused changes in the functioning of the Earth System. For example, when land use and agricultural practices change, the emissions of trace gases such as N_2O, NO, and CH_4 change in highly complex ways, which are extremely difficult to assess at the scales of interest. An even higher level of complexity is encountered when human activities affect the atmospheric levels of some species, something that in turn changes the chemical functioning of the atmosphere and consequently the production rates and lifetimes of aerosol and greenhouse gases. An example of such a mechanism may be the large-scale change of trace gas inputs into the vast photochemical reactor of the tropical troposphere, where most of the photooxidation of long-lived trace gases takes place.

Finally, we must consider feedback loops, in which global change begets global change. Climate change, caused by upsetting the Earth's radiative balance, results in different circulation patterns, changes in water availability at the surface, changes in water vapor content of the atmosphere, and so on, all of which modify the atmospheric

15

budgets of trace gases and aerosols. Thinning of the stratospheric ozone layer yields a higher UV flux into the troposphere, thereby accelerating photooxidation processes.

Understanding the complex interactions between tropospheric chemistry and global change presents a formidable scientific challenge, which can be addressed only by close cooperation between scientific disciplines, tight interaction between observation and modeling, and broad international cooperation.

2.1   Introduction

In the public mind, "global change" has become almost synonymous with "global warming" or "climate change," a narrow reduction of the original meaning. Although there is no doubt that the possibility of climate change is of great concern to the Earth's population, we must not forget that we are living in a period when almost all components of the Earth System are undergoing change. The chemical composition of the atmosphere is being perturbed at a vast scale by human activities. The terrestrial biota are modified by land use change, biomass burning, deforestation, and species extinction. Marine life is impacted by overfishing, eutrophication, and pollution. There is a tendency to see these issues as independent environmental problems, each grabbing the public's attention for some time, and each demanding a specific solution.

This approach obscures the fact that all these phenomena are occurring simultaneously and within the same "Earth System." As a result, they interact with one another, reinforcing or damping each other, or changing each other's temporal evolution. This view is reflected in the "Bretherton" diagram (Figure 2.1), which shows the complex linkages between human activities, physical climate system, and biogeochemical cycles.

It is especially important to examine the Earth System for possible feedbacks, which amplify the effect of perturbations. It is well established that increasing temperatures result in changes in ice albedo, atmospheric water vapor content, and cloudiness, changes that in turn act to increase temperature. If additional positive feedbacks exist, they would add to known feedbacks and, because of the extremely nonlinear behavior at higher gains, could have disproportionately large effects (Lashof et al., 1997).

In this chapter, I explore some of these interactions among human activities, atmospheric chemistry, climate, and ecology, using selected examples or case studies. I proceed from the (relatively) simple to the more complex, keeping in mind that exploring any of the issues addressed here in its full depth and complexity is well beyond the frame of an overview chapter such as this.

2.2   The "Simplest" Case: Anthropogenic Halogenated Hydrocarbons

The clearest evidence of a global change in the Earth System is the changing composition of the atmosphere, particularly the increasing concentrations of some long-lived trace gases emitted by industrial activities. This was first documented for CO_2 by the long-term measurements made by C. D. Keeling at Mauna Loa Station on Hawaii (Bacastow et al., 1985; Keeling et al., 1995) and subsequently for numerous

CONCEPTUAL MODEL of Earth System process operating on timescales of decades to centuries

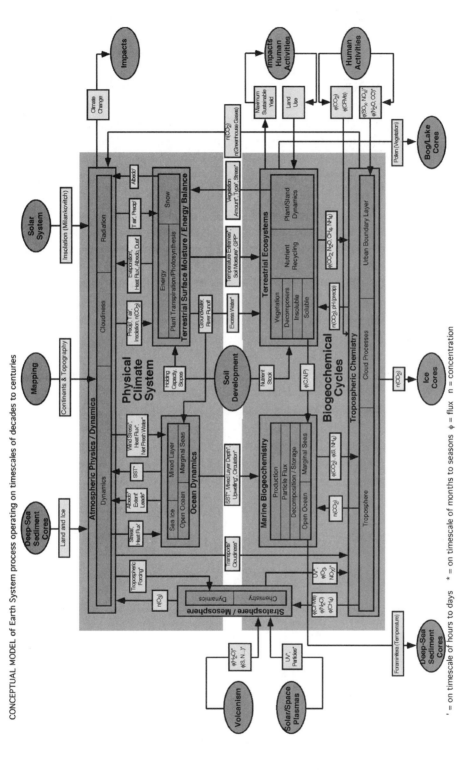

Figure 2.1. The "Bretherton" diagram. This version of the diagram has been produced by the Earth System Science Education Program, Universities Space Research Association, Whitelaw, Wisconsin. See color plate section for color version.

other trace gases such as methane (CH_4), nitrous oxide (N_2O), carbon monoxide (CO), and numerous halogenated hydrocarbons (Houghton et al., 1996, and references therein).

The latter class of compounds includes species such as the chlorofluorocarbons (CFCs, Freons), methyl chloroform (trichloroethane), and the partially halogenated chlorofluorocarbons (HCFCs), which are exclusively human-made and have no natural sources. Because they are produced industrially, often by relatively few manufacturers, accurate records exist of the amounts and times at which they were produced and released into the environment. Most anthropogenic halogenated hydrocarbons have no significant biological sinks and are resistant to hydrolysis in aquatic systems, so their only major sink is photochemical breakdown in the atmosphere. For most substances, this sink follows first-order kinetics, that is, its rate is a linear function of the trace gas concentration.

As a result of the well-characterized and simple source and sink functions of these gases, their concentration in the atmosphere as a function of space and time coordinates can be relatively easily understood and modeled. Figure 2.2 illustrates this behavior with the example of the temporal record of methyl chloroform, a substance that has no known natural sources and is removed almost exclusively by reaction with tropospheric OH. The use of methyl chloroform is severely restricted by the Montreal Protocol, and consequently its production declined sharply around 1990. This resulted in a reversal of its atmospheric trend in 1991, from an average increase of $4.5 \pm 0.1\%$/year to a decline of about 14%/year in 1995/1996. This behavior can be described in an atmospheric model and used to deduce both its weighted-mean atmospheric lifetime (4.8 years) and the global-mean OH concentration (Montzka et al., 1996; Prinn et al., 1995).

Although we have used these compounds as examples of the simplest case, with minimal feedback processes, we should note that even here some complications might arise. If any of these substances could change the stratospheric ozone density to such a degree that it would significantly alter the UV flux into the troposphere, and thus the tropospheric OH abundance, it could influence its own lifetime. Although it has been

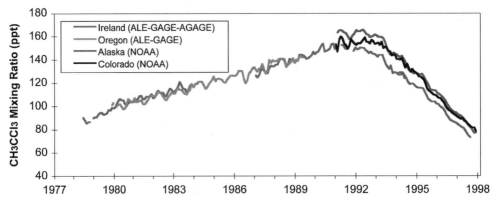

Figure 2.2. Temporal evolution of the atmospheric mixing ratio of methyl chloroform (CH_3CCl_3) in the troposphere over the Northern Hemisphere (from Kurylo et al., 1999). This version of the diagram has been produced by the Earth System Science Education Program, Universities Space Research Association, Whitelaw, Wisconsin. See also color plate section.

argued that significant changes in tropospheric OH have not occurred over the time scale represented in Figure 2.2 (Prinn et al., 1995), this issue is not without controversy (Krol et al., 1998). This issue is discussed in more detail below.

2.3   A More Complex Case: CO_2

In a way, CO_2 represents the opposite extreme from the preceding case: It is subject to strong and complex biogeochemical interactions, and its anthropogenic sources are only minor perturbations on the natural fluxes. The annual fluxes of CO_2 in and out of the terrestrial and marine biota make up some 150 Pg C per year (1 Pg $= 10^{15}$ g), whereas the emission from fossil fuel combustion and cement manufacturing accounts for "only" about 6 Pg year^{-1}. Yet it is this small increment added to the large biogenic fluxes of CO_2 that is responsible for most of the growth of CO_2 concentration in the atmosphere and for about half of the greenhouse gas effect. The rest of the atmospheric CO_2 increase is due to the effect of tropical deforestation, which moves carbon out of the "long-lived terrestrial biomass" reservoir into the atmosphere.

To understand and predict the atmospheric abundance of CO_2, we need a thorough understanding of all the complex biogeochemical interactions that control its transfer between the Earth's compartments, including the deep and shallow oceans, the marine and terrestrial biota, the sediments and soils, and so on. There are a large number of known feedbacks between climate and the cycles of carbon and the "nutrient" elements (N, P, and S), and quite likely an even greater number are still unexplored (see, for example, Lashof et al., 1997). Consequently, CO_2 is probably the most "interesting" trace gas to a biogeochemist. To an atmospheric chemist, however, it is "boring," because it does not undergo any relevant chemical reactions in the atmosphere. Therefore, I do not address the global carbon cycle in any detail in this chapter but restrict myself to these few short remarks.

It may be worthwhile, however, to emphasize one point here, to which we will come back several times in the following sections: the importance of the Tropics in understanding global change. The Tropics are the part of the globe with the most rapidly growing population, the most dramatic industrial expansion, and the most rapid and pervasive change in land use and land cover. The Tropics contain also the largest standing stocks of terrestrial vegetation and have the highest rates of photosynthesis and respiration (Houghton and Skole, 1990; Raich and Potter, 1995). It is therefore likely that changes in tropical land use will have a profound impact on the global carbon cycle in future decades (Houghton et al., 1998).

2.4   Trace Gases with Very Complex Source
and Sink Patterns: CH_4, N_2O

After CO_2, methane is the most important greenhouse gas, and there is unequivocal evidence that its atmospheric concentration is increasing because of human activities. In an effort to understand this increase, a large effort has gone into elucidating the budget of this trace gas.

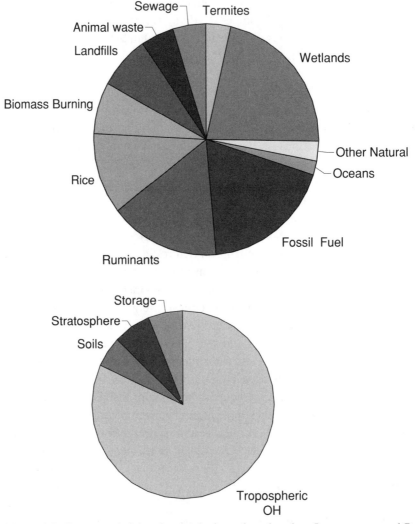

Figure 2.3. Sources and sinks of atmospheric methane based on Intergovernmental Panel on Climate Change (IPCC) 1995 assessment. See also color plate section.

Figure 2.3 shows the sources and sinks of methane in the form of pie diagrams. It is evident from this figure that a large variety of sources contribute to the atmospheric methane budget and that most of these sources are of comparable order of magnitude. The sources that are under human "control" (fossil fuel burning and handling, animal husbandry, rice agriculture, biomass burning, and waste disposal) make up almost three-quarters of the methane sources. In contrast to the halogenated hydrocarbons, however, which are produced industrially, most methane is produced biologically through microbial fermentation (with the exception of the fossil fuel source). Therefore, methane production is dominated by the harnessing of biological processes through human activity. For example, methane production from rice agriculture closely resembles that from natural wetlands, and the emissions from cattle are similar to those from the large herds of wild ruminants that graze on natural grasslands.

The diversity of sources of methane, and their biological nature, results in a complex source pattern, something that makes it quite difficult to assemble an accurate source inventory. There are a considerable number of different rice farming practices, for example, all of which result in substantially different amounts of methane emitted per amount of rice produced, or per hectare farmed. On the other hand, this diversity provides a chance for mitigation by the selection of practices that result in low methane emission factors. As a result, considerable effort has gone into refining our understanding of the processes that control methane emission from the various source types, and into producing more-accurate source estimates.

A very different picture prevails on the sink side of the methane budget. The term labeled "Storage" in Figure 2.2 represents the amount of methane accumulating annually in the atmosphere, a value that can be relatively easily and accurately obtained from the existing measurement network. It appears in the budget as a "sink," in the sense of being a reservoir where some of the methane introduced into the atmosphere ends up, at least for some years. The true sinks – the processes that remove methane from the atmosphere – are dominated by one process, the tropospheric oxidation of methane through its reaction with the hydroxyl radical OH. This sink is so large that even a relatively small uncertainty in its magnitude, some 10%–20%, is as large as many of the individual methane sources.

It may seem, at first glance, that this sink is rather well known and predictable. The global mean OH concentration can be derived from analysis of the distribution in time and space of compounds such as methyl chloroform, as we have discussed above. The reaction rate constant of methane with OH and the global concentration distribution of methane are also well known. Using present-day measured CH_4 and model-predicted OH concentrations, we obtain a photochemical sink that is of the right size to balance the methane budget. Yet how well can we extrapolate this knowledge back into the past, or, more importantly, forward into the future?

Model calculations suggest that tropospheric OH concentrations have decreased by some 25% since the industrial revolution (Crutzen, 1995a). This value is, however, highly uncertain, and other models have predicted much smaller or much larger changes (Thompson, 1992). Furthermore, there is an ongoing controversy about changes in tropospheric OH over the past few decades (Krol et al., 1998; Prinn et al., 1995). The outcome of this discussion has important implications for our understanding of the current methane budget. If the methane lifetime had not been significantly affected by likely changes in OH, the decreasing rate at which methane accumulates in the atmosphere would be consistent with methane sources having reached a plateau, and the atmospheric methane concentration approaching a new steady state (Dlugokencky et al., 1998). If, on the other hand, OH had been growing significantly over the past decade, we would have to assume that methane sources also are currently increasing.

To look into the future, we must remember that most methane oxidation takes place in the Tropics because of the high concentrations of OH resulting from high amounts of water vapor and high UV flux in that region (Andreae and Crutzen, 1997). The Tropics are also the world's most rapidly changing region. Deforestation of the Amazon Basin and subsequent agricultural and industrial development are likely to substantially change the amounts of hydrocarbons and nitrogen oxides released into the tropical atmosphere, resulting in elevated ozone over the region. At the same time, deforestation would change the regional water balance, including the atmospheric water vapor content. Because ozone and water vapor are the precursors of the OH radical, these changes must be expected to have a pronounced influence on OH and consequently on the lifetimes of CH_4, CO, and all the other atmospheric trace gases that are being removed by reaction with OH. Changes in cloudiness resulting from a perturbation of the hydrological cycle in the humid tropics would also impact the OH distribution because of the radiative and chemical effects of clouds on this radical (Mauldin et al., 1997).

The challenge of predicting future CH_4 levels is further complicated by the fact that atmospheric CH_4, CO, and OH are part of a coupled chemical reaction scheme, with complex, nonlinear behavior resulting from simple perturbations (Prather, 1996). Adding CO to this system actually leads to increased CH_4 concentrations, and the couplings and feedbacks in the system result in effects that take longer to decay than the lifetimes of the individual molecules involved. The nonlinearities in this system increase with the concentrations of methane (and CO present in the atmosphere); and at methane source fluxes around three times the present size, runaway growth of methane could occur.

When the possibilities of additional feedbacks with climate and biota are considered, even more complex feedbacks can be expected. The warming resulting from the greenhouse effect may release additional methane from wetlands due to enhanced microbial activity at elevated temperatures, at least initially (Cao et al., 1998; Chapman and Thurlow, 1996; Christensen and Cox, 1995; Lashof et al., 1997; Oechel and Vourlitis, 1994). At longer time scales, effects of changing water table levels and soil moisture content resulting from climate change may reverse the direction of this feedback. Global warming may also liberate methane locked into clathrates in continental slope sediments and permafrost (Harvey and Huang, 1995). This effect may be counteracted in part by stabilization of clathrates due to increased pressure resulting from rising sea levels (Gornitz and Fung, 1994). The overall impact of this feedback process is difficult to assess because of great uncertainties about the amounts of CH_4 present in clathrates, but it is thought to be important mostly in the more distant future (beyond the 21st century) and for high climate sensitivities.

2.5   Indirect Sources and Sinks of Climatically Active Gases: CO, O_3

In the preceding section, we pointed out that gases that are not themselves greenhouse gases may have a climatic effect because they change the rates of production or destruction of greenhouse gases. In this sense, we can attribute a climate forcing and greenhouse warming potential to gases such as CO, which has no significant radiative effect of its own. This is because adding CO to the atmosphere increases the lifetime and abundance of methane, results in the production of ozone, and, following oxidation, adds some CO_2 to the atmosphere. When these effects were simulated in a photochemical model, the cumulative radiative forcing due to CO emissions exceeded at shorter time scales (<15 years) that due to anthropogenic N_2O, one of the important greenhouse gases (Daniel and Solomon, 1998).

Tropospheric ozone, a gas that has no direct emission sources, is the third most important greenhouse gas after CO_2 and CH_4 (Houghton et al., 1996; Portmann et al., 1997; Roelofs et al., 1997; Shine and Forster, 1999; van Dorland et al., 1997). Because the chemical lifetime of ozone in the troposphere is of the same order as the time scales of many atmospheric transport processes (days to weeks), its temporal and spatial distribution is highly inhomogeneous. In the absence of vertically resolved and globally representative data sets on O_3 concentrations, the climatic effect due to this gas must therefore be estimated based on model calculations. The chemical precursors of ozone are hydrocarbons (including methane and NMHC), CO, and the oxides of nitrogen, NO_x. The latter play an especially important role in the ozone budget, because their abundance determines whether the photochemical oxidation of hydrocarbons and CO results in net O_3 production or destruction (Crutzen, 1995b; National Research Council [U.S.] Committee on Tropospheric Ozone Formation and Measurement, 1991).

In many regions of the Earth, especially on the continents, biogenic NMHC emissions are relatively abundant (Fehsenfeld et al., 1992; Guenther et al., 1995), and, in the absence of strong NO_x emissions, their photooxidation results in net O_3 destruction. When NO_x emissions in these regions increase due to development or because deforestation lets NO_x from soil microbial production escape more readily into the troposphere, the system can switch to net O_3 production, strongly enhancing ozone levels (Keller et al., 1991). This is especially critical in the Tropics, where O_3 can be entrained into the intertropical convergence zone (ITCZ) and transported by deep convection into the upper troposphere, where it has the strongest climatic effect. Modeling studies suggest that input of pollutants into convective regions may have strong effects on O_3 levels in the free troposphere (Ellis et al., 1996).

Figure 2.4 shows the sources of nitrogen oxides in the troposphere. Of particular importance to the tropical atmosphere are the emissions from biomass burning, most

NO_x Emissions in Tg N(NO)/yr

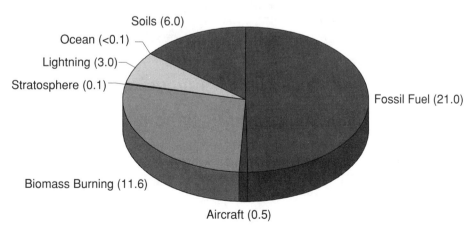

Soils (6.0)
Ocean (<0.1)
Lightning (3.0)
Stratosphere (0.1)
Fossil Fuel (21.0)
Biomass Burning (11.6)
Aircraft (0.5)

Figure 2.4. Sources of nitrogen oxides (NO_x) to the troposphere (data from Wang et al., 1998). See also color plate section.

of which takes place in the Tropics (Andreae, 1993), and the production of NO_x by lightning, which is also abundant in the deep convective thunderstorms of the ITCZ. Because vegetation fires can occur only when the vegetation is dry enough to burn, they are most abundant in the dry season, when the trade wind inversion with its large-scale subsidence prevails over the part of the Tropics in question. Because this inversion prevents convection to heights of more than a few kilometers, it was initially thought that the linkage between dry conditions and subsidence more or less precluded the transport of pyrogenic ozone precursors to the middle and upper troposphere. Recent work has shown, however, that large amounts of smoke can get swept by low-level circulation, such as the trade winds, toward convergent regions over the continents or the ITCZ, and there become subject to deep convection (Andreae et al., 1999; Chatfield et al., 1996; Thompson et al., 1996). This transport pattern can explain the abundance of fire-related O_3 and O_3 precursors observed in the middle and upper troposphere by remote sensing and in situ measurements (Browell et al., 1996; Connors et al., 1996; Olson et al., 1996). Figure 2.5 shows the distribution of O_3 over the tropical South Atlantic during September–October 1992 in comparison with results from earlier studies (DECAFE-88 in the Congo, Andreae et al., 1992; Tropical Atlantic, Kirchhoff et al., 1991) and the ozone climatology over the Pacific Ocean. These results show dramatically the impact that O_3 from biomass burning can have on the entire tropospheric column.

Whether this impact will grow in the future depends both on climate change and on human factors. The amount of fuel available at a given place for burning is a function of ecological factors, such as soil fertility, precipitation, and temperature. It also depends on land use: whether the area has been burned previously, is used for grazing or agriculture, and so on. If climatic variations become more extreme, as climate models have suggested, we can expect a more frequent occurrence of drought years following very wet years. This would result in large amounts of fuel ready to burn in the fire season. Furthermore, in a warmer climate, fire frequency is likely to increase. That would reduce biomass

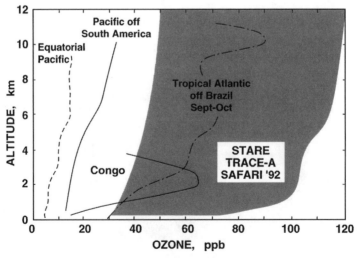

Figure 2.5. Impact of tropical biomass burning on the vertical distribution of ozone over the Equatorial ocean regions. See also color plate section.

carbon storage by changing the age class structure of vegetation as well as cause increased emissions of ozone precursors.

Human activities are of central importance to the frequency and severity of biomass fires. If large parts of the humid Tropics are further deforested, they will transition from a biome essentially free of fires (the tropical rainforest) to biomes with much more frequent fires (grazing lands, agricultural lands, and wastelands). With a higher human population density, the frequency of ignition will also go up. And finally, the amount of biomass burned for cooking and domestic heating, already a major source of emissions in tropical countries, will increase further.

2.6   Aerosols: Complex Spatiotemporal Distributions and Radiative Interactions

Over the past decade, a growing amount of attention has been focused on the climatic effects of atmospheric aerosols. Initially stimulated by a discussion on the climate effects of natural and anthropogenic sulfate aerosols (Charlson et al., 1987; Charlson et al., 1992; Schwartz, 1988), research now spans almost all aerosol types and source mechanisms. Recent reviews can be found in a number of books and articles (Andreae, 1995; Charlson and Heintzenberg, 1995; Houghton et al., 1996; Shine and Forster, 1999). In this chapter, there is not enough space to provide an exhaustive review of all the recent exciting developments in this rapidly expanding field, and I limit the discussion to a few less frequently addressed issues: the role of mineral dust, the function of organic aerosols from biogenic precursors, the link between stratospheric ozone and biogenic sulfate, and the influence of aerosols on smog chemistry.

A few key points must be made to put the interactions among aerosols, climate, and biota into perspective. First, there is no clear distinction between anthropogenic and natural sources. Like O_3, aerosols form in the atmosphere from precursor substances, with the rates of production depending simultaneously on the concentrations of several precursor molecules, most of which could be either biogenic or anthropogenic. Human perturbations can increase or decrease the yields of aerosols from natural precursors, often in surprising ways, as we discuss below.

Second, aerosols interact with climate in much more complex ways than do gaseous molecules. In addition to being able to absorb light (and thereby warm the atmosphere), aerosols can scatter light back into space or enhance the backscattering of light by clouds, something that cools the Earth. Aerosols can also reduce precipitation from clouds, and that enhances clouds' lifetime (a cooling effect). Or, if aerosols absorb radiation and warm an atmospheric layer, that may reduce cloud formation, something that would warm the Earth. Because particles in the atmosphere are created and removed at time scales of days or less, they are very unevenly distributed and cannot be adequately represented by global means, as can the long-lived greenhouse gases. As a consequence of this complex interaction, aerosol effects on climate are usually calculated using three-dimensional climate models, which attempt to include the inhomogeneous distribution of aerosol in time and space. The difficulty both of correctly representing the aerosol distributions in such models and of adequately representing and characterizing the atmospheric physics involved is reflected in the large differences between predictions of aerosol radiative forcing from different models, often as large as a factor of 2 or 3 (Shine and Forster, 1999). This compares with differences of about 7%–10% for the forcing estimates for the well-mixed greenhouse gases. Overall, the cooling effect due to aerosols is considered to be roughly about 50%–100% of the warming effect of the greenhouse gases (Houghton et al., 1996; Shine and Forster, 1999).

Third, aerosols are chemically just as complex as are the gaseous constituents of the atmosphere. This point applies obviously to the organic aerosol, which makes up a substantial fraction of atmospheric particles, but is also true for other aerosol components. It is entirely unrealistic to treat aerosols as simple "pure" compounds, such as ammonium sulfate.

Finally, we must move away from treating aerosols as largely inert products of chemical processes, products that play no important "active" role in atmospheric chemistry. Recent work has shown that reactions in and on aerosols may play an important role in the halogen and sulfur budgets of the atmosphere (Andreae and Crutzen, 1997). Scattering and absorption of UV radiation by aerosols can influence "smog" chemistry in polluted atmospheres (Dickerson et al., 1997). Furthermore, modifications of the optical and chemical characteristics of clouds may have an effect on OH concentrations (Mauldin et al., 1997). In the following paragraphs I illustrate some of these issues.

At first glance, it may be surprising that human perturbations of the atmospheric aerosol load could be significant enough to perturb climate, given that only some 11% of the global aerosol emissions are estimated to come from anthropogenic sources. The production of soil dust and sea spray aerosol, on the other hand, accounts for about 80% of the global source strength (Andreae, 1995). This view, which has been used to argue against a potential influence from anthropogenic aerosols on climate, has several flaws. First, it is not the source strength that is relevant to the climate effect, but rather it is the amounts present in the atmosphere at any given time, the atmospheric burden. Because seasalt aerosol and dust consist mostly of coarse particles, which are rapidly deposited, their share of the burden (68%) is substantially reduced as compared with other sources. Second, it appears that about half of the soil dust aerosol is mobilized as a result of human disturbance of soils and can therefore be considered "anthropogenic" (Tegen and Fung, 1994; Tegen and Fung, 1995; Tegen et al., 1996). When this factor is taken into account, we find that about half of the global aerosol burden is the result of human activity (Figure 2.6), or, in other words, that humans have approximately doubled the aerosol load of the atmosphere.

The case for a substantial effect of anthropogenic aerosols on climate becomes even stronger when we consider the way aerosols interact with the flux of radiation through the atmosphere. We distinguish two basic mechanisms: the scattering and absorption of radiation by the aerosol particles themselves ("direct effect") and the scattering of light by clouds, which can be modified by variations in the concentration of cloud condensation nuclei ("indirect effect"). Light scattering by aerosols is strongly size-dependent, with a maximum effect when the size of the particle and the wavelength of the scattered light are of the same order. For this reason, the submicron sulfate, organic, and smoke aerosols from SO_2 emission and biomass burning have stronger radiative

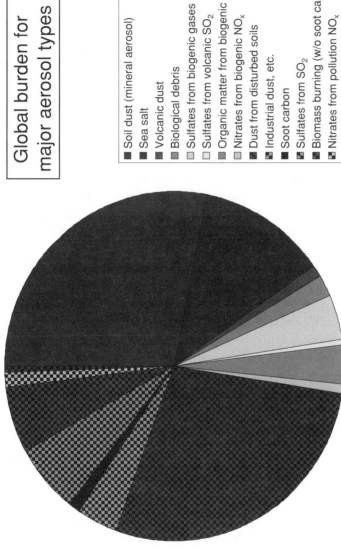

Global burden for major aerosol types

■ Soil dust (mineral aerosol)
■ Sea salt
■ Volcanic dust
■ Biological debris
□ Sulfates from biogenic gases
□ Sulfates from volcanic SO_2
■ Organic matter from biogenic VOC
■ Nitrates from biogenic NO_x
■ Dust from disturbed soils
■ Industrial dust, etc.
■ Soot carbon
■ Sulfates from SO_2
■ Biomass burning (w/o soot carbon)
■ Nitrates from pollution NO_x
■ Organic from anthropogenic VOC

Figure 2.6. Global atmospheric aerosol burden resulting from various sources. See also color plate section.

effects than the more abundant soil dust and seasalt aerosols (Figure 2.7) (Andreae, 1995). As a result, whereas the anthropogenic fraction of the aerosol burden is about 50%, the anthropogenic share of the radiative effect is higher, about 60%.

The fact that shortwave light absorption by dark aerosols (soot) and longwave absorption by silicate minerals act in a warming direction – the opposite of the cooling effect of light scattering – makes the assessment of the net climate effect very difficult. Current estimates suggest that the cooling effect predominates, with a global mean forcing of ~-0.4 W m^{-2}, but with an uncertainty of about ± 0.8 W m^{-2} (Shine and Forster, 1999).

The indirect effect by means of enhancement of cloud albedo resulting from increased cloud condensation nuclei (CCN) concentrations is related to the number rather than the mass concentration of aerosols. Here again, a very rough estimate based on the burden of fine aerosols, which account for most of the CCN burden, suggests that the introduction of anthropogenic particles has more than doubled the amount of CCN in the atmosphere (Andreae, 1995). Because the effect of added CCN is very sensitive to the number of CCN already present and to the type of cloud into which the CCN are introduced (Twomey, 1977), knowledge of the spatiotemporal distribution of CCN sources is critical to an assessment of their effect.

Because the strongest climate effects from increased CCN concentrations are expected for clouds of intermediate optical thickness and low initial (natural) CCN concentrations, the regions of most concern used to be the large areas of marine stratus in the eastern parts of the ocean basins (Charlson et al., 1987). Continental clouds were not thought to be very susceptible to the indirect effect, because it was assumed that continental air had high natural CCN levels. This assumption may have to change in light of recent observations of very low CCN concentrations over Amazonia in the wet season (Roberts et al., 1998). If these measurements prove representative of CCN levels over the tropical continents, they imply that deep convection and rain formation in these regions are occurring naturally at very low cloud droplet number concentrations (CDNC), resulting in very high precipitation probability. Given the low natural CCN concentrations, it would not require very high amounts of anthropogenic emissions to significantly increase CDNC, which would change the rainout efficiency ("overseeding") and could lead to significant changes in the regional water cycle. This could even influence the water vapor content of the tropical troposphere as well as the energy transfer processes in the tropical Hadley cell. The complexity of ice formation in clouds and our sparse knowledge about identity and sources of ice-nucleating particles in clouds further complicate an assessment of the human impact on tropical clouds (Baker, 1997).

So far, we have ignored interactions between human activities and the rate of production of "natural" aerosols. However, this may not be appropriate in a number of instances. Consider the production of sulfate aerosols from marine biogenic dimethyl sulfide (DMS), which has been proposed as the main source for CCN in pristine marine regions (Charlson et al., 1987). An important step in this process is the production of new particles (nucleation) from the gaseous precursor, H_2SO_4. The rate at which this occurs depends on the concentration of gaseous H_2SO_4 molecules, which in turn

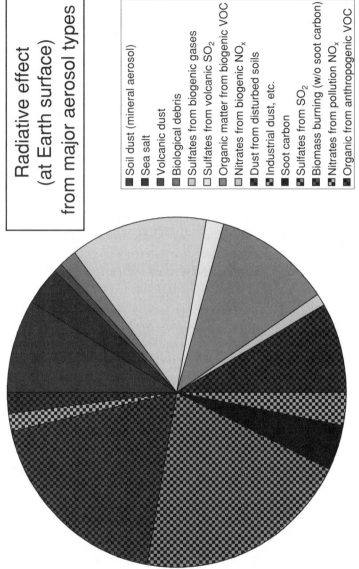

**Radiative effect
(at Earth surface)
from major aerosol types**

- Soil dust (mineral aerosol)
- Sea salt
- Volcanic dust
- Biological debris
- Sulfates from biogenic gases
- Sulfates from volcanic SO_2
- Organic matter from biogenic VOC
- Nitrates from biogenic NO_x
- Dust from disturbed soils
- Industrial dust, etc.
- Soot carbon
- Sulfates from SO_2
- Biomass burning (w/o soot carbon)
- Nitrates from pollution NO_x
- Organic from anthropogenic VOC

Figure 2.7. Extinction of visible radiation at the Earth's surface resulting from the various aerosol types. See also color plate section.

depends on their rate of production from the reaction between SO_2 and OH. Any perturbation of the atmosphere that changes the OH levels in the marine boundary layer may therefore interfere with the rate of new particle and CCN production and have an influence on climate.

One such perturbation may be the increased UV flux that reaches the Earth's surface as a result of the thinning of the stratospheric UV layer (Tang and Madronich, 1995). Toumi et al. (1994) proposed that stratospheric ozone loss from increasing halogen levels may have led to a 3% increase in OH concentration during the 1980s, something that would have resulted in an increase in the production rate of CCN. However, the gaps in our understanding of the mechanisms of CCN production and the climate effects of changing CCN are so large that the uncertainty in the predicted climate effect ranges from an insignificant value to one that would more than compensate for the increased greenhouse forcing over the same period. It may be interesting to explore what changes in the OH concentration in the marine atmosphere have occurred and may still occur as a result of increasing CH_4, CO, and NO_x levels, and what effects those changes may have on CCN concentrations and climate.

An even more striking example of an impact of anthropogenic activities on the production of "natural" aerosols is the oxidation of biogenic volatile organic carbon (VOC) compounds, particularly terpenes, to low-volatility compounds, which condense into aerosol particles (Kanakidou, 1998). The reaction mechanisms and products depend on the chemical environment in which the oxidation reactions occur. In the presence of elevated levels of NO_x, ozone is formed, and it reacts with terpenes very rapidly to form products of low volatility and high aerosol yield (Bowman et al., 1995; Hoffmann et al., 1997; Seinfeld and Pandis, 1998). At low NO_x levels, on the other hand, ozone production is low and terpenes are oxidized predominantly via attack by OH, with lower aerosol yields. Recent estimates suggest that more than 1000 Tg C are emitted annually in the form of biogenic VOC, of which maybe 30% are potential aerosol precursor substances (Guenther et al., 1995). In view of the vast amounts of VOC emitted from vegetation, even small changes in the aerosol production efficiency result in major perturbations of the atmospheric aerosol budget (Andreae and Crutzen, 1997). Kanakidou (1998) has estimated that as much as 80% of the global organic aerosol production may be the indirect result of human impacts on atmospheric chemistry.

The final point I would like to address in this section is the influence of aerosols on gas phase chemistry. The potential of seasalt aerosols to act as a source of reactive gaseous halogen species in the marine boundary layer is now well recognized (Andreae and Crutzen, 1997; Keene et al., 1998; Sander and Crutzen, 1996; Sander et al., 1997; Vogt et al., 1996). This process can be responsible for significant rates of photochemical hydrocarbon oxidation, O_3 destruction, and other reactions in the marine boundary layer (MBL) in addition to OH-based chemistry. Mineral dust in the atmosphere is also far from being an inert substance. It can act as a sink for acidic trace gases, such as SO_2 and HNO_3, and thereby interact with the sulfur and nitrogen cycles (Dentener et al., 1996; Li-Jones and Prospero, 1998; Talbot et al., 1986). Coatings with soluble substances, such as sulfate or nitrate, change the ability of mineral dust aerosols to nucleate cloud droplets (Levin et al., 1996).

In addition to direct chemical interactions, aerosol particles can influence atmospheric chemistry by modifying the UV radiation field. Submicron particles scatter UV radiation very efficiently, and the aerosol loadings present in polluted environments can easily triple or even quadruple the turbidity of the atmosphere at UV wavelengths. This increases the effective path length of the photons in a smoggy environment, and it can thereby enhance the rates of photochemical reactions, including ozone formation (Dickerson et al., 1997). There may be even an analog of the "indirect" climate effect: As variations in CCN change cloud radiative properties, they may also influence the actinic flux in and above clouds and thereby influence the production of the OH radical (Madronich, 1987; Mauldin et al., 1997).

2.7   Climate-Chemistry Feedbacks and the Arctic "Ozone Hole"

In the 1990s, ozone loss in the Arctic stratosphere during the polar sunrise period accelerated dramatically, peaking in 1997 with an ozone loss of $\sim$100–120 Dobson unit (DU) (Müller et al., 1997; Newman et al., 1997). This decrease is comparable to the ozone loss over Antarctica in 1985, at the time when the Antarctic "ozone hole" was first reported (Farman et al., 1985). At the same time, temperatures in the polar vortex dropped to lower temperatures and remained there over longer periods than before. These observations are closely connected: For the development of the chemical reaction sequence that leads to rapid ozone loss, it is necessary that temperatures fall low enough for polar stratospheric clouds (PSC) to form. These clouds must persist long enough for the reactions that regenerate active chlorine species (Cl_2, Cl, ClO) from the inactive reservoir species ($ClONO_2$, HCl), and for HNO_3-rich PSC particles to settle out of the stratosphere. After ozone loss has taken place, the lesser amounts of O_3 present result in less absorption of UV radiation in the stratosphere, and consequently in lower temperatures, which in turn promote ozone loss. Consequently, stratospheric temperature change, PSC formation, and ozone depletion form a feedback system with positive gain, mutually reinforcing one another (Danilin et al., 1998; Portmann et al., 1996).

The situation is further complicated by the fact that the build-up of greenhouse gases in the atmosphere, while it warms the lower atmosphere, actually cools the stratosphere (Ramaswamy and Bowen, 1994). The cooling of the Arctic stratosphere during the past decade, which made rapid ozone loss in the polar region possible, can thus be attributed to three causes: the ozone loss itself through the feedback described in the preceding paragraph; the radiative forcing due to the greenhouse gases; or unrelated fluctuations in the climate system (or a combination of these factors). Which of these mechanisms dominates is crucial to the time scale at which we can anticipate the recovery from low ozone conditions over the Arctic. If the ozone-temperature feedback dominates, recovery will occur approximately at the time scale at which chlorine concentrations in the stratosphere return to levels below those that cause rapid ozone loss. If, on the other hand, cooling was mostly caused by the effect of CO_2 and other greenhouse gases, the temperature/ozone-loss feedback will persist much longer ("slow" recovery). This is because less stratospheric chlorine is required to cause rapid ozone loss at the lower temperatures that would prevail as long as greenhouse gas concentrations remain elevated. If, finally, the observed cooling were mostly caused by unrelated climate fluctuations, recovery would be unpredictable altogether. Injection of large amounts of volcanic aerosol into the stratosphere, another unpredictable "external" forcing, would also act to accelerate O_3 loss and delay recovery (Portmann et al., 1996).

Recent modeling studies suggest that all these mechanisms in fact contributed to the development of an Arctic ozone hole in the 1990s, with the largest effects resulting from the cooling driven by ozone loss interacting with a natural mode of variability (Graf et al., 1998; Shindell et al., 1998). This mode links a strengthened polar night vortex with an enhanced North Atlantic oscillation (Graf et al., 1998). Shindell et al. (1998) have proposed a mechanism by which anthropogenic climate change might be coupled to a strengthened polar vortex. They find that in their model simulations the changes in temperature and winds resulting from increased greenhouse gas concentrations alter the propagation of planetary waves. As a result, planetary waves break up the Arctic polar vortex less frequently, and that leads to significantly colder temperatures existing over longer periods of time in the Arctic stratosphere. These authors estimate that because of this effect the ozone loss over the Arctic by the year 2020 will be double what it would be without greenhouse gas increases, and that recovery from Arctic ozone depletion will be delayed by some 10–15 years.

2.8   Conclusion

This chapter examines the linkages and connections between human perturbations of the Earth System and its chemical, physical, and ecological characteristics. We have seen that it is usually not adequate to consider only the emission of trace gases and aerosols; it is essential also to consider the complex interconnections between any given perturbation and the overall Earth System.

In a few cases, it is relatively easy to assess the impact of anthropogenic emissions on the atmosphere – for example, when the sources of a substance are industrial and its sinks are chemical reactions with first-order kinetics. But in most instances, the emission and removal of climatically active gases and aerosols depend on a multiplicity of human activities and ecological factors, including climate itself.

When land use and agricultural practices change, the emissions of trace gases such as N_2O, NO, and CH_4 change in highly complex ways, which are extremely difficult to assess at the scales of interest. When land use change reaches such vast extent as in the deforestation of the Tropics, it may even cause changes in the climate system, including the hydrological cycle. As a result of these chemical and physical perturbations, the chemical functioning of the atmosphere, and consequently the production rates and lifetimes of aerosols and greenhouse gases, will be modified. The most obvious example for such a mechanism is the large-scale change of trace gas inputs into the tropical troposphere, the vast photochemical reactor where most of the photooxidation of long-lived trace gases takes place. Because of the long time scales involved in ecological change, biogeochemical cycles and climate have a memory of past land use change, and, conversely, current land use change may have long-term consequences reaching far into the future.

In some cases, the human perturbation consists of the release of a precursor compound (e.g., SO_2), which is transformed in the atmosphere to a climatically active substance. In this example, the actual amount of radiatively active sulfate aerosol produced is determined by a complex interplay of atmospheric transport processes, chemical processes in the gas phase, and interactions with other aerosol species. In other cases, such as the production of organic aerosols from biogenic VOCs or sulfate aerosols from DMS, aerosol yields can be modified by anthropogenic changes in atmospheric photooxidation processes.

At longer time scales, we must consider feedback loops in which climate change results in different circulation patterns, changes in water availability at the surface, changes in water vapor content of the atmosphere, and so on. These factors in turn modify the sources, sinks, and atmospheric budgets of trace gases and aerosols, again affecting climate. A dramatic example of this kind of interaction is the coupling between changes in stratospheric temperatures and ozone depletion, something that has shown up over the Arctic during the last decade.

Understanding the complex interactions between tropospheric chemistry and global change presents a formidable scientific challenge. Exciting progress has been made in this area, especially over the past decade, by intensified cooperation between scientific disciplines, close interaction between observation and modeling, and broad international cooperation. However, in our excitement about new conceptual insights into the complexity of the Earth System's workings, we must not lose sight of the fact that the observational database for testing our concepts and models remains rather sparse. High priority must therefore be given to developing new tools and programs for the investigation of our changing planet.

REFERENCES

Andreae, M. O. (1993) The influence of tropical biomass burning on climate and the atmospheric environment. In R. S. Oremland, Ed., Biogeochemistry of Global Change: Radiatively Active Trace Gases, pp. 113–150. Chapman & Hall, New York, NY.

Andreae, M. O. (1995) Climatic effects of changing atmospheric aerosol levels. In A. Henderson-Sellers, Ed. World Survey of Climatology. Vol. 16: Future Climates of the World, pp. 341–392. Elsevier, Amsterdam.

Andreae, M. O., Artaxo, P., Fischer, H., Fortuin, J., Gregoire, J., Hoor, P., Kormann, R., Krejci, R., Lange, L., Lelieveld, J., Longo, K., Peters, W., Reus, M. D., Scheeren, B., Silva Dias, M. D., Strøm, J., and Williams, J. (1999) Transport of biomass burning smoke to the upper troposphere by deep convection in the equatorial region. Nature.

Andreae, M. O., Chapuis, A., Cros, B., Fontan, J., Helas, G., Justice, C., Kaufman, Y. J., Minga, A., and Nganga, D. (1992) Ozone and Aitken nuclei over equatorial Africa: Airborne observations during DECAFE 88. J. Geophys. Res., 97, 6137–6148.

Andreae, M. O., and Crutzen, P. J. (1997) Atmospheric aerosols: Biogeochemical sources and role in atmospheric chemistry. Science, 276, 1052–1056.

Bacastow, R. B., Keeling, C. D., and Whorf, T. P. (1985) Seasonal amplitude increase in atmospheric CO_2 concentration at Mauna Loa, Hawaii, 1959–1982. J. Geophys. Res., 90, 529–540.

Baker, M. B. (1997) Cloud microphysics and climate. Science, 276, 1072–1078.

Bowman, F. M., Pilinis, C., and Seinfeld, J. H. (1995) Ozone and aerosol productivity of reactive organics. Atmos. Environ., 29, 579–589.

Browell, E. V., Fenn, M. A., Butler, C. F., Grant, W. B., Clayton, M. E., Fishman, J., Bachmeier, A. S., Anderson, B. E., Gregory, G. L., Fuelberg, H. E., Bradshaw, J. D., Sandholm, S. T., Blake, D. R., Heikes, B. G., Sachse, G. W., Singh, H. B., and Talbot, R. W. (1996) Ozone and aerosol distributions and air mass characteristics over the South Atlantic basin during the burning season. J. Geophys. Res., 101, 24,043–24,068.

Cao, M.-K., Gregson, K., and Marshall, S. (1998) Global methane emission from wetlands and its sensitivity to climate change. Atmos. Environ., 32, 3293–3299.

Chapman, S. J., and Thurlow, M. (1996) The influence of climate on CO_2 and CH_4 emissions from organic soils. Agricultural and Forest Meteorology, 79, 205–217.

Charlson, R. J., and Heintzenberg, J. (1995) Aerosol Forcing of Climate, p. 416. Wiley, Chichester, New York.

Charlson, R. J., Lovelock, J. E., Andreae, M. O., and Warren, S. G. (1987) Oceanic phytoplankton, atmospheric sulphur, cloud albedo, and climate. Nature, 326, 655–661.

Charlson, R. J., Schwartz, S. E., Hales, J. M., Cess, R. D., Coakley, J. A., Hansen, J. E., and Hofmann, D. J. (1992) Climate forcing by anthropogenic aerosols. Science, 255, 423–430.

Chatfield, R. B., Vastano, J. A., Singh, H. B., and Sachse, G. (1996) A general model of how fire emissions and chemistry produce African/Oceanic plumes (O_3, CO, PAN, smoke) in TRACE-A. J. Geophys. Res., 101, 24,279–24,306.

Christensen, T. R., and Cox, P. (1995) Response of methane emission from arctic tundra to climatic change – Results from a model simulation. Tellus Series B-Chemical and Physical Meteorology, 47, 301–309.

Connors, V. S., Flood, M., Jones, T., Gormsen, B., Nolf, S., and Reichle, H. G., Jr. (1996) Global distribution of biomass burning and carbon monoxide in the middle troposphere during early April and October 1994. In J. S. Levine, Ed. Biomass Burning and Global Change, pp. 99–106. MIT Press, Cambridge, MA.

Crutzen, P. J. (1995a) Overview of tropospheric chemistry: Developments during the past quarter century and a look ahead. Faraday Discuss., 100, 1–21.

Crutzen, P. J. (1995b) Ozone in the troposphere. In H. B. Singh, Ed. Composition, Chemistry, and Climate of the Atmosphere, pp. 349–393. Van Nostrand Reinhold, New York.

Daniel, J. S., and Solomon, S. (1998) On the climate forcing of carbon monoxide. J. Geophys. Res., 103, 13,249–13,260.

Danilin, M. Y., Sze, N. D., Ko, M. K. W., Rodriguez, J. M., and Tabazadeh, A. (1998) Stratospheric cooling and Arctic ozone recovery. Geophys. Res. Lett., 25, 2141–2144.

Dentener, F. J., Carmichael, G. R., Zhang, Y., Lelieveld, J., and Crutzen, P. J. (1996) Role of mineral aerosol as a reactive surface in the global troposphere. J. Geophys. Res., 101, 22,869–22,889.

Dickerson, R. R., Kondragunta, S., Stenchikov, G., Civerolo, K. L., Doddridge, B. G., and Holben, B. N. (1997) The impact of aerosols on solar ultraviolet radiation and photochemical smog. Science, 278, 827–830.

Dlugokencky, E. J., Masarie, K. A., Lang, P. M., and Tans, P. P. (1998) Continuing decline in the growth rate of the atmospheric methane burden. Nature, 393, 447–450.

Ellis, W. G., Jr., Thompson, A. M., Kondragunta, S., Pickering, K. E., Stenchikov, G., Dickerson, R. R., and Tao, W. K. (1996) Potential ozone production following convective transport based on future emission scenarios. Atmos. Environ., 30, 667–672.

Farman, J. C., Gardiner, B. G., and Shanklin, J. D. (1985) Large losses of total ozone in Antarctica reveal seasonal ClO_x/NO_x interaction. Nature, 315, 207–210.

Fehsenfeld, F., Calvert, J., Fall, R., Goldan, P., Guenther, A. B., Hewitt, C. N., Lamb, B., Liu, S., Trainer, M., Westberg, H., and Zimmerman, P. (1992) Emissions of volatile organic compounds from vegetation and the implications for atmospheric chemistry. Global Biogeochem. Cycles, 6, 389–430.

Gornitz, V., and Fung, I. (1994) Potential distribution of methane hydrates in the world's oceans. Global Biogeochem. Cycles, 8, 335–347.

Graf, H. F., Kirchner, I., and Perlwitz, J. (1998) Changing lower stratospheric circulation: The role of ozone and greenhouse gases. J. Geophys. Res., 103, 11,251–11,261.

Guenther, A., Hewitt, C. N., Erickson, D., Fall, R., Geron, C., Graedel, T., Harley, P., Klinger, L., Lerdau, M., McKay, W. A., Pierce, T., Scholes, B., Steinbrecher, R., Tallamraju, R., Taylor, J., and Zimmerman, P. (1995) A global model of natural volatile organic compound emissions. J. Geophys. Res., 100, 8873–8892.

Harvey, L. D. D., and Huang, Z. (1995) Evaluation of the potential impact of methane clathrate destabilization on future global warming. J. Geophys. Res., 100, 2905–2926.

Hoffmann, T., Odum, J. R., Bowman, F., Collins, D., Klockow, D., Flagan, R. C., and Seinfeld, J. H. (1997) Formation of organic aerosols from the oxidation of biogenic hydrocarbons. J. Atmos. Chem., 26, 189–222.

Houghton, J. T., Meira Filho, L. G., Callander, B. A., Harris, N., Kattenberg, A., and Maskell, K. (1996) Climate Change, 1995: The Science of Climate Change, p. 572. Cambridge University Press, Cambridge, UK.

Houghton, R. A., Davidson, E. A., and Woodwell, G. M. (1998) Missing sinks, feedbacks, and understanding the role of terrestrial ecosystems in the global carbon balance. Global Biogeochem. Cycles, 12, 25–34.

Houghton, R. A., and Skole, D. L. (1990) Carbon. In B. L. Turner, W. C. Clark, R. W. Kates, J. F. Richards, J. T. Mathews, and W. B. Meyer, Eds. The Earth as Transformed by Human Action: Global and Regional Changes in the Biosphere over the Past 300 Years, pp. 393–408. Cambridge University Press, New York.

Kanakidou, M. (1998) Impact of the formation of secondary organic aerosols from naturally emitted hydrocarbons on ozone and organic aerosol global budgets: A global three-dimensional study. Symposium of the Commission on Atmospheric Chemistry and Global Pollution (CACGP) and the International Global Atmospheric Chemistry Project (IGAC), Seattle, WA.

Keeling, C. D., Whorf, T. P., Wahlen, M., and Vanderplicht, J. (1995) Interannual extremes in the rate of rise of atmospheric carbon dioxide since 1980. Nature, 375, 666–670.

Keene, W. C., Sander, R., Pszenny, A. A. P., Vogt, R., Crutzen, P. J., and Galloway, J. N. (1998) Aerosol pH in the marine boundary layer: A review and model evaluation. J. Aerosol Sci., 29, 339–356.

Keller, M., Jacob, D. J., Wofsy, S. C., and Harriss, R. C. (1991) Effects of tropical deforestation on global and regional atmospheric chemistry. Clim. Change, 19, 139–158.

Kirchhoff, V. W. J. H., Barnes, R. A., and Torres, A. L. (1991) Ozone climatology at Natal, Brazil, from in situ ozonesonde data. J. Geophys. Res., 96, 10,899–10,909.

Krol, M., van Leeuwen, P. J., and Lelieveld, J. (1998) Global OH trend inferred from methylchloroform measurements. J. Geophys. Res., 103, 10,697–10,711.

Kurylo, M. J., Rodriguez, J. M., Andreae, M. O., Atlas, E. L., Blake, D. R., Butler, J. H., Lal, S., Lary, D. J., Midgley, P. M., Montzka, S. A., Novelli, P. C., Reeves, C. E., Simmonds, P. G., Steele, L. P., Sturges, W. T., Weiss, R. F., and Yokouchi, Y. (1999) Short-Lived Ozone-Related Compounds. In D. L. Albritton, P. J. Aucamp, G. Megie, and R. T. Watson, Eds. Scientific Assessment of Ozone Depletion: 1998, pp. 2.1–2.56. World Meteorological Organization, Geneva.

Lashof, D. A., DeAngelo, B. J., Saleska, S. R., and Harte, J. (1997) Terrestrial ecosystem feedbacks to global climate change. Annu. Rev. Energy Environ., 22, 75–118.

Levin, Z., Ganor, E., and Gladstein, V. (1996) The effect of desert particles coated with sulfate on rain formation in the eastern Mediterranean. J. Appl. Meteor., 35, 1511–1523.

Li-Jones, X., and Prospero, J. M. (1998) Variations in the size distribution of non-sea-salt sulfate aerosol in the marine boundary layer at Barbados: Impact of African dust. J. Geophys. Res., 103, 16,073–16,084.

Madronich, S. (1987) Photodissociation in the atmosphere. 1. Actinic flux and the effects of ground reflections and clouds. J. Geophys. Res., 92, 9740–9752.

Mauldin, R. L., III, Madronich, S., Flocke, S. J., Eisele, F. L., Frost, G. J., and Prevot, A. S. H. (1997) New insights on OH: Measurements around and in clouds. Geophys. Res. Lett., 24, 3033–3036.

Montzka, S. A., Butler, J. H., Myers, R. C., Thompson, T. M., Swanson, T. H., Clarke, A. D., Lock, L. T., and Elkins, J. W. (1996) Decline in the tropospheric abundance of halogen from halocarbons: Implications for stratospheric ozone depletion. Science, 272, 1318–1322.

Müller, R., Crutzen, P. J., Grooss, J.-U., Brühl, C., Russell, J. M., Gernandt, H., McKenna, D. S., and Tuck, A. F. (1997) Severe chemical ozone loss in the Arctic during the winter of 1995–96. Nature, 389, 709–712.

National Research Council (U.S.) Committee on Tropospheric Ozone Formation and Measurement. (1991) Rethinking the Ozone Problem in Urban and Regional Air Pollution. 500 p. National Academy Press, Washington, D.C.

Newman, P. A., Gleason, J. F., McPeters, R. D., and Stolarski, R. S. (1997) Anomalously low ozone over the Arctic. Geophys. Res. Lett., 24, 2689–2692.

Oechel, W. C., and Vourlitis, G. L. (1994) The effects of climate change on land atmosphere feedbacks in arctic tundra regions. Trends in Ecology & Evolution, 9, 324–329.

Olson, J. R., Fishman, J., Kirchhoff, V. W. J. H., Nganga, D., and Cros, B. (1996) Analysis of the distribution of ozone over the southern Atlantic region. J. Geophys. Res., 101, 24,083–24,094.

Portmann, R. W., Solomon, S., Fishman, J., Olson, J. R., Kiehl, J. T., and Briegleb, B. (1997) Radiative forcing of the Earth's climate system due to tropical tropospheric ozone production. J. Geophys. Res., 102, 9409–9417.

Portmann, R. W., Solomon, S., Garcia, R. R., Thomason, L. W., Poole, L. R., and McCormick, M. P. (1996) Role of aerosol variations in anthropogenic ozone depletion in the polar regions. J. Geophys. Res., 101, 22,991–23,006.

Prather, M. J. (1996) Time scales in atmospheric chemistry: Theory, GWPs for CH_4 and CO, and runaway growth. Geophys. Res. Lett., 23, 2597–2600.

Prinn, R. G., Weiss, R. F., Miller, B. R., Huang, J., Alyea, F. N., Cunnold, D. M., Fraser, P. J., Hartley, D. E., and Simmonds, P. G. (1995) Atmospheric trends and lifetime of CH_3CCl_3 and global OH concentrations. Science, 269, 187–192.

Raich, J. W., and Potter, C. S. (1995) Global patterns of carbon dioxide emissions from soils. Global Biogeochem. Cycles, 9, 23–36.

Ramaswamy, V., and Bowen, M. M. (1994) Effect of changes in radiatively active species upon the lower stratospheric temperatures. J. Geophys. Res., 99, 18,909–18,921.

Roberts, G., Andreae, M. O., Maenhaut, W., Artaxo, P., Martins, J. V., Zhou, J., and Swietlicki, E. (1998) Relationships of cloud condensation nuclei to size distribution and aerosol composition in the Amazon Basin. Eos Trans. AGU, 79, F159.

Roelofs, G. J., Lelieveld, J., and van Dorland, R. (1997) A three-dimensional chemistry general circulation model simulation of anthropogenically derived ozone in the troposphere and its radiative climate forcing. J. Geophys. Res., 102, 23,389–23,401.

Sander, R., and Crutzen, P. J. (1996) Model study indicating halogen activation and ozone destruction in polluted air masses transported to the sea. J. Geophys. Res., 101, 9121–9138.

Sander, R., Vogt, R., Harris, G. W., and Crutzen, P. J. (1997) Modeling the chemistry ozone, halogen compounds, and hydrocarbons in the arctic troposphere during spring. Tellus, 49B, 522–532.

Schwartz, S. E. (1988) Are global cloud albedo and climate controlled by marine phytoplankton? Nature, 336, 441–445.

Seinfeld, J. H., and Pandis, S. N. (1998) Atmospheric chemistry and physics: From air pollution to climate change. 1326 p. John Wiley, New York.

Shindell, D. T., Rind, D., and Lonergan, P. (1998) Increased polar stratospheric ozone losses and delayed eventual recovery owing to increasing greenhouse-gas concentrations. Nature, 392, 589–592.

Shine, K. P., and Forster, P. M. D. F. (1999) The effect of human activity on radiative forcing of climate change: A review of recent developments. Global and Planetary Change, 205–225.

Talbot, R. W., Harriss, R. C., Browell, E. V., Gregory, G. L., Sebacher, D. I., and Beck, S. M. (1986) Distribution and geochemistry of aerosols in the tropical North Atlantic troposphere – Relationship to Saharan dust. J. Geophys. Res., 91, 5173–5182.

Tang, X. Y., and Madronich, S. (1995) Effects of increased solar ultraviolet radiation on tropospheric composition and air quality. Ambio, 24, 188–190.

Tegen, I., and Fung, I. (1994) Modeling of mineral dust in the atmosphere: Sources, transport, and optical thickness. J. Geophys. Res., 99, 22,897–22,914.

Tegen, I., and Fung, I. (1995) Contribution to the atmospheric mineral aerosol load from land surface modification. J. Geophys. Res., 100, 18,707–18,726.

Tegen, I., Lacis, A. A., and Fung, I. (1996) The influence on climate forcing of mineral aerosols from disturbed soils. Nature, 380, 419–422.

Thompson, A. M. (1992) The oxidizing capacity of the earth's atmosphere: Probable past and future changes. Science, 256, 1157–1165.

Thompson, A. M., Pickering, K. E., McNamara, D. P., Schoeberl, M. R., Hudson, R. D., Kim, J.-H., Browell, E. V., Kirchhoff, V. W. J. H., and Nganga, D. (1996) Where did tropospheric ozone over southern Africa and the tropical Atlantic come from in October 1992? Insights from TOMS, GTE TRACE-A, and SAFARI 1992. J. Geophys. Res., 101, 24,251–24,278.

Toumi, R., Bekki, S., and Law, K. S. (1994) Indirect influence of ozone depletion on climate forcing by clouds. Nature, 372, 348–351.

Twomey, S. (1977) The influence of pollution on the short-wave albedo of clouds. J. Atmos. Sci., 34, 1149–1152.

van Dorland, R., Dentener, F. J., and Lelieveld, J. (1997) Radiative forcing due to tropospheric ozone and sulfate aerosols. J. Geophys. Res., 102, 28,079–28,100.

Vogt, R., Crutzen, P. J., and Sander, R. (1996) A mechanism for halogen release from seasalt aerosol in the remote marine boundary layer. Nature, 383, 327–330.

Wang, Y. H., Jacob, D. J., and Logan, J. A. (1998) Global simulation of tropospheric O_3-NO_x-hydrocarbon chemistry, 1. Model formulation. J. Geophys. Res., 103, 10,713–10,725.

3 Atmospheric CO_2 Variations

Response to Natural and Anthropogenic Earth System Forcings

INEZ FUNG

ABSTRACT

In this chapter, we focus on the land sink of anthropogenic CO_2, because humans have a history of using the terrestrial biosphere for our purpose and because efforts to control atmospheric CO_2 levels involve deliberate manipulation of the biosphere. We present atmospheric evidence for the land sink and use information about its interannual variations to infer its stability.

3.1 Introduction

The Mauna Loa CO_2 record is a clear documentation of the increasing concentration of CO_2 in the atmosphere as a result of anthropogenic activities. By 1999, the atmospheric CO_2 abundance had increased by 25% since the beginning of the preindustrial era. The cumulative increase, together with the concomitant increase in CH_4, N_2O, CFCs, and other greenhouse gases, presents a total radiative forcing of $\sim$2–3 W/m^2 to the climate system in the 1990s. This forcing is countered to some degree by the increase in sulphate and other aerosols in the atmosphere.

The decreasing $^{14}C/^{12}C$ ratio in tree rings (Suess, 1955) proves that the atmospheric CO_2 increase is due to the addition of fossil (^{14}C-free) carbon. However, the CO_2 increase rate, as determined from the atmospheric record, is only 50%–60% that emitted by fossil fuel combustion (Figure 3.1). Thus, the land and oceans have absorbed the remainder of the fossil fuel CO_2 as well as the CO_2 released due to land use modification. It is important to figure out the partitioning of the carbon sink between the land and sea inasmuch as the residence time of carbon is shorter on land than that in the oceans. Terrestrial storage may be transient and easy to destabilize with climate warming.

3.2 Evidence for a Land Sink

The existence of a land sink for anthropogenic CO_2 has been inferred since the early carbon budget calculations using one-dimensional ocean models (e.g., Oeschger et al., 1975). Because of the heterogeneity of the land surface and the lack of terrestrial flux measurements of long duration, direct detection of the land sink has been difficult. The most uncontroversial evidence for the land sink is now found in the changing ratio of

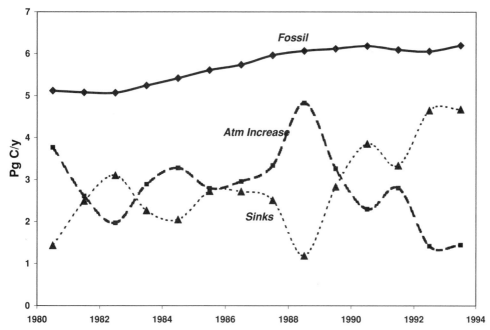

Figure 3.1. The global carbon budget for 1980–1994. The deforestation source is not included, and its magnitude must be added to the displayed sink strength to yield the actual sink for anthropogenic CO_2.

O_2/N_2 in the atmosphere (Keeling and Shertz, 1992). Atmospheric O_2 concentration (and hence the O_2/N_2 ratio) decreases with the combustion of fossil fuels and with the net uptake of carbon by the terrestrial biosphere. The O_2/N_2 to CO_2 ratios are different for the two processes, but O_2/N_2 ratios are not altered by atmosphere-ocean carbon exchange. Using the high-precision measuremants of O_2/N_2 in the atmosphere, Keeling and Shertz (1992) infer, for the period 1991–1994, a net carbon uptake of 2.0 ± 0.9 PgC/y by the terrestrial biosphere. This translates into an anthropogenic carbon sink of 3.6 PgC/y for a deforestation source of 1.6 PgC/y. As a residual, the ocean sink for the period is 1.7 ± 0.9 PgC/y.

Atmospheric $\delta^{13}C$ variations could also potentially distinguish between the terrestrial and oceanic sinks (Keeling et al., 1995; Francey et al., 1996). Atmospheric $\delta^{13}C$ has been decreasing from −6.4‰ in the preindustrial era to −8‰ in the 1990s with the addition of fossil fuel carbon, which has an isotopic value of −25‰, that of the plant material from which fossil fuels are derived. The observed atmospheric $\delta^{13}C$ decrease is less than expected if all the fossil fuel CO_2 remained airborne, further supporting the need for terrestrial and oceanic uptake, which remove "lighter" carbon and leave an atmosphere with a "heavier" $^{13}C/^{12}C$ ratio. However, constraining the ocean-land partitioning of the carbon sink requires information about (1) the isotopic disequilibrium associated with the gross fluxes and (2) the discrimination of the land uptake processes. These are discussed below.

Gross fluxes, assumed to cancel in CO_2, leave a $\delta^{13}C$ signature in the atmosphere. Fluxes from the atmosphere to the land or ocean carry today's "light" $^{13}C/^{12}C$ ratios, whereas the returning fluxes are "heavier," with the difference determined by the age of the donor pools. There are considerable uncertainties in the magnitudes and distributions of the isotopic disequilibrium associated with gross terrestrial and oceanic exchanges (Fung et al., 1997). The oceanic isotopic disequilibrium is governed principally by the oceanic circulation. The isotopic disequilibirum associated with the biosphere depends on the age of the respired carbon, to which the old, slowly decomposing soil carbon pools contribute a small percentage of the total flux but contribute a significant weighting of the $\delta^{13}C$ signature. In tropical forests, the disequilibrium is large even though carbon turnover is fast in the soil, because trees there live for 30–40 years. In the tundra, the disequilibrium is also large, because decomposition is slow in the cold climate. Fung et al. (1997) used a comprehensive biogeochemistry model to study factors contributing to isotopic disequilibrium associated with terrestrial fluxes. They obtained a value of 24‰ PgC/y that is larger than those used in previous studies (Tans et al., 1993; Keeling et al., 1995; Francey et al., 1996). With their estimate, the isotopic disequilibrium associated with the land and ocean gross fluxes explains about half the discrepancy between the observed atmospheric decrease and the source contribution (Figure 3.2). A greater degree of isotopic disequilibrium would contribute to a "heavy" atmospheric $\delta^{13}C$ and imply a smaller net uptake by the biosphere.

Because ^{13}C discrimination by terrestrial uptake is $\sim$10 times greater than by oceanic uptake, it is not unreasonable to anticipate a land carbon sink to contribute to the observed atmospheric $\delta^{13}C$ trend. However, if one assumes at this point that terrestrial photosynthesis has a discrimination of 18‰, one obtains a smaller oceanic uptake than obtained from the O_2/N_2 analysis. The dilemma stems from the uncertainty about the type of vegetation responsible for the carbon uptake. C_3 and C_4 vegetation have different photosynthetic pathways and hence different degrees of discrimination against ^{13}C during photosynthesis. $\delta^{13}C$ of C_3 plants varies with ambient temperature and humidity and CO_2 levels but is around -25‰ whereas that for C_4 plants is around -12‰. C_4 photosynthesis thus imparts an isotopic signature in the atmosphere that is similar to oceanic uptake. By combining their $\delta^{13}C$ isotopic disequilibria with the ocean-land sink partitioning derived from O_2/N_2 by Keeling and Shertz (1992), Fung et al. (1997) estimate that $\sim$15% of the terrestrial uptake is by C_4 plants. Corn, legumes, and some grasses are C_4 plants.

In summary, the O_2/N_2 and $\delta^{13}C$ records show that the terrestrial biosphere and oceans have removed the anthropogenic CO_2 from the atmosphere. Not all the terrestrial uptake is by forests (C_3 vegetation); a small fraction may have occurred in grasslands and agricultural lands.

3.3   Global CO_2 Variations

In this section, we explore the interannual variations of the contemporary carbon budget, and we use the variations to deduce how the sinks have responded to climate perturbations. The globally integrated budget of atmospheric CO_2 can be written as follows:

$$\frac{\partial M}{\partial t} = FF + DEF - S \tag{1}$$

Its variation from 1980 to 1994 is shown in Figure 3.1. Fossil fuel emission of CO_2 is estimated from United Nations statistics (Andres et al., 1996), and the atmospheric

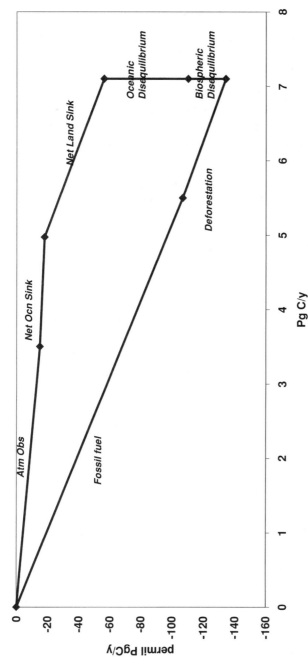

Figure 3.2. The carbon budget for the 1980s, as determined simultaneously by the variations of the ^{12}C and ^{13}C fluxes. The slope of each line is given by the $\delta^{13}C$ (relative to the atmosphere) of each flux. Note that the isotopic disequilibria associated with gross biospheric and oceanic exchange explain approximately half the ^{13}C discrepancy between the sources and observations. The slope of the "net land sink" line is given by the photosynthetic discrimination of the vegetation responsible for the sink and will vary with the C_3:C_4 fraction of the vegetation types involved.

CO_2 growth rate is calculated as the time derivative of the globally averaged CO_2 concentration. The residual is the total sink, removal by land and ocean. The total sink is larger and must include all the CO_2 released by deforestation, estimated to be 1.6 ± 1 PgC/y (Schimel et al., 1996).

Fossil fuel emissions were about 5.2 PgC/y in the early 1980s and reached 6.1 PgC/y by 1994. Atmospheric CO_2 growth rate shows large interannual variations, despite the relatively flat nature of the fossil fuel source. The growth rate was as high as 4.5 PgC/y in 1988 and decreased rapidly to 2.5 PgC/y in 1990, and to 1.8 PgC/y in 1993. The sink, derived as the residual, shows a factor of 3 difference between these years. The period 1989–1990 has been noted as an anomalous period (Enting et al., 1995) because the global $\delta^{13}C$ was nearly constant (Francey et al., 1996).

These growth rate variations are the result of the sensitive response of the land and ocean sinks to interannual climate perturbations. They warn that future growth rates of atmospheric CO_2 will not only follow anthropogenic inputs but will also be influenced by terrestrial and oceanic carbon dynamics in a changing climate.

3.4   Hemispheric CO_2 Variations

CO_2 concentrations measured at a large number of sites in the remote marine atmosphere show that the concentrations are higher in the Northern than in the Southern Hemisphere. More than 90% of the fossil fuel CO_2 source is in the Northern Hemisphere. The concentration difference averaged ~ 3 ppmv for the period 1980–1994. This hemispheric gradient is less steep than would be expected if all the fossil fuel CO_2 remained airborne, thus suggesting that the Northern hemispheric CO_2 sink is greater than the Southern Hemisphere sink. Combining the CO_2 concentration gradients with the compilation of air-sea flux of CO_2 from shipboard measurements, Tans et al. (1990) argued that the terrestrial biosphere in the Northern Hemisphere acted as a substantial sink of anthropogenic CO_2 in the decade of the 1980s. Ciais et al. (1995), using as well the gradient of $\delta^{13}C$ in the atmosphere, concluded that a large fraction of the NH land sink was at middle to high latitudes from 1992 to 1993.

The interannual variations in this hemispheric gradient, as captured by the concentrations at Mauna Loa (MLO) and the South Pole (SPO), are shown in Figure 3.3. The MLO-SPO concentration difference varies from 1.6 ppmv in 1983 to 3.3 ppmv in 1990. Between 1988 and 1990, there continued to be a positive trend in the hemispheric gradient while the global growth rate started to decrease. After 1990, CO_2 growth rate at MLO slowed with the global growth rate.

The atmospheric gradient variations have been used in global three-dimensional models to infer the year-to-year variations of the land and ocean sinks (e.g., Rayner et al., 1999). Here we illustrate the interannual variations in the hemispheric sinks using a simple two-box model.

$$\frac{\partial Mn}{\partial t} = -\frac{Mn - Ms}{\tau} + FF - Sn \tag{2}$$

$$\frac{\partial Ms}{\partial t} = +\frac{Mn - Ms}{\tau} - Ss \tag{3}$$

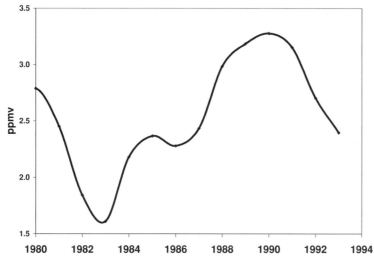

Figure 3.3. The difference between the annual mean CO$_2$ concentrations for Mauna Loa, Hawaii (MLO), and the South Pole (SPO) for 1980–1994.

In Equations 2 and 3, *Mn* and *Ms* are the CO$_2$ mass in the Northern and Southern Hemispheres, respectively, τ is the interhemispheric exchange time, taken to be 1.2 years, and *Sn* and *Ss* are the sink in the Northern and Southern Hemispheres, respectively. We have assumed that all the fossil fuel CO$_2$ is released in the Northern Hemisphere.

We have investigated the interannual variations of τ, using the Goddard Institute for Space Studies global three-dimensional climate model forced by the observed sea surface temperatures, volcanic optical depths, greenhouse gases, and ozone variations for 1980–1994. An inert tracer with a time-invariant Northern Hemisphere industrial source pattern is transported in the climate model. When the model results were reduced to a two-box representation as in Equations 2 and 3, the effective τ was found to vary by a factor of 20%. This magnitude of variation turns out to be unimportant for the CO$_2$ source/sink problem, because the MLO-SPO difference varies by a factor of 2 over the same period.

Equations 2 and 3 together with the requirement that *Sn* + *Ss* = *Sink* yield the interannual variations of the Northern Hemisphere and Southern Hemisphere sinks. These are shown in Figure 3.4, together with the southern oscillation index (SOI), a measure of the state of the El Niño/southern oscillation. The SOI displayed are annual means for each calendar year, consistent with the way the CO$_2$ data have been averaged. We have not included the deforestation source explicitly, because the contemporary deforestation source is located in the Tropics and makes no contribution to the hemispheric gradient. In this way, the hemispheric sinks described below do not include half the deforestation source ($\sim$0.8 PgC/y).

Averaged 1980–1988, the NH sink was 2.5 $\pm$ 1 PgC/y, whereas the SH sink was 0.5 $\pm$ 0.5 PgC/y. From 1988 to 1990, the total sink increased by 2 PgC/y (compare Figure 3.1) with approximately equal contributions from the NH and SH.

Figure 3.4 shows that *variations* in the NH sink are larger than the *variations* in the SH sink. This is not unexpected, because interannual climate perturbations are more

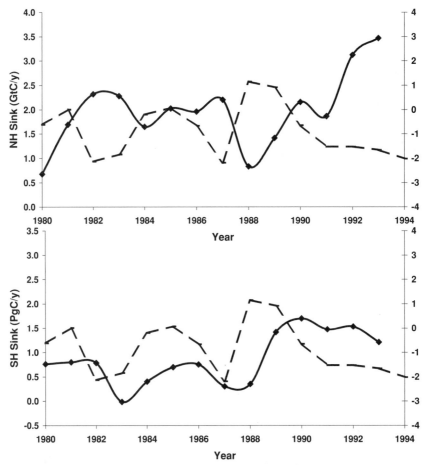

Figure 3.4. Variations of the Northern Hemisphere (top panel) and Southern Hemisphere (bottom panel) sinks and the southern oscillation index (SOI, dashed line) for 1980–1994, as determined by a two-box model. Note that the calculation does not include deforestation, and so half the deforestation release should be added to the displayed sink strengths to obtain the actual sink. Negative values of the SOI indicate El Niño years.

pronounced in the Northern Hemisphere. The dominant mode of interannual climate variations is the El Niño/southern oscillation, captured in the SOI, the normalized anomalous pressure difference between Tahiti and Darwin, Australia. During an El Niño, there is weaker east-west pressure gradient in the equatorial Pacific (negative SOI) leading to weaker easterly winds. Anomalously warm sea surface temperatures during the El Niño lead to global-scale perturbations in temperature and precipitation. The NH CO_2 sink is out of phase with the SOI; the NH sink is strong during strong El Niños (SOI < 0) and is weak during La Niñas (SOI > 0). The SH sink has the opposite phase relationship with the SOI; the sink lags the SOI by a year (the resolution of the investigation). The phase lag suggests alteration of total carbon concentrations resulting from circulation changes.

There was a steep increase in the global sink strength from 1988 to 1990. In our calculation, half of this sink was in the NH. This NH sink was predominantly terrestrial, because atmospheric δ^{13}C did not decrease despite the continued increase in fossil fuel combustion (see Francey et al., 1996). After 1990, the NH sink continued to increase to 4 PgC/y in 1993, whereas the SH sink leveled off at 1.5 PgC/y. The NH sink is greater than would have been expected from the SOI. This NH sink also cannot be dominated by the land, inasmuch as atmospheric δ^{13}C resumed its decline after 1990. Francey et al. (1996) inferred that the global ocean sink dominated after the 1990s. We hypothesize here that the North Atlantic will become a significant sink after the 1990s. The hypothesis is suggested by an analysis of the surface marine conditions in the Atlantic. It turns out that the decadal North Atlantic oscillation (NAO) index was very positive starting in 1990 (Hurrell, 1995). The NAO index represents the pressure difference between Azores high- and Icelandic low-pressure systems. Positive NAO index is accompanied by strong westerly winds and storminess, which are conducive to CO$_2$ uptake by the ocean.

3.5   Seasonal CO$_2$ Variations

A terrestrial carbon sink is an annual imbalance between uptake and release, each of which is seasonal. Together, they impart a signature in the seasonal oscillations of CO$_2$ in the atmosphere. The annual CO$_2$ fluxes due to photosynthetic uptake and decomposition release are largest ($\sim$1 kg/m^2/y) in the Tropics and decrease poleward. However, the tropical fluxes are nearly synchronous and cancel, and so they leave little signature in the atmospheric CO$_2$ record. When averaged over the globe, the growing season net flux is $\sim$15%–20% of the annual net primary production. The largest contribution to the growing season net flux, and hence the atmospheric CO$_2$ oscillations, is from the high-latitude biosphere, where the growing season is shorter than the decomposition season (Fung et al., 1987; D'Arrigo et al., 1987; Randerson et al., 1997).

Like the annual atmospheric CO$_2$ growth rate, the CO$_2$ seasonal cycles have not remained uniform through time. The amplitudes of the oscillations at the Northern Hemisphere observing sites have been increasing (Keeling et al., 1996; Randerson et al., 1997, 1999), suggesting changing dynamics of the terrestrial biosphere. The increasing amplitudes may be signatures of net uptake as well as net release, as illustrated below.

The seasonal cycle at Mauna Loa captures the dynamics of the Northern Hemisphere biosphere. The annual minima occur at the end of the growing season and result mainly from the net photosynthetic uptake (which is not compensated by decomposition release) integrated over the growing season. Similarly, the annual maxima reflect the release integrated through the rest of the year. The anomalous (departures from 1980 to 1994) mean maxima and minima are shown in Figure 3.5. Photosynthetic uptake (CO$_2$ seasonal minima) was enhanced in 1982 and reduced in 1988, consistent with the variations found in the global and NH sinks (compare Figures 3.1 and 3.4). The photosynthetic increase from 1988 to 1992 matched the sink trend but is smaller in magnitude, thus supporting the suggestion of the northern oceans as a significant sink for this period. The variations in respiration (CO$_2$ seasonal maxima) also vary in time, so the CO$_2$ amplitude variations at MLO cannot be used as an indicator of net biospheric uptake.

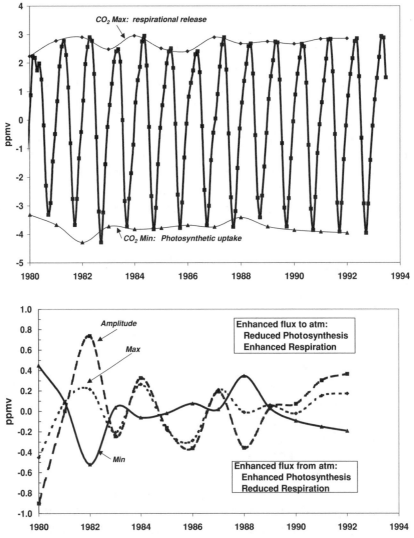

Figure 3.5. Top panel: Detrended CO_2 seasonal cycle at Mauna Loa, Hawaii (MLO) for 1980–1994. The variations in the minima indicate variations in photosynthetic rates, whereas variations in the maximum indicate variations in respiration rates. Bottom panel: Anomalous concentration maxima, minima, and peak-trough amplitude of the seasonal cycle at MLO. The anomalies are defined relative to the mean for the period. Positive anomalies denote enhanced flux to the atmosphere.

The phasing of the CO_2 cycles also has not remained constant. Analysis of the monthly growth rate of the seasonal cycle at Point Barrow and other high-latitude sites shows that CO_2 drawdown has been occurring earlier in the season (Keeling et al., 1996; Randerson et al., 1997). Randerson et al. (1997) show that only early net primary productivity by the high-latitude biosphere can explain the CO_2 amplitude increase, net uptake, normalized difference vegetation index (NDVI) trend, and change in the shape of the CO_2 cycle. This argues, at least for the high-latitude sites, that the increasing amplitudes signify net uptake and that climate change is the driving mechanism.

3.6   Processes Associated with Terrestrial Carbon Uptake Variability

The rates of photosynthesis and decomposition are sensitive to climatic perturbations. Dai and Fung (1993) suggested that climate-induced interdecadal variations in the terrestrial net flux may be half as much as that of the anthropogenic sink. This large natural fluctuation presents a challenge to the unambiguous identification and detection of the anthropogenic carbon sink, especially in records with durations shorter than the climatic fluctuations.

Many processes have been cited as candidates for the anthropogenic carbon sink: CO_2 and nitrogen fertilization (Holland et al., 1997), global warming, forest recovery following agricultural abandonment, storage as forestry products (Winjum et al., 1998), and many others. It is difficult to distinguish among these processes from the "top-down" atmospheric approach. Direct information is being gathered to document the contribution of each of these processes, and no single process is expected to dominate the land sink. Furthermore, it is unlikely that these processes vary interannually by 1–2 Pg C/y, and so climate perturbation appears the principal cause for the interannual variations in the land sink.

CO_2 exchange processes at high latitudes are limited by temperature and are hence very sensitive to temperature perturbations. Temperature, however, affects not only photosynthesis and decomposition but also thaw and the subsequent alteration of the water budget and mobilization of nutrients.

A positive trend in photosynthesis at high latitudes has been reported in the >15-year record of the NDVI, the satellite index of photosynthetic activity (Myneni et al. 1997). The magnitude of the trend is uncertain because of inadequate calibration and atmospheric correction for the multisatellite, multisensor record (Fung, 1997); nevertheless, the sign of the trend appears robust. Like Keeling et al. (1996), Myneni et al. also infer an earlier growing season from the satellite data. This is consistent with thermometer records of surface air temperature, which show the greatest warming rate in the springtime at high latitudes.

The sensitivity of respiration on net ecosystem exchange is illustrated by the flux tower measurements at a boreal forest site in Saskatchewan, Canada (Goulden et al., 1998). There, both uptake and release increased with time, and the integrated net ecosystem exchange at the top of the canopy suggests that this site is a net source, and not a sink, of atmospheric CO_2 for 1991–1994. At this boreal site, respiration has a stronger dependence on temperature than would be expected from decomposition rates, suggesting that the volume of thaw has increased. Although this site may not be representative of the entire boreal zone, inasmuch as there is not widespread increase in thaw, this site warns that the current high-latitude carbon sink may not be sustained with global warming.

The competing effect of temperature on high-latitude photosynthesis and respiration is further illustrated by Randerson et al. (1999). Monthly CO_2 fluxes since 1980, as inferred from the high-latitude atmospheric CO_2 oscillations and a transport model, have different relationships with temperature in different seasons. The springtime flux anomalies (Figure 3.6) are negatively correlated with surface air temperature anomalies,

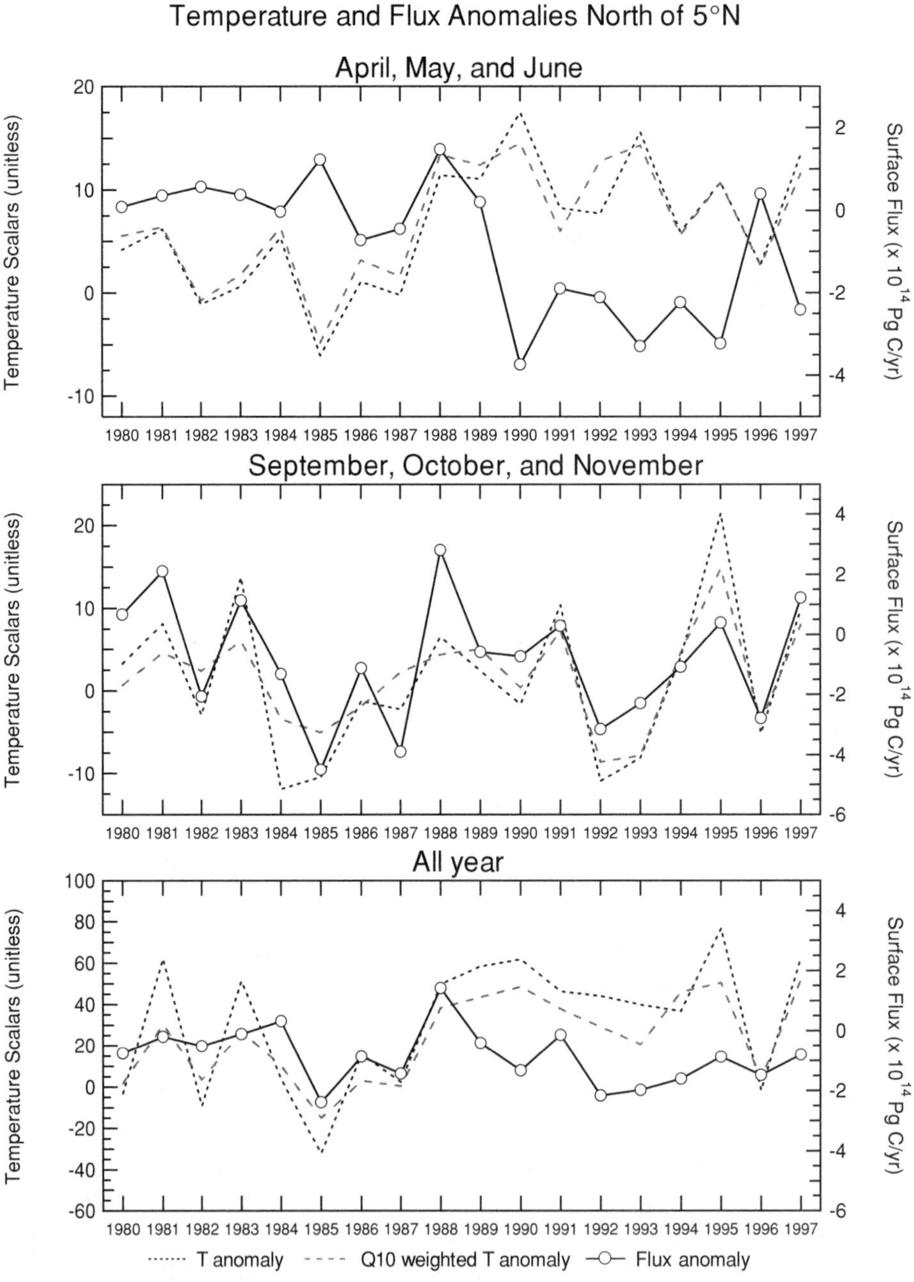

Figure 3.6. Relationship between inferred CO_2 flux anomalies and temperature anomalies north of 5°N for 1980–1997. The temperature anomalies and the Q_{10}-weighted temperature anomalies have been weighted by the net primary productivity for the region. Q_{10} is a measure of the temperature sensitivity of decomposition rates. (Taken from Randerson et al., 1999.)

a result consistent with Myneni et al. (1997), whereas those in the fall are positively correlated with temperature, consistent with Goulden et al. (1998). As has been argued above, although the competition favors net uptake in the present climate, further increases in temperature may change the dynamics of the competition.

At middle and low latitudes, the variations in precipitation are likely to be as important as the variations in temperature in causing interannual variations in photosynthesis and respiration. This is demonstrated by Goulden et al. (1996), who suggest that the NH land sink may vary interannually by ± 1 PgC/y because of seasonal climate perturbations.

3.7   The Human Dimension

In addition to burning fossil fuels, humans affect the carbon cycle via direct modification of the terrestrial biosphere. Deforestation, occurring now in the Tropics, generally replaces high-density biomass (forests) with low-density biomass (crops and pastures). If the mode of deforestation includes burning, a fraction of the aboveground biomass carbon is released immediately to the atmosphere. The fraction released depends on the characteristics of the fuel – its size and moisture content – as well as on the repetition of the burns, and it ranges from <5% for large logs to 35% for multiple burns (Carvalho et al., 1998). The detritus left on site adds to the decomposition pool, and exposed soils are subject to rapid oxidation. The turnover time of carbon in the tropics is 5–10 years, and so deforestation contributes to a sustained release of CO$_2$ several years after the initial clearing.

Deforestation does not occur uniformly across a landscape but rather is concentrated in small areas (Skole and Tucker, 1993). In Brazil, the distribution of the cleared areas shows that >50% of the deforested plots are >100 ha and <15% of the deforested plots are >1000 ha (INPE, 1999). These plot sizes are large and represent the interests of commercial and large-scale enterprises. The analysis also shows that the deforestation rate in Brazil varies by a factor of 2 from year to year. The fluctuation is related to the Brazilian economy and land and timber prices. Hence, coarse-resolution ($1°$ x $1°$ $\sim 10^6$ ha) satellite imagery will likely overestimate the deforestation rate, and statistical extrapolation of a small-area survey to an ecosystem or a country is subject to bias. Our preliminary reanalysis of the deforestation source suggests that the annual CO$_2$ release is smaller, by a factor of 2, than the Schimel et al. (1996) estimate; however, a firmer estimate cannot be made until the high-resolution satellite data for Africa, South Asia and Southeast Asia have been analyzed.

Analysis of the contemporary deforestation focuses on the Tropics. Over the past 100 years, the expansion of agriculture, mainly in the middle latitudes, has contributed a cumulative carbon source that is half the fossil fuel source. The cleared areas are now under management or are abandoned, so year-by-year carbon balance is now close to neutral in some areas or is showing slow accumulation in above- and belowground carbon in other areas (Post et al., 1997). The recovery is incomplete, so the present-day carbon inventory in mid-latitudes is still less than that during the preindustrial era (DeFries et al., 1999). This recovery, when coupled with the inadvertent stimulation of photosynthesis via elevated CO$_2$ levels and nitrogen deposition, results in a net carbon uptake in the middle latitudes.

The preceding discussion highlights the fact that the magnitude and direction of the carbon flux are tied to the history of land use, which is driven by a variety of forces in addition to population pressures. Recent deforestation leads to a transient carbon source to the atmosphere, and abandonment of past clearings leads to transient carbon sequestration by the biosphere. Viewing the contemporary carbon budgets in short 5- or 10-year snapshots may thus obfuscate the causes of and thus the mitigation strategies for the atmospheric CO_2 increase.

3.8   Implications for the Future

The natural carbon cycle involves large fluxes ($\sim 10^2$ PgC/y) into and out of the biosphere, and large fluxes ($\sim 10^2$ PgC/y) into and out of the oceans. At equilibrium, these fluxes cancel. The contemporary carbon sink ($\sim 10^2$ PgC/y) is a result of the incomplete cancellation of these opposing fluxes and represents a $\sim 1\%$ enhancement of the CO_2 fluxes out of the atmosphere. The large opposing fluxes are regulated by climate and are hence sensitive to climate perturbations.

In this chapter, we have examined the interannual variations of atmospheric CO_2 concentrations and have shown that the CO_2 sinks are climatically sensitive. There is currently carbon absorption by the high-latitude biosphere because photosynthesis enhancement is greater than the respiration enhancement. It is not likely that this sink will continue with global warming. Our reliance on the biosphere can lead to carbon sources at one time and carbon sinks at another. How atmospheric CO_2 levels, and climate, will evolve will depend on the delicate competition between direct human-induced and indirect climate-induced alterations of the biosphere. Strategies for deliberate manipulation of the carbon cycle must take into account the evolving climatic envelope.

REFERENCES

Andres, R., Marland, G., Fung, I., and Matthews, E. (1996). A 1×1 distribution of carbon dioxide emissions from fossil fuel consumption and cement manufacture, 1950–1990. Global Biogeochem. Cycles, 10, 419–430.

Carvalho, J. A., Higuchi, N., Araujo, T. M., and Santos, J. C. (1998). Combustion completeness in a rainforest clearing experiment in Manaus, Brazil. J. Geophys. Res., 103, 13195–13199.

Ciais, P., Tans, P. P., Trolier, M., White, J. W. C., and Francey, R. J. (1995). A large northern hemisphere terrestrial CO_2 sink indicated by the $^{13}C/^{12}C$ ratio of atmospheric CO_2. Science, 269, 1098–1102.

Dai, A. G., and Fung, I. (1993). Can climate variability contribute to the missing CO_2 sink? Global Biogeochemical Cycles, 7, 599–609.

D'Arrigo, R., Jacoby, G., and Fung, I. (1987). Boreal forests and atmosphere-biosphere exchange of carbon dioxide. Nature, 329, 321–323.

DeFries, R. S., Field, C. B., Fung, I., Collatz, G. J., and Bounoua, L. (1999). Combining satellite data and biogeochemical models to estimate global effects of human-induced land cover change on carbon emissions and primary productivity. Global Biogeochem. Cycles, 13, 803–815.

Enting, I. G., Trudinger, C. M., and Francey, R. J. (1995). A synthesis inversion of the concentration and $\delta^{13}C$ of atmospheric CO_2. Tellus, 47B, 35–52.

Francey, R. J., Tans, P. P., Allison, C., Enting, I. G., White, J. W. C., and Trolier, M. (1996). Changes in oceanic and terrestrial carbon uptake since 1982. Nature, 324, 237–238.

Fung, I.Y., Tucker, C. J., and Prentice, K. C. (1987). On the variability of atmospheric-biosphere exchange of CO_2. Advances in Space Research, 7, (11)175–(11)180.

Fung, I. (1997). A greener north. Nature, 386, 659–660.

Fung, I., Field, C. B., Berry, J. A., Thompson, M. V., Randerson, J. T., Malmstrom, C., Vitousek, P., Collatz, G. J., Sellers, P. J., Randall, D. A., Denning, A. S., Badeck, F., and John, J. (1997). Carbon-13 exchanges between the atmosphere and biosphere. Global Biogeochem. Cycles, 11, 507–533.

Goulden, M., Munger, J., Fan, S.-M., Daube, B., and Wofsy, S. C. (1996). Exchange of carbon dioxide by a deciduous forest: Response to interannual climate variability. Science, 271, 1576–1578.

Goulden, M., Wofsy, S. C., Harden, J., Trumbore, S., Crill, P., Gower, S., Fries, T., Daube, B., Fan, S.-M., Sutton, D., Bazzaz, A., and Munger, J. (1998). Sensitivity of boreal forest carbon balance to soil thaw. Science, 279, 214–217.

Holland, E. A., Braswell, R., Lamarque, J., Townsend, A., Sulzman, J., Muller, J.-F., Detener, F., Brasseur, G., Levy, H., Penner, J., and Roelofs, G. (1997). Variations in the predicted spatial distribution of atmospheric nitrogen deposition and their impacts on carbon uptake by terrestrial ecosystems. J. Geophys. Res., 102, 15849–15866.

Hurrell, J. W. (1995). Decadal trends in the North Atlantic Oscillation – regional temperatures and precipitation. Science, 269, 676–679.

Instituto Nacional de Pesquisas Espaciais (INPE) (1999). Amazonia: Deforestation 1995–1997. 23 pp. IBAMA MMA, INPE.

Keeling, R., and Shertz, S. (1992). Seasonal and interannual variations in atmospheric oxygen and implications for the global carbon cycle. Nature, 358, 723–727.

Keeling, C. D., Whorf, T., Whalen, M., and van der Plicht, J. (1995). Interannual extremes in the rate of rise of atmospheric carbon dioxide since 1980. Nature, 375, 666–668.

Keeling, C. D., Chin, J., and Whorf, T. (1996). Increased activity of northern vegetation inferred from atmospheric CO$_2$ measurements. Nature, 382, 146–149.

Myneni, R., Keeling, C. D., Tucker, C. J., Asrar, G., and Nemani, R. (1997). Increased plant growth in the northern high latitudes from 1981–1991. Nature, 386, 698–702.

Oeschger, H., Siegenthaler, U., Schotterer, U., and Gugelman, A. (1975). A box diffusion model to study the carbon dioxide exchange in Nature. Tellus, 27, 168–192.

Post, W. M., King, A., and Wullschleger, S. (1997). Historical variations in terrestrial biospheric carbon storage. Global Biogeochem. Cycles, 11, 99–109.

Randerson, J. T., Thompson, M. V., Conway, T., Fung, I., and Field, C. B. (1997). The contribution of terrestrial sources and sinks to trends in the seasonal cycle of atmospheric carbon dioxide. Global Biogeochem. Cycles, 11, 535–560.

Randerson, J. T., Field, C. B., Fung, I., and Tans, P. P. (1999). Increases in early season ecosystem uptake explain changes in the seasonal cycle of atmospheric CO$_2$ at high northern latitudes. Geophys. Research Lett., 26, 2765–2768.

Rayner, P. J., Enting, I. G., Francey, R. J., and Langenfelds, R. (1999). Reconstructing the recent carbon cycle from atmospheric CO$_2$, δ^{13}C and O$_2$/N$_2$ observations. Tellus, 51B, 213–232.

Schimel, D. S. et al. (1996). Radiative forcing of climate change, in Climate Change 1995: The Science of Climate Change, edited by J. T. Houghton, L. G. Meira Filho, B. Callendar, N. Harris, A. Kattenberg, and K. Maskell, chap. 2, pp. 65–131, Cambridge University Press, New York.

Skole, D., and Tucker, C. J. (1993). Tropical deforestation and habitat fragmentation in the Amazon: satellite data from 1978 to 1988. Science, 260, 1905–1910.

Suess, H. (1955). Radiocarbon concentration in modern wood. Science, 122, 415.

Tans, P. P., Fung, I., and Takahashi, T. (1990). Observational constraints on the global atmospheric CO$_2$ budget. Science, 247, 1431–1438.

Tans, P. P., Berry, J. A., and Keeling, R. (1993). Oceanic ^{13}C/^{12}C observations: A new window on oceanic CO$_2$ uptake. Global Biogeochem. Cycles, 7, 353–368.

Winjum, J., Brown, S., and Schlamadinger, B. (1998). Forest harvests and wood products: sources and sinks of atmospheric carbon dioxide. Forest Science, 44, 272–284.

4 Modeling and Evaluating Terrestrial Biospheric Exchanges of Water, Carbon Dioxide, and Oxygen in the Global Climate System

MARTIN HEIMANN

ABSTRACT

The climate of the Earth is, to a considerable degree, controlled by the terrestrial biosphere. Its surface properties (albedo and roughness) are important parameters for the surface exchanges of energy and momentum. Even more important are the tightly coupled exchange fluxes of water, carbon dioxide, and oxygen both by diffusion through the stomata of leaves during photosynthesis and through soils and organic matter by respiration. By means of process models of terrestrial biogeochemistry and coupling these to general circulation models in various ways, one can attempt to estimate the degree to which this terrestrial biospheric control is effective. Furthermore, the exchanges of these gases are reflected in spatiotemporal variations of their atmospheric distribution. Modeling of these patterns and comparing them to observations constitutes a powerful tool to evaluate the performance of terrestrial biogeochemical models.

A recent sensitivity model simulation with the European Centre Hamburg Atmosphere Model (ECHAM) general circulation model demonstrates that the presence or absence of terrestrial vegetation induces near surface land temperature changes of as much as $8\,°C$, doubled precipitation, and nearly threefold changes in evapotranspiration. The largest contributions to these changes are found to be caused by the enhanced surface water recycling in the presence of a "green world," as compared with a global desert. The simulation sets upper bounds on possible climate modifications induced by anthropogenic changes in land use. Terrestrial CO_2 exchanges play a crucial role in the global carbon cycle, not only because of the never-ending quest for the "missing sink" but also for the quantification of climatic feedbacks, in particular in the context of global warming. Atmospheric observations clearly demonstrate that climatic fluctuations significantly reduce or enhance the growth rate of atmospheric CO_2, at least on interannual to decadal time scales. It is also clear that the terrestrial biosphere contributes significantly to these variations. Parallel to H_2O and CO_2, oxygen is also exchanged with the terrestrial biosphere. Oxygen is not a greenhouse gas, but it provides an important diagnostic for terrestrial and oceanic biogeochemical processes. Furthermore, oxygen is composed of three different stable isotopes, which, through fractionation effects, provide additional information on the global biogeochemical cycles of H_2O, CO_2, and oxygen.

4.1   Introduction

There is no doubt that the biosphere has the potential to considerably influence the climatic environment on the surface of the Earth, as witnessed, for example, by a walk on a hot summer day from an open field into a neighboring forest. During recent years it has become increasingly evident that this influence is not restricted only to local modifications of the environment but that there exist controls of the biosphere on the climate system that have the potential to significantly modify the climate on a global scale. There exist three different fundamental interaction pathways between the physical climate system and the biosphere, operating on different temporal and spatial scales. They include the following:

1. The biophysiological controls by which the vegetation on land actively controls evapotranspiration and thus the surface balances of energy and water;
2. A global, biogeochemical control through emissions and uptake of trace gases, such as carbon dioxide, methane, nitrous oxide, and others, which influence the radiation balance of the atmosphere;
3. A biogeographical feedback exerted by the presence of vegetation regimes with different physical properties (albedo, roughness, conductivity, etc.) determining the surface exchanges of energy, momentum, and water.

The quantification of these biospheric controls on the Earth's climate and the determination of global-scale potential impacts induced by anthropogenic modifications require that these feedback processes be explicitly incorporated in comprehensive models of the Earth System. Traditionally, various scientific communities working on different temporal and spatial scales (Figure 4.1A) have explored this. Accordingly, the fast interactions are being explored with soil-vegetation-atmosphere-transport (SVAT) models; terrestrial biogeochemical models (TBMs) are used to describe the exchanges of the trace gases with the atmosphere, and biome models have been developed to investigate the biogeographical feedbacks (Prentice et al., 1992). Recently, first attempts have been made to consistently combine these three approaches in the form of dynamical global vegetation models (DGVMs), as shown schematically in Figure 4.1B.

The extent to which present-day DGVMs are able to describe biosphere-climate interactions under present, past, or future conditions is being described by Prentice (Chapter 11, this volume). Here I illustrate the state of the art in modeling biospheric exchanges of water, carbon dioxide, and oxygen.

4.2   Water Cycle

Evapotranspiration plays a crucial role in the surface energy balance over land. Biological controls on evapotranspiration are foremost the stomata in plant leaves, which determine the conductivity. A second important factor is the amount of soil water that is available for evapotranspiration, an amount that is determined to a considerable degree by the rooting depth of plants (Shukla and Mintz, 1982; Nepstad et al., 1994; Kleidon and Heimann, 1998).

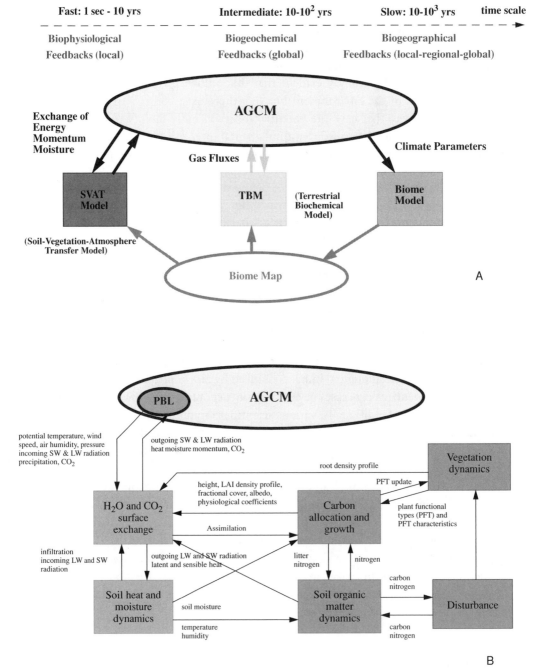

Figure 4.1. A: Schematic depiction of the traditional approach to investigate biospheric interactions with the physical climate system. B: Schematic of the processes represented in a comprehensive dynamical global vegetation model. See color plate section for a color version.

What is the overall effect of vegetation on the climate of the Earth? A recent sensitivity modeling study (Kleidon et al., 1999; Fraedrich et al., 1999) demonstrates this control. In two experiments with the climate model ECHAM4 (Roeckner et al., 1996), the standard surface parameters in nonglaciated regions have been replaced either by values representative of a desert or by an evergreen forest that is optimally adapted to the prevailing climate. The two model experiments explore extremes of future anthropogenic changes in land use: a hypothetical scenario of global desertification and a hypothetical scenario of global afforestation. Apart from the different land surface parameterizations, the two simulation experiments included identical, simple boundary conditions: present-day climatological annual cycles of sea surface temperatures and present-day land ice cover. The model simulations were run for 11 years, and the first year was discarded to avoid spin-up effects. The simulations were performed in T42 horizontal model resolution (corresponding to approximately 2.8° latitude by 2.8° longitude) with 19 model layers in the vertical dimension.

Figure 4.2A shows the difference in the 10-year mean annual 2m temperature, and Figure 4.2B shows the ratio of the 10-year mean annual precipitation of the two simulations. The 2m-temperature difference demonstrates the fundamental role of evapotranspiration in the green planet simulation, lowering the temperature by as much as 8° in the Tropics over land. This happens despite the generally lower land surface albedo in the green planet simulation. The effect of the lower surface reflectance in the green planet simulation is dominant only over parts of the Sahara desert, where limited precipitation doesn't provide enough water for sustained evaporative cooling. Over the northern temperate latitudes, temperatures tend to be higher over the Asian continent, which is related to the vegetation reducing the surface albedo under low snow cover. Because of the prescribed sea surface temperatures, there are no significant temperature differences over the oceans between the two simulations. Based on Figure 4.2A, one could argue that one sees only a local effect, caused by the differently specified surface parameterization in the two model runs. However, an inspection of the other meteorological fields in the simulations shows that this is not the case. Indeed, the climate of the green planet simulation is characterized by a much more vigorous water cycle, with doubled precipitation (Figure 4.2B) and tripled evapotranspiration over land as compared with the desert world simulation (Kleidon et al., 1999). The relatively larger enhancement of evapotranspiration versus precipitation implies that global river runoff is reduced in the green planet simulation, and that is caused by the large water-holding capacity of vegetation. The enhanced water cycle in the green planet simulation leads to increased convection in the Tropics and thus a stronger Hadley circulation, with higher upper troposphere temperatures and decreased precipitation in the descending branches of the Hadley cell over the Oceans. The green planet simulation also exhibits a more zonal circulation structure in mid-latitudes (Fraedrich et al., 1999). Overall, the dominant effect of the vegetation is found to be the strongly enhanced evapotranspiration, which tends to turn the Earth almost into a pure aquaplanet.

Clearly, the two climate simulations can demonstrate only a gross picture of the maximal control of the vegetation on the climate of the Earth, because they are subject to at least two major limitations: They do not consider any oceanic feedbacks, which

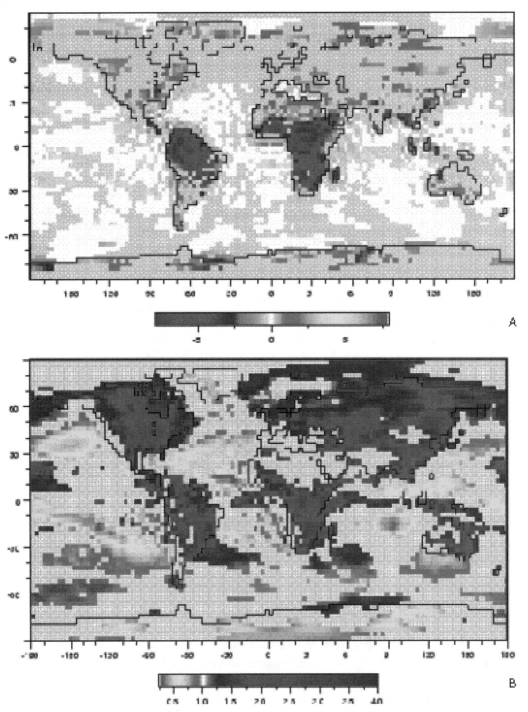

Figure 4.2. A: Difference in annual mean surface temperature (2m temperature) between the green planet and the desert world simulation. Colored areas indicate regions with significant (Student t-test, p > 0.99), and gray areas indicate regions with insignificant, temperature differences. B: Ratio of precipitation in the green planet to precipitation in the desert world simulation. Colored areas indicate regions with significant (Student t-test, p > 0.99), and gray areas indicate regions with insignificant, precipitation differences. See color plate section for color version.

might amplify or lessen the vegetation impact to a yet unknown degree. Furthermore, the prescribed surface vegetation in the green planet simulation might not be ecologically sustainable in every part of the world. Nevertheless, the simulations serve to illustrate that indeed, vegetation must be considered an important integral part in any credible Earth System Model.

The drastic changes investigated here are not likely to occur within the next century. Nevertheless, there exist several more-subtle feedback mechanisms in the vegetation-climate system that might significantly affect the response of the climate system to the anticipated future change in greenhouse gas forcing. For example, a recent simulation study has demonstrated the effects of possible changes in stomata control on evapotranspiration and thus on the surface temperature (Sellers et al., 1996). Plants regulate their stomata conductivity to make maximum use of the available leaf water. The stomata, however, are also the pathway for the CO_2 that is needed for photosynthesis. From empirical studies it is well established that the conductivity is not only inversely related to water stress but also directly proportional to the rate of leaf photosynthesis (Ball, 1988; Collatz et al., 1991). Hence, photosynthesis of atmospheric CO_2 can directly control leaf conductance and thus evapotranspiration, and this process provides a tight coupling of the cycles of carbon and water. The consequences of this tight coupling have only recently been investigated in atmospheric general circulation models that, necessarily, include a description of the terrestrial carbon cycle as part of the land surface package. Using such a model it has been demonstrated (Sellers et al., 1996; Bounoua et al., 1999) that this coupling might considerably enhance the land surface warming under a doubled atmospheric CO_2 concentration. If the latter is rising, plants may profit by increased assimilation; however, they may also "decide" to downregulate photosynthesis to present-day levels – for example, by developing fewer stomata per leaf area (Woodward, 1987) or by more intensive but shorter leaf photosynthesis periods (Hattenschwiler and Koerner, 1996; Koerner, 1998). Whether downregulation indeed occurs in a world with enhanced CO_2 is questionable, but it reflects a possibility. If so, evapotranspiration would be significantly reduced, yielding an additional surface warming beyond the direct CO_2-induced radiative effect. According to the present simulations (Sellers et al., 1996; Bounoua et al., 1999), this additional warming might be on the order of 30% of the direct response.

4.3   Exchanges of Carbon Dioxide

It is well known that in addition to the ocean, the terrestrial biosphere plays an important role in the global carbon cycle and the regulation of the atmospheric CO_2 concentration. There exists the fundamental question of the fate of the excess CO_2 that is directly injected into the atmosphere by anthropogenic activities (burning of fossil fuels and from changes in land use). This question traditionally concerns the atmospheric budget of CO_2 and how this changes with time as a function of the increasing CO_2 concentration (Schimel et al., 1995, 1996; Heimann, 1997; Prentice et al., 2000; Prentice, Chapter 11, this volume). There exists also a scientifically challenging second question, regarding the climatic feedbacks on the global carbon cycle.

Clearly, as witnessed by the changes in CO_2 during the glacial cycles (Petit et al., 1999) or during the Holocene (Indermühle et al., 1999), there exists considerable scope in the Earth System for climate-induced shifts in carbon storage between the different reservoirs on different time scales.

As a climate forcing agent, only the mean atmospheric CO_2 concentration is of any significance, because the spatiotemporal concentration variations induced by sources and sinks at the surface of the Earth are too small to have any radiative effects. Neverthe-less, an accurate observation of these variations all over the globe provides an important means to evaluate the surface exchange fluxes simulated by carbon cycle models. As part of the international Carbon Cycle Model Linkage Project (CCMLP), funded in part by the U.S. Electric Power Research Institute (EPRI) and with support from the Global Analysis, Interpretation and Modelling (GAIM) task force of the IGBP, several global terrestrial carbon cycle models have been scrutinized using this approach (Heimann et al., 1998; Kicklighter et al., 1999).

Most pronounced is the seasonal cycle with atmospheric variations of as much as 20 ppmv in high latitudes in the north, which gradually becomes smaller in mid-latitudes and toward the equator, vanishes south of the equator, and reappears in the deep Southern Hemisphere with opposite phase. Figure 4.3 shows the result of a simulation

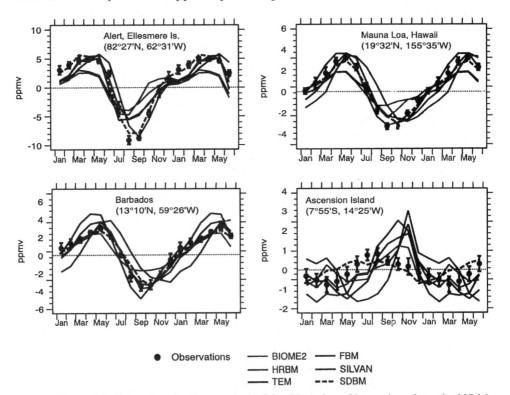

Figure 4.3. Seasonal cycle of atmospheric CO_2. Black dots: Observations from the NOAA-CMDL database (Conway et al., 1994), with error bars reflecting interannual variability. Colored solid lines: simulation results from five prognostic terrestrial carbon cycle models (Heimann et al., 1998). Dashed black line: simulation by a diagnostic terrestrial model driven by remote sensing and climate data (Knorr and Heimann, 1995). See also color plate section.

study, in which five geographically explicit, process-based terrestrial models are evaluated by injecting the simulated surface exchange fluxes into a global atmospheric transport model to calculate the concentration variations at the monitoring stations (Heimann et al., 1998). The terrestrial models included in the study are of intermediate complexity. They describe the major processes of the cycling of carbon (photosynthesis, leaf development and abscission, allocation of carbon to different parts of the plants, litter fall, soil carbon, heterotrophic respiration) on a global grid of 0.5° latitude and longitude in a prognostic fashion, but with a prescribed, fixed vegetation distribution. For comparison, the results from a sixth model are also shown (Simple Diagnostic Biosphere Model [SDBM], Knorr and Heimann, 1995). This is not a prognostic model but uses satellite remote sensing information ("greenness index" NDVI) to describe the photosynthetic activity of the vegetation.

Shown are the simulated seasonal cycles of atmospheric CO_2 at four stations in the Northern Hemisphere and the Tropics: Alert in northern Canada, Mauna Loa, Hawaii, Barbados in the Antilles, and Ascension Island in the southern Atlantic Ocean. The observations are from the global monitoring network of the Climate Monitoring and Diagnostic Laboratory of the U.S. Oceanographic and Atmospheric Administration (NOAA-CMDL) (Conway et al. 1994); the error bars reflect both instrumental errors and interannual variability. The modeled signals also include an oceanic component derived from a global ocean carbon cycle model (Six and Maier-Reimer, 1996), which, however, is very small at the Northern Hemisphere stations.

The simulations show that the agreement with the observations is not perfect. Indeed, by comparing the simulations of the prognostic models with the diagnostic SDBM, it is seen that the latter matches the atmospheric observations significantly better (Heimann et al., 1998). This indicates that the modeled processes leading to the leaf area still require refinement. The simulation results at Ascension Island also show that all the prognostic models overestimate the observed seasonal cycle, an error that can be traced to defects in the modeled water cycle, in particular to an assumed too shallow rooting depth in the Tropics.

Modeling the seasonal cycle in this way has become a standard test for the evaluation of terrestrial carbon cycle models (Nemry et al., 1999). Clearly, a test such as shown here addresses only particular aspects of a terrestrial model and cannot be conclusive.

An additional atmospheric signature of variability in the carbon cycle is provided by the interannual fluctuations in the growth rate of atmospheric CO_2. After subtraction of the seasonal cycle, the globally averaged CO_2 concentration exhibits significant variations that have been shown to correlate with large-scale climate fluctuations, such as the El Niño-southern oscillation (Bacastow, 1976; Bacastow et al., 1980) or the climate anomaly induced by the Mt. Pinatubo volcanic eruption after 1992 (Sarmiento, 1993). These variations are substantial, showing year-to-year variations of as much as 2–$4 \, \mathrm{PgC} \, a^{-1}$ or more than 50% of the annual emissions from the burning of fossil fuels. They are larger than the documented interannual changes in the anthropogenic sources and hence clearly represent the response of the global carbon cycle to short-term climatic fluctuations. Unfortunately, both the ocean and the terrestrial carbon cycle contribute to these fluctuations, and although concurrent measurements of the $^{13}C/^{12}C$ isotopic composition of atmospheric CO_2 could in principle be used to discriminate between the two components, the existing isotope ratio observations are contradictory (Keeling et al., 1995; Francey et al., 1995). Recent modeling and observational studies neverthe-less point to a relatively smaller contribution of the ocean to the interannual variability of atmospheric CO_2.

These interannual variations are a prime target for the assessment of the short-term climate sensitivity of terrestrial carbon cycle models. Within CCMLP a series of model simulations have been carried out in which the carbon cycle models have been driven by the observed changes in temperature and precipitation compiled by the Climate Research Unit of the University of East Anglia (updated from Jones, 1994; Hulme et al., 1998). Interestingly, the terrestrial models were able to reproduce a sig-nificant fraction of the observed interannual variability generated in the Tropics, both in magnitude and in phase. They were less successful, however, in the simulation of the extratropical fluctuations, in particular the Mt. Pinatubo anomaly after 1992 (Heimann et al., unpublished]. This indicates a too-weak modeled climate sensitivity in temper-ate latitudes. As an example, Figure 4.4 shows the global response of one of the more successful prognostic models compared to the observations. From the latter, the an-thropogenic long-term trend has been removed. The agreement between simulation and observations is fair, that is, significant at the 90% confidence level ($r^2 = \sim 0.7$), but the modeled amplitude of the variations is too large. Also shown, for compari-son, is the globally averaged atmospheric growth rate as modeled by an oceanic carbon cycle model driven with the observed surface fluxes of wind stress, temperature, and freshwater fluxes from the European Centre for Medium Range Weather Forecasts

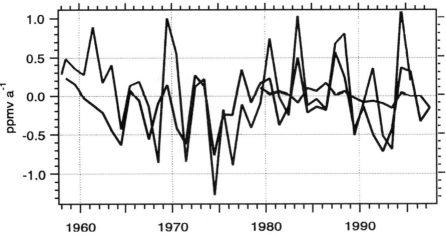

Figure 4.4. Interannual, climate-driven variability in the global carbon cycle. Red line: observed growth rate of globally averaged atmospheric CO_2 concentration after subtraction of the an-thropogenic trend (Heimann, 1997). Green line: growth rate of atmospheric CO_2 simulated by a process-based terrestrial model driven by observed changes in temperature and precipitation (Sitch et al., in preparation). Blue line: growth rate of atmospheric CO_2 simulated by a coupled ocean general circulation–carbon cycle model forced with observed anomalies of surface wind stress, temperatures, and freshwater fluxes (LeQuéré, in press). See also color plate section.

(ECMWF) analyses over the period 1979–1997 (LeQuéré et al., in press). Clearly, the oceanic contribution is dwarfed by the terrestrial fluctuations.

4.4   Exchanges of Oxygen and Its Isotopes

Parallel to the exchanges of water vapor and carbon dioxide, oxygen is also exchanged by the terrestrial biosphere. This exchange proceeds in relatively well-known, fixed stoichiometric ratios of about 1.1 mole of oxygen released for 1 mole of CO_2 assimilated during photosynthesis and reversed during oxidation of organic matter. On geological time scales this exchange is crucial for the habitability of the planet. The present amounts of oxygen in the atmosphere, however, are so large that the fluctuations induced by biotic and abiotic processes are small (a few ppmv on a background concentration of 200,000 ppmv). As a result, their measurement has only very recently become feasible. Variations in the ratio of molecular oxygen to nitrogen (which is the quantity that is measured) provide a means to separate terrestrial from oceanic carbon exchange fluxes (Keeling and Shertz, 1992; Keeling et al., 1993; Keeling et al., 1996; Bender et al., 1996). This is because CO_2 and molecular O_2 exchanges with the ocean exhibit quite different stoichiometric relations compared with the exchanges on land. Because of the relatively well-known, fixed stoichiometric relations for terrestrial exchanges, atmospheric oxygen measurements primarily have been used to evaluate oceanic models, such as through the seasonal cycle of CO_2 and O_2 (Keeling et al., 1998) or the mean annual meridional gradient in CO_2 and O_2 (Stephens et al., 1998).

Oxygen consists of three stable isotopes (^{16}O, ^{17}O, and ^{18}O) with characteristic isotopic ratios in the different molecular forms (H_2O, CO_2, and O_2). Chemical, physical, and biological transformation processes induce fractionations, which lead to different isotopic ratios in the various reservoirs of these compounds (e.g., Keeling, 1995). As described below, until now primarily the $^{18}O/^{16}O$ ratio has been of interest in observational and modeling studies. Because almost all fractionation processes are mass-dependent, the behavior of ^{17}O can be predicted from ^{18}O and hence appears to be redundant. However, it has recently been found that exchange processes with the ozone chemistry in the stratosphere induce a mass-independent fractionation that is not yet completely understood (Thiemens et al., 1991; Mauersberger et al., 1993, 1999). This effect creates a ^{17}O anomaly that may be traced through the various exchange processes of CO_2 and O_2 within the entire biogeochemical system.

To model these isotopes, the cycles of water, carbon, and oxygen all must be described in a consistent way. This requires an atmospheric general circulation model that includes a full description of the oxygen isotopes in the hydrological cycle (Hoffmann et al., 1998). This provides the background isotopic composition of precipitation, groundwater, and water vapor all over the globe, which, together with the energy and water balance of the canopy, determines the isotopic composition of leaf water. This in turn is the anchor point that determines the isotopic composition of molecular oxygen released to the atmosphere during photosynthesis, and it also imprints its isotopic signature to the CO_2 that is exchanged through the stomata of the leaves with the atmosphere. Conversely, during respiration the CO_2 released carries the isotopic signature of the leaf and soil water, while the strongly fractionated accompanying oxygen consumption induces an isotopic signal in the atmospheric oxygen. The oxygen isotopic composition of atmospheric CO_2 and O_2 is also influenced by oceanic exchanges and by the (small) impacts from the anthropogenic burning of fossil fuels.

In the case of molecular oxygen, these exchange processes lead to an atmosphere enriched in ^{18}O compared with mean ocean water, known as the Dole effect (Dole, 1935). Variations in this quantity in the past reflect not only the relative strength of terrestrial and oceanic productivity (Bender et al., 1994) but also the geographical location of terrestrial oxygen exchanges, which may change with the displacements of the vegetation zones during the glacial–interglacial cycles. The isotopic composition of atmospheric O_2 is also not uniform but is expected to exhibit spatiotemporal variations. These variations are relatively small (∼0.02%, Seibt, 1997), but it is expected that they will become measurable within the near future. The fractionation effects associated with the exchanges of CO_2, however, generate easily detectable atmospheric signatures (Francey and Tans, 1987).

All the isotopic fractionation processes are relatively well understood and can be described in models (Ciais et al., 1997a; Ciais et al., 1997b; Seibt, 1997). In these modeling studies, however, the isotopic tracers are not yet treated in a fully consistent way but are simulated using an off-line model hierarchy. That is, an atmospheric general circulation model computes the climate fields and the isotopic composition of precipitation, groundwater, and near surface water vapor. These output fields are then used to drive a terrestrial carbon cycle model that computes the exchange fluxes of CO_2 and O_2 and their isotopic compositions. In a third step, these exchange fluxes are given to an atmospheric transport model for the computation of the atmospheric signatures in the concentration and isotopic ratio fields at the observing stations. Clearly, with the advent of comprehensive DGVMs that include the full biogeochemical cycles and can be coupled to climate models, simulation studies of the oxygen isotopic tracers provide a promising approach for the evaluation of the modeled exchanges on a regional and global scale.

4.5   Conclusion

Terrestrial biospheric exchanges of water, carbon dioxide, and oxygen are controlled essentially by the same set of processes. Hence, a realistic representation in comprehensive biogeochemical models necessitates a consistent treatment of all three species. The existing models have not yet reached the stage depicted in Figure 4.1B, in that they still are being used mostly off-line in an off-line mode, that is, driven by the output of climate models. Nevertheless, the process knowledge exists to represent the various exchange processes in a realistic manner as an interactive component within comprehensive Earth System Models, and, as shown here, appropriate techniques exist that allow an evaluation of the model performances on regional and global scales.

Through exchanges of H_2O and CO_2, the biosphere can significantly modify the climate of the Earth. Hence, a consistent implementation of the terrestrial carbon cycle and its coupling to the hydrological cycle on land represent an immediate next model development step. Eventually, the coupling to existing comprehensive models of the oceanic carbon cycle will permit a consistent treatment of the carbon cycle in the next-generation atmosphere-ocean climate models. Such models are needed to investigate the history of climate, and they are also an indispensable tool for the assessment of anthropogenically induced climate change in the near future.

As outlined here, the cycle of oxygen and its isotopes is tightly coupled to the carbon cycle. Hence, variations of atmospheric molecular O_2 and of the stable oxygen isotope ratios in H_2O, CO_2, and O_2 provide an additional tool to evaluate the modeled biospheric exchange processes. Variations of these tracers are being or will be monitored soon, and their temporal variations on longer time scales have been observed in ice cores. It is hoped that the inclusion of the cycles of oxygen and its isotopes in the next generation of coupled climate–biogeochemistry models will receive high priority.

REFERENCES

Bacastow, R. B. (1976). Modulation of atmospheric carbon dioxide by the southern oscillation. Nature, 261, 116–118.

Bacastow, R. B., Adams, J. A., Keeling, C. D., Moss, D. J., Whorf, T. P., and Wong, C. S. (1980). Atmospheric carbon-dioxide, the Southern Oscillation, and the weak 1975 El-Nino. Science, 210, 66–68.

Ball, J. T. (1988). An analysis of stomatal conductance. Ph.D. Thesis, Stanford University, Stanford, 89 pp.

Bender, M., Ellis, T., Tans, P., Francey, R., and Lowe, D. (1996). Variability in the O_2/N_2 ratio of southern hemisphere air, 1991–1994 – Implications for the carbon cycle. Global Biogeochemical Cycles, 10, 9–21.

Bender, M., Sowers, T., and Labeyrie, L. (1994). The Dole effect and its variations during the last 130,000 years as measured in the Vostok ice core. Global Biogeochemical Cycles, 8, 363–376.

Bounoua, L., Collatz, G. J., Sellers, P. J., Randall, D. A., Dazlich, D. A., Los, S. O., Berry, J. A., Fung, I., Tucker, C. J., Field, C. B., and Iensen, T. G. (1999). Interactions between vegetation and climate: Radiative and physiological effects of doubled atmospheric CO_2. Journal of Climate, 12, 309–324.

Ciais, P., Denning, A. S., Tans, P. P., Berry, J. A., Randall, D. A., Collatz, G. J., Sellers, P. J., White, J. W. C., Trolier, M., Meijer, H. A. J., Francey, R. J., Moufray, P., and Heimann, M. (1997a). A three-dimensional synthesis study of $\delta^{18}O$ in atmospheric CO_2. 1. Surface fluxes. Journal of Geophysical Research-Atmospheres, 102, 5857–5872.

Ciais, P., Tans, P. P., Denning, A. S., Francey, R. J., Trolier, M., Meijer, H. A. J., White, J. W. C., Berry, J. A., Randall, D. A., Collatz, G. J., Sellers, P. J., Moufray, P., and Heimann, M. (1997b). A three-dimensional synthesis study of $\delta^{18}O$ in atmospheric CO_2. 2. Simulations with the TM2 transport model. Journal of Geophysical Research-Atmospheres 102, 5873–5883.

Collatz, G. J., Ball, J. T., Grivet, C., and Berry, J. A. (1991). Physiological and environmental-regulation of stomatal conductance, photosynthesis and transpiration – a model that includes a laminar boundary-layer. Agricultural and Forest Meteorology, 54, 107–136.

Conway, T. J., Tans, P. P., Waterman, L. S., and Thoning, K. W. (1994). Evidence for interannual variability of the carbon cycle from the National Oceanic and Atmospheric Administration Climate Monitoring and Diagnostics Laboratory global air sampling network. Journal of Geophysical Research-Atmospheres, 99, 22831–22855.

Dole, M. (1935). The relative atomic weight of oxygen in water and air. J. Am. Chem. Soc., 57, 2731.

Fraedrich, K., Kleidon, A., and Lunkeit, F. (1999). A green planet versus a desert world: Estimating the effects of vegetation extremes on the atmosphere. J. of Climate, in press.

Francey, R. J., and Tans, P. P. (1987). Latitudinal variation in ^{18}O of atmospheric CO_2. Nature, 327, 495–497.

Francey, R. J., Tans, P. P., Allison, C. E., Euting, I. G., White, J. W. C., and Trolier, M. (1995). Changes in oceanic and terrestrial carbon uptake since 1982. Nature, 373, 326–330.

Hattenschwiler, S., and Korner, C. (1996). System-level adjustments to elevated CO_2 in model spruce ecosystems. Global Change Biology, 2, 377–387.

Heimann, M. (1997). A review of the contemporary global carbon cycle and as seen a century ago by Arrhenius and Högbom. Ambio, 26, 17–24.

Heimann, M., Esser, G., Haxeltine, A., Kaduk, J., Kicklighter, D. W., Knorr, W., Kohlmaier, G. H., McGuire, A. D., Melillo, J., Moore, B., Otto, R. D., Prentice, I. C., Sauf, W., Schloss, A., Sitch, S., Wittenberg, U., and Wurth, G. (1998). Evaluation of terrestrial Carbon Cycle models through simulations of the seasonal cycle of atmospheric CO_2: First results of a model intercomparison study. Global Biogeochemical Cycles, 12, 1–24.

Hoffmann, G., Werner, M., and Heimann, M. (1998). Water isotope module of the ECHAM atmospheric general circulation model – a study on timescales from days to several years. Journal of Geophysical Research-Atmospheres, 103, 16871–16896.

Hulme, M., Osborn, T. J., and Johns, T. C. (1998). Precipitation sensitivity to global warming: Comparison of observations with HadCM2 simulations. Geophys. Res. Letts., 25, 3379–3382.

Indermühle, A., Stocker, T. F., Joos, F., Fischer, A., Smith, H. J., Wahlen, M., Deck, B., Mestroianni, D., Tschumi, J., Bluuier, T., Meyer. R., and Stauffer, B. (1999). Holocene carbon-cycle dynamics based on CO_2 trapped in ice at Taylor Dome, Antarctica. Nature, 398, 121–126.

Jones, P. D. (1994). Hemispheric surface air temperature variations: a reanalysis and an update to 1993. J. Clim., 7, 1794–1802.

Keeling, C. D., Whorf, T. P., Wahlen, M., and Vanderplicht, J. (1995). Interannual extremes in the rate of rise of atmospheric carbon dioxide since 1980. Nature, 375, 666–670.

Keeling, R. F., Stephens, B. B., Najjar, R. G., Doney, S. C., Archer, D., and Heimann, M. (1998). Seasonal variations in the atmospheric O_2/N_2 ratio in relation to the kinetics of air-sea gas exchange. Global Biogeochemical Cycles, 12, 141–163.

Keeling, R. F., Piper, S. C., and Heimann, M. (1996). Global and hemispheric CO_2 sinks deduced from changes in atmospheric O_2 concentration. Nature, 381, 218–221.

Keeling, R. F. (1995). The atmospheric oxygen cycle – the oxygen isotopes of atmospheric CO_2 and O_2 and the O_2/N_2 ratio. Reviews of Geophysics, 33, 1253–1262.

Keeling, R. F., Najjar, R. P., Bender, M. L., and Tans, P. P. (1993). What atmospheric oxygen measurements can tell us about the global carbon-cycle. Global Biogeochemical Cycles, 7, 37–67.

Keeling, R. F., and Shertz, S. R. (1992). Seasonal and interannual variations in atmospheric oxygen and implications for the global carbon-cycle. Nature, 358, 723–727.

Kleidon, A., Fraedrich, K., and Heimann, M. (1999). A green planet versus a desert world: Estimating the maximum effect of vegetation on the land surface climate. Climatic Change, in press.

Kleidon, A., and Heimann, M. (1998). Optimised rooting depth and its impacts on the simulated climate of an atmospheric general circulation model. Geophysical Research Letters, 25, 345–348.

Kicklighter, D. W., Bruno, M., Donges, S., Esser, G., Heimann, M., Helfrich, J., Ift, F., Joos, F., Kaduk, J., Kohlmaier, G. H., McGuire, A. D., Melillo, J. M., Meyer, R., Moore, B., Nadler, A., Prentice, I. C., Sauf, W., Schloss, A. L., Sitch, S., Wittenberg, U., and Wurth, G. (1999). A first-order analysis of the potential role of CO_2 fertilization to affect the global carbon budget: a comparison of four terrestrial biosphere models. Tellus, B51, 343–366.

Knorr, W., and Heimann, M. (1995). Impact of drought stress and other factors on seasonal land biosphere CO_2 exchange studied through an atmospheric tracer transport model. Tellus, B47, 471–489.

Koerner, C. (1998). Tropical forests in a CO_2-rich world. Climatic Change, 39, 297–315.

Le Quéré, C., Orr, J. C., Monfray, P., Aumont, O., and Madec, G. (1999). Interannual variability of the oceanic sink of CO_2 from 1979 through 1997. Global Biogeochemical Cycles, in press.

Mauersberger, K., Erbacher, B., Krankowsky, D., Gunther, J., and Nickel, R. (1999). Ozone isotope enrichment: Isotopomer-specific rate coefficients. Science, 283, 370–372.

Mauersberger, K., Morton, J., Schueler, B., Stehr, J., and Anderson, S. M. (1993). Multi-isotope study of ozone – implications for the heavy ozone anomaly. Geophysical Research Letters, 20, 1031–1034.

Nemry, B., Francois, L., Gerard, J. C., Bondeau, A., and Heimann, M. (1999). Comparing global models of terrestrial net primary productivity (NPP): Analysis of the seasonal atmospheric CO_2 signal. Global Change Biology, 5(Suppl 1), 65–76.

Nepstad, D. C. et al. (1994). The role of deep roots in the hydrological and carbon cycles of amazonian forests and pastures. Nature, 372, 666–669.

Petit, J. R., Jouzel, J., Raynaud, D., Barkov, N. I., Barnola, J. M., Basile, I., Bender, M., Chappellaz, J., Davis, M., Delaygue, G., Delmotte, M., Kotlyakov, V. M., Legrand, M., Lipenkov, V. Y., Lorius, C., Pepin, L., Ritz, C., Saltzman, E., and Stievenard, M. (1999). Climate and atmospheric history of the past 420,000 years from the Vostok ice core, Antarctica. Nature, 399, 429–436.

Prentice, I. C., Cramer, W., Harrison, S. P., Leemans, R., Monserud, R. A., and Solomon, A. M. (1992). A global biome model based on plant physiology and dominance, soil properties and climate. Journal of Biogeography, 19, 117–134.

Prentice, I. C., Heimann, M., and Sitch, S. (2000). The carbon balance of the terrestrial biosphere: Ecosystem models and atmospheric observations. Global Ecology and Biogeography Letters, in press.

Roeckner, E., Oberhuber, J. M., Bacher, A., Christoph, M., and Kirchner, I. (1996). ENSO variability and atmospheric response in a global coupled atmosphere-ocean GCM. Climate Dynamics, 12, 737–754.

Sarmiento, J. L. (1993). Carbon-cycle–atmospheric CO_2 stalled. Nature, 365, 697–698.

Schimel, D. S., Alves, D., Enting, I., Heimann, M., Joos, F., Raynaud, D., Wigley, T., Prather, M., Derwent, R., Ehhalt, D., Fraser, P., Sanhueza, E., Zhou, X., Jonas, P., Charlson, R., Rodhe, H., Sadasivan, S., Shine, K. P., Fouquart, Y., Ramaswamy, V., Solomon, S., Srinivasan, J., Albritton, D., Isaksen, I., Lal, M., and Wuebbles, D. (1996). Radiative forcing of climate change. In: J. T. Houghton, L. G. M. Filho, B. A. Callander, N. Harris, A. Kattenberg, and K. Maskell (Editors), Climate Change 1995: The Science of Climate Change. Cambridge University Press, Cambridge, pp. 65–132.

Schimel, D. S., Enting, I. G., Heimann, M., Wigley, T. M., Raynaud, D., Alves, D., and Siegenthaler, U. (1995). CO_2 and the carbon cycle. In: J. T. Houghton, L. G. M. Filho, J. Bruce, H. Lee, B. A. Callander, E. Haites, N. Harris, and K. Maskell (Editors), Climate Change 1994: Radiative Forcing of Climate Change and an Evaluation of the IPCC IS92 Emission Scenarios. Cambridge University Press, Cambridge, pp. 35–71.

Seibt, U. (1997). Simulation der $^{18}O/^{16}O$-Zusammensetzung von atmosphärischem Sauerstoff (Simulation of the $^{18}O/^{16}O$-composition of atmospheric oxygen), Diploma thesis, University of Tübingen (in German). Published as Examensarbeit Nr. 48, Max-Planck-Institut für Meteorologie, Hamburg, Germany.

Sellers, P. J., Bounoua, L., Collatz, G. J., Randall, D. A., Dazlich, D. A., Los, S. O., Berry, J. A., Fung, I., Tucker, C. J., Field, C. B., and Jensen, T. G. (1996). Comparison of radiative and physiological effects of doubled atmospheric CO_2 on climate. Science, 271, 1402–1406.

Shukla, J., and Mintz, Y. (1982). Influence of land-surface evapo-transpiration on the earth's climate. Science, 215, 1498–1501.

Six, K. D., and Maier-Reimer, E. (1996). Effects of plankton dynamics on seasonal carbon fluxes in an ocean general circulation model. Global Biogeochemical Cycles, 10, 559–583.

Stephens, B. B. et al. (1998). Testing global ocean carbon cycle models using measurements of atmospheric O_2 and CO_2 concentration. Global Biogeochemical Cycles, 12, 213–230.

Thiemens, M. H., Jackson, T., Mauersberger, K., Schueler, B., and Morton, J. (1991). Oxygen isotope fractionation in stratospheric CO_2. Geophysical Research Letters, 18, 669–672.

Woodward, F. I. (1987). Stomatal numbers are sensitive to increases in CO_2 from preindustrial levels. Nature, 327, 617–618.

5   Carbon Futures

WALLACE S. BROECKER

ABSTRACT

In this chapter I summarize my views as to where we stand with regard to our knowledge of the way in which the fossil fuel CO_2 we release has distributed and will distribute itself among the atmosphere, ocean, and terrestrial biosphere reservoirs. I also make the case that during the next two decades, we must carefully explore all the aspects of purposeful disposal of fossil fuel-derived CO_2 both on the sea floor and in deep continental reservoirs.

5.1   Introduction

The Kyoto Protocols have enhanced the interest of researchers regarding the fate of the CO_2 produced as a result of fossil fuel burning. In particular, attention has been focused on the forces driving excess storage of carbon in the terrestrial biosphere. This is indeed an important research thrust, but in my opinion, these protocols direct too much attention to the next few decades when, instead, we should be thinking in terms of a century or more, for it is on this time scale that the most serious threats posed by the ongoing build-up of greenhouse gases are likely to be manifested. What will matter 70 to 130 years from now will be the CO_2, CH_4, and N_2O contents of the atmosphere rather than the shape of the path by which these levels are achieved. In the scramble to meet the Kyoto limits, countries will be tempted to resort to promises of expanded storage of carbon in forests. The availability of this loophole tends to downplay the need to launch an all-out effort to create the wherewithal to decarbonize our energy system through the development of new alternate energy systems and the means to capture and store the CO_2 released from the existing systems. Several decades from now, critical decisions regarding the extent and rapidity of deployment of these systems will have to be made. These decisions will presumably be based on the magnitude of the greenhouse impacts observed during the next two or three decades.

In any case, let us look at the large-scale picture of carbon sources and sinks. In Figure 5.1, I summarize the sizes of the important reservoirs. Although not everyone will agree with my choices, the conclusions would be much the same no matter whose values I were to adopt. The first point is that the build-up of CO_2 will not be limited by availability of fossil fuels; even when petroleum and natural gas become scarce, abundant

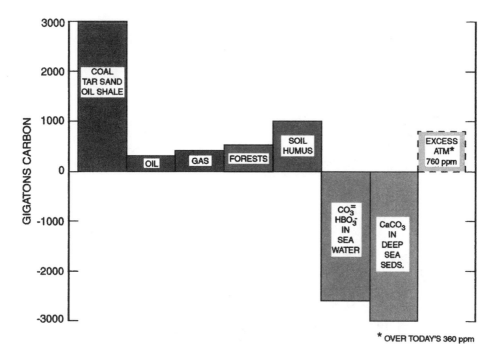

Figure 5.1. Potential carbon sources and sinks. The first three bars show estimates of the sizes of the as yet untapped reservoirs of fossil fuel carbon. The next two bars show the amounts of carbon currently stored in forests and soil humus. The next bar shows the total buffer capacity of the sea, that is,

$$CO_2 + CO_3^= + H_2O \rightarrow 2HCO_3^-$$

and

$$CO_2 + HBO_3^- + H_2O \rightarrow HCO_3^- + H_2BO_3$$

The next bar shows the amount of $CaCO_3$ in the upper meter of deep sea sediments (i.e., that available for dissolution on the time scale of several thousand years):

$$CO_2 + CaCO_3 + H_2O \rightarrow 2HCO_3^- + Ca^{++}$$

Finally, the last bar shows the amount of extra CO_2 required to raise its content in the atmosphere by 400 ppm.

coal, oil shale, and tar sand reserves will remain. Second, there is, of course, no limit to how much CO_2 the ocean-atmosphere-biosphere system can hold. Rather, the question is how the CO_2 we release will distribute itself among these reservoirs.

5.2   Atmosphere-Ocean Partitioning

Were the rate of CO_2 exchange between air and sea and the rate of mixing within the sea infinitely rapid, then five-sixths of the CO_2 we have produced to date would have taken up residence in the sea. Only one-sixth would remain airborne. But based on the observed distribution of anthropogenic tracers (^{14}C, ^3H ...) in the sea, it has been shown that only about 15% of the sea's capacity for CO_2 uptake is being utilized. The reason is that the resistance to uptake is dominated by the finite rate of transfer of water from the surface to the interior of the sea. Because of this limitation, only about

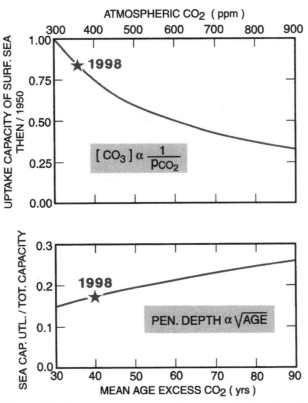

Figure 5.2. The upper panel approximates the decline of the buffer capacity (and hence the capacity for excess CO_2 uptake) of surface ocean water as a function of atmospheric CO_2 content. The lower panel approximates the change in the fraction of the sea's buffer capacity utilized as a function of the mean age of fossil fuel CO_2 molecules. The square root relationship comes from the observation based on natural radiocarbon that the entire sea is ventilated on the time scale of 10^3 years and the observation that on the time scale of a decade, bomb ^{14}C and ^{3}H were spread through about 10% of the ocean's volume.

two-sixths of the CO_2 we have produced has been taken up by the sea. Most of the other four-sixths remains in the atmosphere.

The partitioning between air and sea will evolve with time. Two opposing tendencies will influence this evolution. First, CO_2 is taken up mainly because it reacts with $CO_3^=$ and HBO_3^- present in the sea to form HCO_3^- and H_2BO_3. Because the amounts of $CO_3^=$ and HBO_3^- in the ocean are finite, the uptake of CO_2 is driving down their concentrations and hence also the sea's capacity for further uptake (see Figure 5.2). The uptake capacity of surface seawater (per ppm of atm CO_2 rise) can be approximated as inversely proportional to the CO_2 partial pressure in the overlying air. Thus, were the atmosphere's CO_2 content to reach twice today's (i.e., 720 ppm), then, at that time, the uptake potential for excess CO_2 by surface waters would drop to about half of today's value. As can be seen by comparison with the exact calculations listed in Table 5.1, although not exact, this approximation is useful for first-order thinking.

The second aspect of the evolution of the air-sea partitioning has to do with the fraction of the sea's capacity that is utilized. This fraction depends on the time available

Table 5.1. CO_2 Uptake Capacity of Ocean Water as a Function of Temperature and Carbon Dioxide Partial Pressure. As can be Seen, a Tripling of the Atmosphere's CO_2 Content Leads to a 2.5-Fold Reduction in the $CO_3^=$ Ion Content and a Fourfold Reduction in the Uptake Capacity Per Unit Rise in Atmospheric CO_2 Content for Thermocline and Deep Ocean Water Source Waters. These Calculations were Carried out Using the Accepted Equilibrium Constants for Carbonate and Borate. The Starting Point was the Composition for Three Arbitrarily Selected Samples Measured as Part of the GEOSECS Program.

T °C	Alk μeq/kg	PCO_2 μatm	ΣCO_2 μmol/kg	$CO_3^=$ μmol/kg	$\Delta\Sigma CO_2/\Delta pCO_2$ μmol/kg/μatm
Warm Surface Ocean					
26.7	2375	300	1967.1	279	0.78
		500	2082.2	206	0.42
		700	2150.1	164	0.27
		900	2196.4	137	0.20
Thermocline Outcrops					
11.3	2295	300	2053.3	167	0.61
		500	2140.7	114	0.30
		700	2190.2	87	0.20
		900	2223.9	71	0.15
Deep Ocean Outcrops					
1.0	2339	300	2158.6	123	0.54
		500	2235.8	82	0.28
		700	2280.4	62	0.18
		900	2311.9	49	0.14

for the CO_2 molecules to be carried from the sea surface into its interior (see Figure 5.2). The longer the time available, the deeper the penetration. Between the end of World War II and the Arab oil boycott in 1974, the fossil fuel CO_2 production had accelerated at a pace averaging close to 3% per year. Were this pace continued, the mean age of fossil fuel molecules would approach 69 ln 2/3, or 33 years. However, as in the decades preceding World War II and during the past two decades, the rate of increase has been less steep, so the mean age is closer to 40 years.

Taken together, the distributions of natural and bomb-test radiocarbon suggest that the fraction of the ocean's capacity utilized increases roughly as the square root of the penetration time (see Figure 5.3). So if, for example, the rate of rise in emissions were to be reduced to 1% per year, then at the end of the 21st century, the mean age of CO_2 molecules would approach 70 years (i.e., ln 2 yr.). The fraction of the ocean ventilated would rise by a factor approaching $\sqrt{70/40}$, or 1.3. In the extreme, if fossil fuel use were to cease in the year 2000, the mean age of the excess CO_2 molecules in the year 2100 would be 140 years and the fraction of the sea's capacity to be utilized would increase by $\sqrt{140/40}$, or by about a factor approaching 2. Hence, if these rules apply, no matter what scenario is envisioned for the next century, no more than 30% of the ocean's capacity for partitioning will be utilized by the year 2100.

But how reliable is the square root relationship? Although the distribution of natural radiocarbon constrains the rate at which CO_2 is carried into the deep sea by way

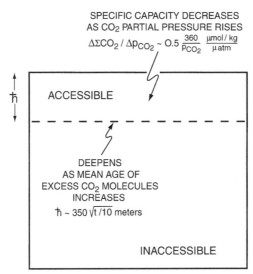

Figure 5.3. First-order model of ocean uptake of fossil fuel CO_2. The specific capacity of seawater to absorb excess CO_2 is a fixed function of temperature and chemical composition. For today's thermocline and deep sea, the ΣCO_2 content increases by about 0.5 μmol/kg per μatm increase in pCO_2. However, the uptake capacity drops as the CO_2 content of the atmosphere rises. Based on the distributions of natural radiocarbon and bomb radiocarbon and tritium, the volume of the ocean accessible to fossil fuel CO_2 varies roughly as the square root of time. Based on the mean penetration depth of bomb ^{14}C (i.e., 350 meters) one decade after the implementation of the ban on atmospheric testing, the penetration depth as a function mean age of fossil fuel molecules can be estimated.

of the thermo–haline circulation (deeper than 1750 meters) and the distribution of bomb ^{14}C (see Broecker et al., 1985) constrains the rate at which CO_2 is being stored in the thermocline (at depths less than 500 meters), we have as yet no strong constraint on the rate at which the ocean's intermediate waters (500–1750 meters) are being ventilated. Thus, we must either put our faith in the interpolation between the time scales for ventilation of the deep sea (centuries) and the time scales for ventilation of the thermocline (decades), or we must accept the as-yet somewhat shaky results of simulations made using ocean general circulation models (OGCMs) to estimate the storage in this all-critical mid-depth portion of the ocean. My impression is that we are still a long way from adequately modeling what goes on in this critical depth range.

When these two influences on the distribution of CO_2 between the ocean and atmosphere are considered together, for most emission scenarios, the reduction in capacity will roughly balance the expected increase in the penetration depth of CO_2. Thus, the current roughly 2 to 1 split between air and sea is likely to apply, at least for the next several decades.

In light of the large number of calculations carried out in OGCMs, one might conclude that the simplified calculations of the ocean presented here are unnecessary. However, one must keep in mind that although the models have been adjusted to yield reasonably good representations of the distributions of both natural and bomb radiocarbon, there currently is no way to confirm that their modes of ventilation of the intermediate depth range ocean are correct. Inasmuch as this reservoir will become

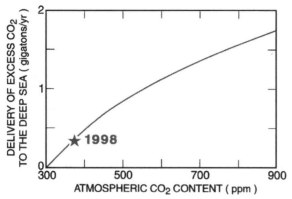

Figure 5.4. Estimates of the delivery of excess CO_2 to the deep sea via thermo-haline ventilation. The assumptions are (1) that 30 Sverdrups (i.e., 30×10^6 m^3/sec) of water descend to the deep sea from the polar outcrops, (2) that these descending waters have achieved CO_2 partial pressure close to that in the atmosphere (see Table 5.1), and (3) that on the time scale of a century, little of the excess CO_2-charged deep water is recirculated. Because the rate of deep-sea ventilation may decrease as the globe warms and because newly formed deep waters generally do not achieve equilibrium with the overlying atmosphere, the results of this calculation are likely to be upper limits.

ever more important during the course of the next century, the current model-based projections could well be flawed.

One aspect of the entry of fossil fuel CO_2 into the ocean can be usefully isolated from the rest. Deep waters formed in the northern Atlantic and in the Southern Ocean carry excess CO_2 into the deep sea. Because of the very long residence time of deep water in the ocean, over the course of the next century little of the CO_2 carried away from the surface in this way will return to the atmosphere. Based on an estimate of 30 Sverdrups for the average deep water ventilation flux (Broecker et al., 1998) and using the information listed in Table 5.1, a rough estimate of the amounts of CO_2 delivered to the deep can be made (see Figure 5.4).

One other factor must be considered. To the extent that the planet warms, the rate of ventilation of the ocean's interior is likely to decrease, reducing the fraction of uptake by the sea. I will return to this point.

5.3   Atmosphere-Terrestrial Biosphere Partitioning

The processes governing the partitioning of CO_2 between the atmosphere and the terrestrial biosphere are complex and not well understood. Most carbon cycle specialists agree that during the 1980s the terrestrial carbon inventory remained nearly constant, that is, greening roughly matched forest cutting. But, before 1989 there were no O_2/N_2 measurements, so this conclusion is perhaps open to challenge. Monitoring of the rate of O_2 decline (Keeling et al., 1996) makes clear that during the early 1990s greening outstripped forest cutting. The net uptake of CO_2 by the biosphere during this period balanced roughly 30% of the production of CO_2 by fossil fuel burning. Based on the geographic distribution of excess CO_2 in the atmosphere, most of this storage appears to have occurred in the North Temperate Zone. In a recent paper, based on a deconvolution of atmospheric CO_2 data Fan et al. (1998) suggest that the excess storage occurred mainly in North America. However, other treatments of the same data set do not support this conclusion.

What drives this excess storage? Four possibilities exist:

1. Unusually favorable growth conditions during the early 1990s allowed photosynthesis to temporarily outstrip respiration.
2. Regrowth of forests on land previously used for farming (this regrowth is confined mainly to temperate North America and Europe).
3. Fertilization by the excess atmospheric CO_2. (Because CO_2 fertilization is global, to explain the observation that storage has occurred mainly in the North Temperate Zone, one would have to conclude that deforestation in the Tropics has roughly balanced the CO_2 growth enhancement.)
4. Fertilization with fixed nitrogen released from automobiles and farms (these releases have occurred mainly in temperate North America and Europe).

With regard to future expectations for biospheric storage, it makes a significant difference which combination of these drivers is responsible. Most ecologists attribute enhanced storage to some combination of CO_2 and fixed nitrogen fertilization. This explanation is consistent with a north temperate sink but not with dominance by a sink in North America. Rather, the storage should be more nearly equally split between Eurasia and North America. Direct estimates of forest regrowth suggest that it can account for no more than 20% of the required excess storage. The fact that a dramatic increase in the excess storage term appears to have commenced at the end of the 1980s suggests a meteorologic explanation. If so, then according to the analysis by Fan et al. (1998), this impact was centered over North America. If indeed the explanation is meteorological, then these deconvolutions of atmospheric CO_2 data do not provide us with the information we need if we are to predict the century-long changes in terrestrial storage.

My guess is that fixed nitrogen is a more important driver than most ecologists believe it to be. Because the release and deposition of fixed nitrogen occurs mainly at the north temperate latitudes, this would fit the constraint based on the distribution of excess CO_2 in the atmosphere. I am suspicious that by thinking in terms of terrestrial "Redfield" ratios, we may be underestimating the power of nitrogen forcing. An analogy to an auto factory is useful in this regard. Fixed nitrogen may be more akin to assembly lines than to fenders. Thus, by increasing the supply of fixed nitrogen to forests, we are doing more than permitting carbon sequestration calculated as some set multiple of the number of moles of nitrogen. Rather, we are strengthening the apparatus that produces wood and humus. If so, we are elevating the steady-state carbon storage to a new plateau. But, as has been pointed out, the extent of this elevation has limits. Once the nitrogen supply saturates, as it has in most European forests, other ingredients (light, water, soil-derived nutrients, etc.) will become limiting. Hence, further addition of fixed nitrogen will no longer be effective.

All estimates of future storage are subject to question because of the potential influence of global warming on the humus inventory in soils. Everyone agrees that the warmer the temperature, the shorter the lifetime of soil humic compounds. But because both the magnitude of the expected global warming and the coefficient relating temperature and humus turnover times are uncertain, it is not clear to me whether, taken together, greening and warming will lead to a net increase or a net decrease in the carbon inventory of soils.

So what can be said about the evolution of biospheric storage over the next century? Can any useful limits be set? With regard to CO_2 fertilization, let us assume that CO_2 growth enhancement can increase wood stocks by no more than $\sqrt{p_{CO_2}^{then}}/\sqrt{p_{CO_2}^{now}}$. Were CO_2 concentration in air to rise to 650 ppm by the year 2100, then the maximum increase in standing biomass would be 500 ($\sqrt{650/360} - 1$), or 170 GtC. Of course, if there is a significant synergistic impact of fixed nitrogen, this number could be larger. But, on the other hand, the temptation to harvest tropical forests and the impact of rising soil temperatures loom large. Thus, I find it hard to believe that at the end of the next century storage in the biosphere will have increased by more than 200 Gt. Indeed the increase may even prove to be much smaller than this.

5.4   Ocean Research

It is my opinion that estimates of CO_2 uptake using existing OGCMs are no more reliable than that obtained from the tracer calibrated box-diffusion model of Siegenthaler and Oeschger (1978). One has only to compare the plots of excess CO_2 versus bomb ^{14}C obtained by the existing models to realize that modelers are a long way from producing simulations that adequately replicate thermocline ventilation, let alone intermediate water formation. Because this latter reservoir will assume ever greater importance in the uptake of CO_2 during the course of the next century, there is a way to go before estimates of future ocean uptake more accurate than $\pm 20\%$ can be obtained. Hence, my agenda for improvement of our ability to predict the uptake by the ocean of excess atmospheric CO_2 would have as its core the improvement of ocean general circulation models. Key to this exercise is the development of models that are capable of duplicating not only the 3D distribution of density but also the spatial and temporal evolution of the transient tracer distributions (i.e., ^3H, ^3He, ^{14}C, CFCs, etc.). Of course, to strengthen the constraints placed by tracers, we must periodically update the global distributions of these substances. As time passes, the residence time of the bomb-produced tracers approaches that of the fossil fuel CO_2 molecules. Hence, the World Ocean Circulation Experiment (WOCE) surveys conducted during 1990 (i.e., 30 years after the major atmospheric H-bomb tests were conducted) will greatly strengthen the constraints placed by the earlier Geochemical Ocean Sections Study (GEOSECS) surveys (1972–1977).

This approach will be adequate only as long as the ocean operates in its preindustrial state. As global warming progresses, there will be a tendency for the ventilation of the ocean's interior to slow. The reason is that both the warming and the "wetting" of the temperate latitude outcrops from which the thermocline and intermediate waters descend will reduce the density of these waters. We need a set of observations that will indicate that this slowdown is under way. This will prove to be a difficult task because tracers provide integral measures of ocean operation, and as a result their distributions are not particularly sensitive to gradual changes in ventilation rate.

Keeling et al. (1996) have suggested a window into this problem. It involves the measurement of the ratios of O_2 to N_2 and Ar to N_2 in the atmosphere. The secular trend and hemispheric seasonality in the O_2 to N_2 ratio carries information regarding the global plant productivity. The secular trend and seasonal swings in atmospheric CO_2 can be used to constrain the terrestrial contribution to global plant productivity. The residual O_2 change is then a measure of the strength of the oceanic life cycles. Were ventilation of the thermocline to weaken, fewer nutrients would be brought to the surface and presumably the warm-season production of O_2 would drop. By keeping track of the secular trend and the amplitude of the annual O_2/N_2 swing, one can perhaps obtain a measure of the strength of thermocline ventilation.

The Ar to N_2 ratio in the air changes as the result of breathing by the ocean. For each calorie of heat taken up by the sea, both Ar and N_2 are released to the atmosphere. The important point is that the ratio of Ar to N_2 in the gas so released is not the same as that in the atmosphere. Hence, both the secular warming of the ocean and the seasonal heat exchange between ocean and atmosphere are recorded by the Ar to N_2 ratio in the atmosphere. By monitoring the secular trend in this ratio for the globe as a whole, one can obtain a measure of the rate of warming of the ocean. Were ventilation to slow, the rate of heat uptake would also slow. Seasonal swings in Ar/N_2 will also be of interest. The more swamp-like the upper ocean becomes, the larger should be the seasonal cycle in its surface temperature and hence in the hemispheric Ar to N_2 ratio.

Many of those who do research on the global carbon cycle push for surveys of surface water pCO_2 and upper-ocean ΣCO_2 as a means of establishing the magnitude of an ocean's uptake. I deem both of these strategies to be flawed. The problem with the ΔpCO_2 (air-sea) is that the accuracy required to yield an uptake estimate accurate to 10 percent is ± 1 μatm. Not only would the seasonal swing in pCO_2 have to be accurately established everywhere in the world ocean, but also corrections would have to be made for the sea-to-land transport of CO_2 supplying the carbon carried by rivers to the sea and for the ΔT between the ocean's "skin" and the interior of its mixed layer. Estimates suggest that together these two effects may have maintained a preindustrial ocean mixed-layer CO_2 partial pressure averaging as much as 4 μatm higher than the atmosphere. If these corrections are neglected, the uptake of excess CO_2 by the ocean would be underestimated by 1 GtC/yr. I doubt whether sufficient accuracy can ever be achieved by this approach, and even if it could, the cost would be enormous.

This is not to say that surveys of surface ocean pCO_2 are not useful. Indeed, those who seek to deconvolve terrestrial sources and sinks from the distribution of CO_2 in the atmosphere must have this information in order to correct for the oceanic distribution of CO_2 sources and sinks. This task is less demanding; the level of required accuracy is easily attainable.

The ΣCO_2 approach has more promise. A repeat of the WOCE survey 20 years from now will allow the integral uptake of CO_2 over this period to be accurately assessed. Furthermore, the 3D distribution of the ΣCO_2 excess will provide yet another powerful constraint on ocean circulation models. In combination with the distributions of bomb ^{14}C and the CFCs, these results will allow the influence of differing atmospheric equilibration times ($\sim$10 years for ^{14}C, $\sim$0.5 years for CO_2, and $\sim$0.1 year for CFCs) to be addressed in model simulations. But, by itself, this integral uptake estimate has no predictive power. We must, as I've already said, emphasize model development.

5.5   Terrestrial Research

As already stated, developing the means to predict how storage of carbon on the continents will evolve is a very tall order – so tall that reliable predictions may prove impossible. In the absence of predictive power, the best we can do is to document the magnitude of the annual changes in storage and to better constrain the locations at which this storage is occurring. The key measurements to be made are those that document the secular trends in the atmospheric CO_2 content and in O_2 to N_2 ratio and their interhemispheric gradients. To make the most effective use of this atmospheric data, it is important that vertical profiles of CO_2 be obtained. Modeling studies (Denning et al., 1996a,b) clearly show that without such data, rectification effects related to the dynamics of the atmospheric boundary layer can lead to significant biases in the interhemispheric gradient. In this regard, it is essential to quantify all the transports that generated gradients in atmospheric CO_2 content before the onset of fossil fuel burning. If these gradients are not properly isolated, they could easily lead to false conclusions regarding current terrestrial sources and sinks for fossil fuel CO_2. There is one aspect of this problem with which I am quite familiar. It has to do with the interhemispheric transport of CO_2 via the Atlantic Ocean's conveyor circulation (Broecker and Peng, 1992). Peng and I estimated that this transport is on the order of 0.6 GtC per year (see Figure 5.5). Keeling and Peng (1995) confirmed the importance of this transport mode and demonstrated that it affects the distribution of O_2 as well. To date no ocean model has duplicated the strong natural uptake of CO_2 that Broecker and Peng demonstrated to be occurring in the northern Atlantic. Hence, this phenomenon is not included in inversions of atmospheric CO_2 distributions. To evaluate the possible impact of this northern Atlantic sink, Fan et al. (1998) carried out simulations that suggest that were the uptake of CO_2 by the northern Atlantic to be increased by 0.6 GtC/yr, then the model-derived terrestrial storage in North America would drop by 0.9 GtC/yr.

Much emphasis is being placed on eddy correlation–based CO_2 fluxes determined using towers (Wofsy et al., 1993). For the first time, it is possible to estimate the net uptake (summer) and release (rest of year) of CO_2 by forests. Although this is a major advance, one must keep in mind that the terrestrial biosphere is a crazy quilt consisting of thousands upon thousands of small patches, each with its own vegetation history. A majority of sites are likely to be storing carbon, for releases are in part catastrophic because of fire, blowdown, or cutting. Only rarely has a true steady state been achieved. Thus, although the tower studies will help to elucidate the factors influencing photosynthesis and respiration, taken alone they are not likely to yield the answers we seek in our quest to predict carbon storage.

In my estimation, if we are to develop the needed predictability, then we will have to carry out a host of purposeful manipulations involving both fixed nitrogen and CO_2. For nitrogen alone, purposeful fertilization of checkerboard plots within natural forests

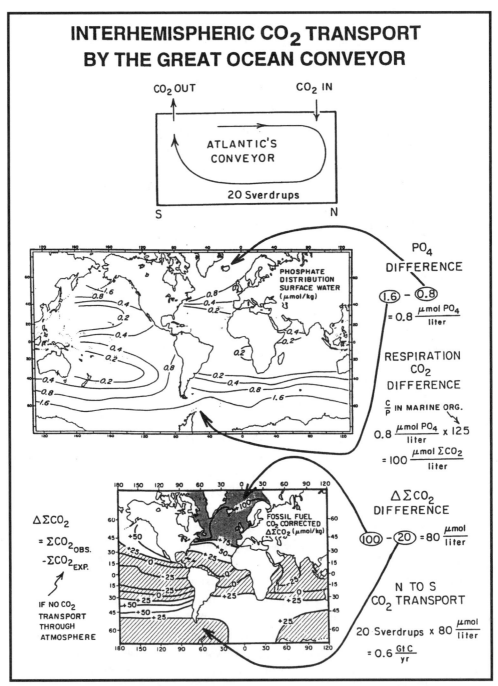

Figure 5.5. Based on measurements of salinity, ΣCO_2, alkalinity, and phosphate, Broecker and Peng (1992) have calculated the amount of CO_2 taken up from or given off to the atmosphere by surface ocean waters across the globe. Because of their relatively low PO_4 (and hence respiration CO_2) content, cold waters in the northern Atlantic take up far more atmospheric CO_2 than do their high-phosphate Southern Ocean counterparts. This excess CO_2 is transported to the Southern Hemisphere by the lower limb of the Atlantic's conveyor circulation. To date, no model has succeeded in duplicating these observations. Hence, the attempts to deconvolve continental sources and sinks are contaminated by the preindustrial ocean source and sink pattern.

throughout the world should be conducted. These experiments should be continued for a decade or more. As compared with almost all other aspects of carbon cycle research, these experiments would be modest in cost. To me, they are a must!

CO_2 and CO_2 plus fixed nitrogen experiments are more challenging because they require either CO_2 injection rings (see, for example, Ellsworth et al., 1995), open top chambers (see, for example, Turnbull et al., 1998, and Tissue et al., 1997), or total enclosures. A modest number of experiments of the first two types are now under way. But because of their short duration, none is above criticism. What is needed as a supplement are structures large enough to permit growth from seed to full canopy under a set of controlled CO_2 levels. The only facility currently available and equipped for such an experiment is the triple-domed portion of Biosphere 2.

5.6   The Next Century

Today, as the result of fossil fuel burning we release each year an amount of CO_2 containing 6.8 Gt of carbon to the atmosphere, and as the result of deforestation roughly 1.0 Gt more. With global population expected to rise to about 9 billion and per capita energy consumption in developing countries expected to dramatically increase, unless a major change occurs in our energy habits, this annual total is bound to rise significantly. Let us take as a base case that over the next 100 years 1000 GtC of carbon from fossil fuel burning and deforestation will be released as CO_2 to the atmosphere (i.e., an average of 10 GtC/yr).

As summarized in Figure 5.6, if there were no change in biomass, this scenario would cause the CO_2 content of the atmosphere to rise to about 800 ppm. On the other hand, if the biomass were to increase by 200 Gt (scenario 2), then the atmospheric CO_2 content would rise to only 640 ppm. For both scenarios the total ocean uptake is assumed to be about 200 GtC.

Inasmuch as most environmentalists are convinced that even a CO_2 rise to 640 ppm is unacceptable, either we must cut deeply into our use of fossil fuels as an energy source or we must capture and bury a large portion of the CO_2 produced when we burn them. Because this is a chapter on CO_2 futures, rather than energy futures, I will concentrate on the latter option. As outlined by Brookhaven's Meyer Steinberg in a report to the Department of Energy (Steinberg et al., 1984), it is feasible to remove CO_2 from stack gases at fossil fuel-fired power plants. Its implementation would raise the cost of electricity by about a factor of 2. Using current technology, the effluent would be passed through a tower, where the CO_2 would be absorbed into an organic solvent. The solvent would then be heated and passed through a second tower, where the pure CO_2 would be released. This CO_2 would be compressed to liquid form (at $\sim$30 atm pressure) and sent through a pipeline to the disposal site. Disposal could be either at sea or on land (Marchetti, 1977) depending on local circumstance (see Herzog et al., 1997). For example, in Japan, where earthquake hazards loom large, disposal at sea would be the preferred option. By contrast, in the central United States, disposal in salty aquifers or spent petroleum and natural gas reservoirs would be the most economical option.

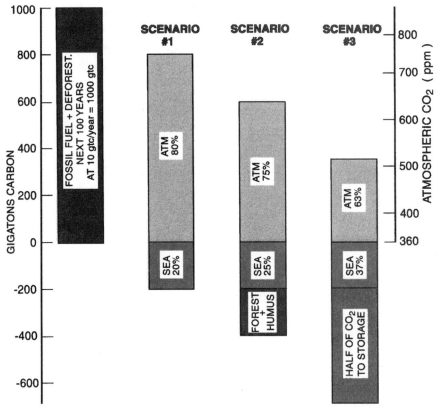

Figure 5.6. If over the next century 1000 GtC are released to the atmosphere as the result of fossil fuel burning and deforestation, then in the absence of significant greening of the terrestrial biosphere roughly 80% will remain in the atmosphere, raising its CO_2 content to about 750 ppm. If we are very lucky and greening driven by fixed nitrogen and carbon dioxide increases storage in the terrestrial biosphere by as much as 200 GtC, CO_2 would increase to about 635 ppm. The only way by which the CO_2 content could be held below 500 ppm would be to capture and store on the order of half of the CO_2 generated.

Experiments conducted by MBARI's Peter Brewer et al. (1999) using a remotely controlled submersible confirm that because of its higher compressibility, liquid CO_2 becomes more dense than seawater at depths below 3 kilometers. Furthermore, the Brewer et al. (1999) experiments show that when in contact with seawater, liquid CO_2 is rapidly converted to solid hydrate form (~6 H_2O to 1 CO_2). This solid is more dense than either seawater or liquid CO_2. Hence, if the liquid CO_2 were to be discharged into the deep sea, it would fall to the sea floor, creating a pile of solid hydrate. This hydrate would gradually dissolve into the bottom water and from there would be spread into the vast deep-sea reservoir. In this way, the route to the deep sea could be short-circuited, allowing the capacity for CO_2 uptake by this vast reservoir to be tapped on a time scale far shorter than the several hundred years required if the CO_2 enters the sea at the surface. Because the time scale for return to the surface is also measured in hundreds of years, this type of disposal would prevent the return of the CO_2 to the surface until well after the greenhouse crisis was behind us.

One might ask how we can be certain that no short circuits exist via which the CO_2 transferred to the sea floor could escape to the atmosphere on a much shorter time scale. First of all, natural radiocarbon contained in the sea's dissolved inorganic carbon acts as a clock, recording the isolation time of the waters that have descended into the abyss from the high-latitude source regions. This clock tells us that the isolation times average several hundred years and range up to a millennium in the remote deep northern Pacific. In addition, we have an even better indicator, namely, the ^{3}He released from the crests of mid-ocean ridges (mean depth 2500 meters). The distribution of this helium isotope in the deep sea tells us two things. First of all, the ridge-crest ^{3}He becomes rather well mixed throughout the vast volume of deep seawater before it reaches the surface. This observation suggests that the CO_2 will very likely do likewise (Farley et al., 1995). If so, it will be largely neutralized by reaction with the carbonate and borate ion contained in the deep sea before reaching the surface. Second, from the inventory of ^{3}He in the deep sea and estimates of the rate at which it is being released from ridge crests, it appears that the average helium atom remains in the deep sea for many hundreds of years before reaching the surface and escaping to the atmosphere (and from there to outer space).

With regard to disposal on the continents, the volume of pore space in the 1- to 2-kilometer depth range is vast. Most of these pores are currently filled with water. Much of this water is too salty for use in agriculture. A much smaller portion of the pore space is filled with petroleum and natural gas. The liquid CO_2 pumped into such reservoirs would displace the less dense resident water. CO_2 disposed of in this way would remain trapped for many thousands of years.

Were we to sequester, let's say, 25% of the 1000 GtC likely to be converted to CO_2 during the next century in deep continental reservoirs, the volume of liquid CO_2 would be about 1000 cubic kilometers. Taking the reservoir porosity to be 20%, the volume of reservoir rock required would be 5000 cubic kilometers. Were these reservoirs to average 1 kilometer in thickness, they would cover only 5000 square kilometers (i.e., if all in one place, a square only 70 kilometers on a side). Hence, there is plenty of space available for CO_2 storage on the continents.

5.7   Summary

Unless there is a dramatic reduction in the fraction of the energy produced from fossil fuels over the course of the next century, the only way to prevent the atmosphere's CO_2 content from rising well above 500 ppm will be to implement large-scale CO_2 sequestration. Although not negligible in their potential contribution, taken together, energy conservation and terrestrial carbon storage certainly won't do the job. Rather, we must launch a serious effort to develop the means to sequester CO_2 and the political mechanism necessary to make sequestration economically competitive.

ACKNOWLEDGMENTS

Abhijit Sanyal provided the calculations listed in Table 5.1. Jorge Sarmiento thoughtfully criticized an earlier draft. Support was provided by Department of Energy grant DE-FG02-97ER62441.

REFERENCES

Brewer, P. G., Friederich, G., Peltzer, E. T., and Orr, Jr., F. M. (1999). Direct experiments on the ocean disposal of fossil fuel CO_2. Science, 284, 943–45.

Broecker, W. S., Peng, T.-H., Ostlund, H. G., and Stuvier, M. (1985). The distribution of bomb radiocarbon. J. Geophys. Res., 90, 6953–70.

Broecker, W. S., and Peng, T.-H. (1992). Interhemispheric transport of CO_2 by ocean circulation. Nature, 356, 587–89.

Broecker, W. S., Peacock, S., Walker, S., Weiss, R., Fahrbach, E., Schroeder, M., Mikolajewicz, U., Heinze, C., Key, R., Peng, T.-H., and Rubin, S. (1998). How much deep water is formed in the Southern Ocean? J. Geophys. Res., 103, 15,833–43.

Denning, A. S., Collatz, G. J., Zhang, C., Randall, D. A., Berry, J. A., Sellers, P. J., Colello, G. D., and Dazlich, D. A. (1996a). Simulations of terrestrial carbon metabolism and atmospheric CO_2 in a general circulation model. Tellus, 48B, 521–42.

Denning, A. S., Randall, D. A., Collatz, G. J., and Sellers, P. J. (1996b). Simulations of terrestrial carbon metabolism and atmospheric CO_2 in a general circulation model, Part 2: Simulated CO_2 concentrations. Tellus, 48B, 543–67.

Ellsworth, D., Oren, R., Huang, C., Phillips, N., and Hendry, G. (1995). Leaf and canopy responses to elevated CO_2 in a pine forest under free-air CO_2 enrichment. Oecologia, 104, 139–46.

Fan, S., Gloor, M., Mahlman, J., Pacala, S., Sarmiento, J., Takahashi, T., and Tans, P. (1998). A large terrestrial carbon sink in North America implied by atmospheric and oceanic carbon dioxide data and models. Science, 282, 442–46.

Farley, K. A., Maier-Reimer, E., Schlosser, P., Broecker, W. S., and Bonani, G. (1995). Constraints on mantle ^3He fluxes and deep-sea circulation from an oceanic general circulation model. J. Geophys. Res., 100, 3829–39.

Herzog, H., Drake, E., and Adams, E. (1997). CO_2 capture, reuse, and storage technologies for mitigating global climate change. A White Paper Final Report, Mass. Inst. Technology, Cambridge, MA, 66 pp.

Keeling, R. F., and Peng, T.-H. (1995). Transport of heat, CO_2 and O_2 by the Atlantic's thermo-haline circulation. Phil. Trans. R. Soc. Lond., 348, 133–42.

Keeling, R. F., Piper, S. C., and Heimann, M. (1996). Global and hemispheric CO_2 sinks deduced from changes in atmospheric O_2 concentration. Nature, 381, 218–21.

Marchetti, C. (1977). On geo-engineering and the CO_2 problem. Climate Change, 1, 59–68.

Siegenthaler, U., and Oeschger, H. (1978). Predicting future atmospheric carbon dioxide levels. Science, 199, 388–95.

Steinberg, M., Cheng, H. C., and Horn, F. (1984). A systems study for the removal, recovery and disposal of carbon dioxide from fossil fuel power plants in the U.S. Report for U.S. Department of Energy, 76 pp.

Tissue, D. T., Thomas, R. B., and Strain, B. R. (1997). Atmospheric CO_2 enrichment increases growth and photosynthesis of Pinus taeda: a 4-year experiment in the field. Plant Cell and Environment, 20, 1123–34.

Turnbull, M. H., Tissue, D. T., Griffin, K. L., Rogers, G. N. D., and Whitehead, D. (1998). Photosynthetic acclimation to long-term exposure to elevated CO_2 concentration in Pinus radiata D. Don. is related to age of needles. Plant Cell and Environment, 21, 1019–28.

Wofsy, S. C., Goulden, M. L., Munger, J. W., Fan, S.-M., Bakwin, P. S., Daube, B. C., Bassow, S. L., and Bazzaz, F. A. (1993). Net exchange of CO_2 in a mid-latitude forest. Science, 260, 1314–17.

PART TWO

THE HUMAN PERSPECTIVE

6 Global Climate Change in the Human Perspective

STEPHEN H. SCHNEIDER

ABSTRACT

What does it mean to consider climate change in the human perspective? Throughout human history, climate has both promoted and constrained human activity. In fact, humans only very recently have been able to reduce their dependence on climate through advances in technology and organization. The other consideration in the discussion of climate change in the human perspective is how human action affects climate. Are our actions causing the climate to change in ways or at rates that will threaten natural systems or make human adaptations difficult? What actions can we or should we take to alleviate the effects of human action on climate change? To approach these questions, we often use mathematical modeling and computer simulation general circulation models (GCM) to aid our understanding of the relationship between human action and global climate change. Integrated assessment models (IAM) are important tools to study the impacts of climate change on the environment and society as well as the costs and benefits of various policy options and decisions.

I present a brief overview of the climate debate, modeling, and the current understanding of the climate processes. I discuss how IAMs evaluate the effects of human-induced climate change and the implications of policy options. Finally, I suggest areas for further consideration.

6.1 Can a Forecast Climate Signal Be Detected in the Climate Record?

Twenty thousand years ago, a mere blink in geologic time, a visitor to the now-productive U.S. Corn Belt would not have seen the heart of one of the world's foremost granaries, but rather would have seen open spruce parkland forest, where many of the tree species were the same kinds that are found today 500 to 1000 kilometers north in the boreal forests of Canada (e.g., Wright et al., 1993). Similarly, if we could somehow have been flying over the Great Basin in the western United States, we would have seen massive fossil lakes. Some of them stretched hundreds of miles, such as the former Lake Bonneville in Utah. We would also have seen the now-fossil beaches (currently visible flying into the Salt Lake City, Utah, Airport or over Mono Lake, California) from high-water stands that date back ten to fifteen thousand years ago. The Ice Age, which at its maximum some twenty thousand years ago was about $5\,^\circ C$ to $7\,^\circ C$ colder than our current global climate, disappeared in what is, to nature, a relatively rapid period of about five to ten thousand years. The average rate of temperature change from the Ice Age to the current 10,000-year period of relative climate stability, our so-called Holocene Interglacial, is about $1\,^\circ C$ change for every one thousand years. Of course, there were more-rapid periods embedded within this time frame (e.g., Broecker, 1997), but for the moment, let's consider only the sustained average rates.

Such changes not only correspond with radical alterations in the ecosystems of the Earth, but also have been implicated in the extinction of what is known as the charismatic megafauna (woolly mammoths, saber tooth tigers, etc.). Fossil pollen evidence tells us that the vegetation habitats during the more "rapid" parts of the transition from ice age to interglacial around ten to twelve thousand years ago saw what paleoclimatologists call "no analog habitats," that is, combinations of pollen abundances that do not exist on Earth today (Overpeck et al., 1992). All this change was natural, of course, and there are two reasons for mentioning it in our context of a human perspective. First, it reminds us that the climate and ecosystems change by themselves, without need of humans (the latter is what we call anthropogenic causation). Second, climate change of about several degrees on a global average basis is a very significant change from the point of view of natural systems.

Explanations of the Ice Age vary, the most popular one being a change in the amount of sunlight coming in between (a) winter and summer and (b) the poles and the equator. These changes in the distribution of seasonal or latitudinal sunshine are caused by slow variations in the tilt of the Earth's axis and other orbital elements, but these astronomical variations alone cannot totally explain the climatic cycles (e.g., Crowley and North, 1991). If these orbital variations and other factors (such as the increased reflectivity of the Earth associated with more ice) are combined, our best climate theories (embodied through mathematical models that comprise the physical laws of conservation of mass, energy, and momentum) suggest that the Ice Age should have been several degrees warmer than it actually was, especially in the Southern Hemisphere. What could account for this extra cold? Perhaps the models are not sensitive enough; they do not respond sufficiently to a change in so-called "radiative climate forcing," which is the change in the amount of radiant energy coming to the Earth from external factors such as orbital variations or extra ice. Another possibility (more likely, I think) is that something else also changed at the same time.

These theories can be better reconciled with what happened between ice ages and interglacials if one assumes that several watts of energy over every square meter of the Earth were taken away in the ice age by some other mechanism at a global scale. But what could be such a mechanism? The obvious candidate would be a change in the composition of the Earth's atmosphere, a change that affects both its reflectivity and its heat trapping capacity (e.g., decreases in the well-known greenhouse effect or increases in atmospheric dust). But what evidence is there that greenhouse gases – for example, carbon dioxide, methane, nitrous oxide, or water vapor – had lower concentrations twenty thousand years ago than in the interglacial? About fifteen years ago that evidence came through loud and clear from the ice caps of the world. Air trapped in these glaciers provides a library of the history of the Earth's atmosphere back some 200,000 years. It shows that during the past two ice ages carbon dioxide concentration was about 40% less and methane half of the average value during the current and penultimate interglacials (Eddy and Oeschger, 1993). It also shows that since the industrial revolution carbon dioxide has increased beyond any levels experienced in the past 150,000 years (at least) by nearly 30%, and methane by 150% – two figures that virtually no knowledgeable scientist disputes (e.g., IPCC, 1996a). Moreover, nearly all climate scientists agree that these documented increases in greenhouse gas concentrations are a result of so-called anthropogenic emissions, which are driven by increasing numbers of people pursuing higher standards of living and clearing land and using technology to achieve these growth-oriented goals.

If the carbon dioxide and methane decreases in the last ice age helped to explain the ice age coldness, can they tell us something about how the anthropogenic increase of these gases due to human activities might cause climate change in the future? The answer is, "Not directly," because it is possible that there are other factors we have not accounted for in the ice age story that could well have been involved, and there are still many unanswered questions associated with the ice age cycles. It is simply a circumstantial bit of evidence that suggests that the estimated levels of carbon dioxide and methane gases during the ice ages are consistent with the predictions of the greenhouse effect (e.g., Hoffert and Covey, 1992). During the ice ages, when surface temperatures were lower by about $5\,°C$ to $7\,°C$, the estimated levels of greenhouse gases were about half of current levels. From this and other information about ice caps and the distribution of sunlight, we infer that a doubling of CO_2 would raise surface temperatures by about $3\,°C \pm 1.5\,°C$. This is known as the "climate sensitivity range." The magnitude of climate sensitivity that best helps to explain the ice age coldness is $2\,°C$–$3\,°C$. If the best estimate of the temperature change associated with a doubling of CO_2 were $10\,°$ warming, which is twice the value at the high end of the climate sensitivity range accepted by a mainstream climate scientist today (e.g., IPCC, 1996a), then the ice ages should have been even colder than they were. On the other hand, if the Earth would warm only by half a degree or less if CO_2 doubled, then it would be tougher to explain the magnitude of the ice ages without finding some other mechanism not yet identified. Of course, the latter is possible. So what other lines of circumstantial evidence or direct evidence do we have for estimating the sensitivity of the climate to greenhouse gas increases?

From thousands of laboratory experiments and direct measurements, millions of balloon observations, and trillions of satellite data bits, we know that the basic structure of the energy flows in and out of the Earth's atmosphere are relatively well understood. We know that water vapor, carbon dioxide, or methane traps enough energy on the Earth to warm the surface about $33\,°C$ relative to that which would occur in their absence.

This well-known natural greenhouse effect has been known for a century and a half and is not under dispute. Nor do many climatologists dispute that there has been about a $0.5\,°C$ ($\pm 0.2\,°C$) globally averaged warming trend at the Earth's surface over the past century, nor that 1998 was by several tenths of a degree the warmest year globally in the instrumental record. In much greater dispute is whether a small increment in this envelope of greenhouse gases since the industrial revolution would produce a noticeable response (i.e., a "climate signal"). It is difficult to detect a small climate signal (less than 0.5 °C) because the natural variability of global surface temperature is several tenths of a degree Celsius from year to year. Also, century-long 0.5 °C global warming trends are not common; they may have occurred about every one thousand years or so. However, as Mann et al. (1999) show, the latter half of the 20th century stands out remarkably as above the climatic noise of the millennium as the warmest period.

The debate over whether that signal has been detected and can be attributed to human activities has been intense lately. This intensity has been based on significant new pieces of evidence (e.g., Santer et al., 1996; Wigley et al., 1998) – albeit each piece is circumstantial – and a few loud, well-publicized denials (e.g., Robinson and Robinson, 1997) that the totality of evidence has any meaning (e.g., see the review by Edwards and Schneider, 1997). In the absence of clear, direct empirical evidence, one often must use either circumstantial evidence or incomplete bits of direct evidence with uncertainties attached. When the preponderance of such evidence gets strong enough, then most scientists begin to accept, tentatively of course, the likelihood of causal connections (e.g., Chapter 8 of IPCC, 1996a). Some people shed their skepticism at different levels than others, so naturally there will be a cacophonous debate over whether a climate signal has been "detected," let alone whether it could be attributed to human activities. One can always find some scientist who will want a 999 out of 1000 probability of certainty, and others who will accept the proposition at 8 or 9 chances out of 10. This is not exact science. Instead, it is a value judgment about the acceptability and meaning of a significant, but not conclusive, body of evidence. The scientific job is to assess (1) what can happen and (2) what the odds are of its happening (see, for example, this discussion in Chapter 6 of Schneider, 1997a). Let me discuss this process further.

I have mentioned the ice ages because this is a "natural experiment" that we use, not to forecast the future, but to build understanding of climate processes and to validate the tools that we do use to forecast the future (e.g., Schneider, 1993) – that is, our climate theories embodied in mathematical models. Are there any other such natural experiments? There are several, the two most prominent being (1) episodic volcanic eruptions that throw dust in the stratosphere that reflects for a few years a few watts per square meter of solar energy that otherwise would have reached the lower atmosphere and (2) the seasonal cycle. Let's consider volcanic eruptions first.

Volcanic dust veils should cool the planet. In fact, a number of climate modeling groups forecast that the last major eruption, Mt. Pinatubo in 1991, would cool the Earth's lower atmosphere for a few years on the order of several tenths of a degree – in advance of the actual data to confirm. Indeed, that is roughly what happened. However, it could be argued that a few tenths of a degree cooling, or warming for that matter, might be a natural internal fluctuation in the Earth's climate system, and indeed, as noted earlier, fluctuations of that magnitude are a part of the natural background "climatic noise." How then might we distinguish the climatic signal of the volcanic eruption from the noise of the natural internal variability? In any one eruption it is difficult to do so because the signal to noise ratio is about 1 – that is, the magnitude of the cooling expected is about equal to the magnitude of the natural internal fluctuations in non-volcanic years. Therefore, for any one volcanic dust event we cannot have very much confidence that a signal has been observed. So the fact that the Pinatubo results showed up about as predicted doesn't, by itself, give a lot of confidence, although as a circumstantial bit of evidence it is quite useful. However, another volcanic eruption in 1983, El Chichón, was also followed by several tenths of a degree cooling, as was the effect after Mt. Agung in 1963 and Mt. Krakatoa in 1883.

In other words, by looking at the results of several volcanic eruptions, a number of scientists (including Mass and Schneider, 1977) discovered that indeed there was a clear and obvious correlation. This suggests that when a few watts of energy over every square meter of the Earth is removed for a few years by volcanic dust veils in the stratosphere, the lower atmosphere will cool by a few tenths of degrees – the very magnitude predicted by the same computer models that we use to forecast the effects of a few watts per square meter of sustained (i.e., over a century or more) heating from global greenhouse gas increases.

What other natural experiments might we have to test climate sensitivity? Another one that happens every year is the seasons. Winter predictably follows summer, being some $15°$ colder in the Northern Hemisphere and $5°$ colder than summer in the Southern Hemisphere. The reason the Southern Hemisphere has a smaller seasonal cycle is because it has much more ocean than land, and water has a higher heat-retaining capacity than does land or air. Because a season is not long enough for the planet to reach an equilibrium temperature change, the more land-dominated Northern Hemisphere has lower heat storage capacity and thus a larger seasonal cycle of surface temperature. How well do the climate models reproduce this change? Extraordinarily well. Although the absolute temperatures that may be simulated by models can be off by as much as $5°$ or $6°$ in some regions of the world for some seasons, the models' capacity to reproduce the amplitude of the seasonal cycle of surface air temperatures, by and large, is quite good. (It is less good for some other variables, however, particularly for the hydrological systems. See Chapter 5 of IPCC, 1996a.) Now, if we were making a factor of 10 error by either overestimating or underestimating the sensitivity of the climate to radiative forcing, it would be difficult for the models to reproduce the different seasonal cycle surface temperature amplitudes over land and oceans as well as they do. This is another piece of circumstantial evidence suggesting that current estimate of climate sensitivity is not off by a factor of 10, as some contrarians assert. Indeed, indirect evidence such as ice ages, volcanic eruptions, and the seasonal cycle simulation skills of models are prime reasons why many of us in the scientific community have for more than twenty years expected that clear signs of anthropogenic climate change were not unlikely by the 21st century (e.g., see p. 11 of Schneider and Mesirow, 1976, in which I projected that "demonstrable climatic changes could occur by the end of this century").

In summary, then, in my opinion it is unlikely that natural variability is the explanation of all recent climate change, especially that which has been documented in the last half of the 20th century. However, because much of the debate over detection and attribution of human-caused climate change hinges on the projections of climatic models, it is necessary to have at least a cursory understanding of how they work. Although it is impossible to treat more than the highlights of the nature and use of climatic models in only a few pages or so, I nonetheless offer the following section in the hopes of reducing somewhat the confusion that may exist in many people's minds after listening to the often acrimonious and technically complex debate over climatic models and their credibility.

6.2   Overview of Climate Modeling Fundamentals

Engineers and scientists build models – either mathematical or physical ones – primarily to perform tests that are too dangerous, too expensive, or perhaps impossible to perform with the real thing. To simulate the climate, a modeler must decide which components of the climatic system to include and which variables to involve. For example, if we choose to simulate the long-term sequence of glacials and interglacials (the period between successive ice ages), our model must include explicitly the effects of all the important interacting components of the climate system operating over the past million years or so. These include the atmosphere, oceans, sea ice/glaciers (cryosphere), land surface (including biota), land subsurface, and chemical processes (including terrestrial and marine biogeochemical cycles), as well as the external or "boundary forcing" conditions such as input of solar radiant energy (e.g., see IPCC, 1996a).

The problem for climate scientists is quantitatively to separate out cause and effect linkages from among the many factors that interact within the climate system. It is a controversial effort because there are so many subsystems, so many forcings, and so many interacting complex sets of processes operating at the same time that debates about the adequacy of models often erupt. These difficulties are compounded because it is sometimes difficult to determine a trend when there is a large variation around the trend, let alone the possibility that there can be trends in that variability as well.

6.2.1   Modeling the Climate System

How are climate models constructed? First, scientists look at observations of changes in temperatures, ozone levels, and so forth. This allows us to identify correlations among variables. Correlation is not necessarily cause and effect; just because event B tracks event A doesn't mean event B was caused by event A. One must actually prove that relationship is causal and explain how it happened. Especially when unprecedented events are being considered, a first-principles approach, rather than a purely empirical-statistical approach, is desirable. However, observations can lead to a hypothesis of cause and effect – "laws" – that can be tested (for example, see Root and Schneider, 1995). The testing is often based on simulations with mathematical models run on a computer. The models, in turn, must be tested against a variety of observations – present and paleoclimatic. That is how the scientific method is typically applied. When a model or set of linked models appears plausible, it can be fed unprecedented changes, such as projected human global climate change forcings: changes to atmospheric composition or land surfaces that are not known to have happened before. Then the model can be asked to make projections of future climate, ozone levels, forests, species extinction rates, and so on.

The most comprehensive weather simulation models produce three-dimensional details of temperature, winds, humidity, and rainfall all over the globe. A weather map generated by such a computer model – known as a general circulation model – often looks quite realistic, but it is never faithful in every detail. To make a weather map generated by computer we must solve six partial differential equations that describe the fluid motions in the atmosphere. It sounds in principle as if there's no problem: we know that those equations work in the laboratory; we know that they describe fluid motions and energy and mass relationships (e.g., Washington and Parkinson, 1986). Why then aren't the models perfect simulations of the atmospheric behavior?

One answer is that the evolution of weather from some starting weather map (known as the initial condition) cannot be uniquely determined beyond about 10 days, even in principle. A weather event on one day cannot be said to determine an event 30 days in the future, all those commercial "long-range" weather forecasts notwithstanding. But the inherent unpredictability of weather details much beyond 10 days (owing to the chaotic internal dynamics of the atmosphere) doesn't preclude accurate forecasts of long-term averages (climate rather than weather). The seasonal cycle is absolute proof of such deterministic predictability, inasmuch as winter reliably follows summer and the cause and effect are known with certainty. Unfortunately, this distinction between the in-principle unpredictability of long-term weather and the possibility of long-term climatic projections is often missed in the public debate, especially by non-climate scientist authors (e.g., Robinson and Robinson, 1997).

6.2.2 Grids and Parameterization

The other answer to the imperfection of GCM simulations, even for long-term averages, is that nobody knows how to solve those six complex mathematical equations exactly. It's not like an algebraic equation, in which one can get the exact solution by a series of simple operations. There isn't any known mathematical technique to solve such coupled, nonlinear partial differential equations exactly. We approximate the solutions by taking the equations, which are continuous, and breaking them into discrete chunks, which we call *grid boxes*. A typical GCM grid size for a "low-resolution" model is about the size of Germany horizontally, and that of a "high-resolution" GCM is about the size of Belgium. In the vertical dimension there are two (low-resolution) up to about twenty (high-resolution) vertical layers that are typically spanning the lowest 10 (up to 40) kilometers of the atmosphere.

Clouds are very important to the energy balance of the earth-atmosphere system because they reflect sunlight away and trap infrared heat. But because none of us has ever seen a single cloud the size of Belgium, let alone Germany, we have a problem of scale; how can we treat processes that occur in nature at a smaller scale than we can resolve by our approximation technique of using large grid boxes? For example, we cannot calculate clouds explicitly because individual clouds are typically the size of a dot in this grid box. But we can put forward a few reasonable propositions on cloud physics: if it's a humid day, for example, it's more likely to be cloudy. If the air is rising, it's also more likely to be cloudy.

These climate models can predict the average humidity in the grid box, and whether the air is rising or sinking on average. So then we can write what we call a *parametric representation*, or *parameterization*, to connect large-scale variables that are resolved by the grid box (such as humidity) to unresolved small-scale processes (individual clouds). Then we get a prediction of grid box-averaged cloudiness through this parameterization. So-called "cumulus parameterization" is one of the important – and controversial – elements of GCMs that occupy a great deal of effort in the climate modeling community. Therefore, the models are not ignoring cloudiness, but neither are they explicitly resolving individual clouds. Instead, modelers try to get the average effect of processes that can't be resolved explicitly at smaller scales than the smallest resolved scale (the grid box) in the GCM (e.g., Trenberth, 1992). Developing, testing, and validating many such parameterizations are the most important tasks of the modelers because these parameterizations determine critically important issues such as climate sensitivity. The *climate sensitivity* is the degree of response of the climate system to a unit change in some forcing factor: typically, in our context, the change in globally averaged surface air temperature to a fixed doubling of the concentration of atmospheric carbon dioxide above preindustrial levels. This brings us to one of the most profound controversies in climate science, and one of the best examples of the usefulness, and fragility, of computer modeling.

6.2.3   The Greenhouse Effect

If the Earth only absorbed radiation from the sun without giving an equal amount of heat back to space by some means, the planet would continue to warm up until the oceans boiled. We know that the oceans are not boiling, and surface thermometers plus satellites have shown that the Earth's temperature remains roughly constant from year to year (notwithstanding the interannual globally averaged variability of about $0.2\,°C$ or the $0.5\,°C$ warming trend in the 20th century). This near constancy requires that about as much radiant energy (within a few watts per square meter) leaves the planet each year in some form as is coming in. In other words, a near equilibrium, or energy balance, has been established. The components of this energy balance are crucial to the climate.

All bodies with temperature give off radiant energy. The Earth gives off a total amount of radiant energy equivalent to that of a *black body* – a fictional structure that represents an ideal radiator – with a temperature of roughly $-18\,°C$ ($255\,°K$). The mean global surface air temperature is about $15\,°C$ ($288\,°K$), some $33\,°C$ warmer than the Earth's black body temperature. The difference is caused by the well-established natural greenhouse effect.

The term *greenhouse effect* arises from the classic analogy to a greenhouse, in which the glass allows the solar radiation in and traps much of the heat inside. However, the moniker is a bit of a misnomer because the mechanisms are different. In a greenhouse the glass primarily prevents convection currents of air from taking heat away from the interior. Greenhouse glass is not keeping the enclosure warm primarily by its blocking or reradiating infrared radiation; rather, it is constraining the physical transport of heat by air motion.

Although most of the Earth's surface and all clouds (except thin, wispy clouds) are reasonably close approximations to a black body, the atmospheric gases are not. When the nearly black body radiation emitted by the Earth's surface travels upward into the atmosphere, it encounters air molecules and aerosol particles. Water vapor, carbon dioxide, methane, nitrous oxide, ozone, and many other trace gases in the Earth's gaseous envelope tend to be highly selective – but often highly effective – absorbers of terrestrial infrared radiation. Furthermore, clouds (except for thin cirrus) absorb nearly all the infrared radiation that hits them, and then they reradiate energy almost like a black body at the temperature of the cloud surface – colder than the Earth's surface most of the time.

The atmosphere is more opaque to terrestrial infrared radiation than it is to incoming solar radiation, simply because the physical properties of atmospheric molecules, cloud particles and dust particles tend on average to be more transparent to solar radiation wavelengths than to terrestrial radiation. These properties create the large surface heating that characterizes the greenhouse effect, by means of which the atmosphere allows a considerable fraction of solar radiation to penetrate to the Earth's surface and then traps (more precisely, intercepts and reradiates) much of the upward terrestrial infrared radiation from the surface and lower atmosphere. The downward reradiation further enhances surface warming and is the prime process causing the greenhouse effect.

This is not a speculative theory but a well-understood and validated phenomenon of nature (e.g., Raval and Ramanathan, 1989). The most important greenhouse gas is water vapor, because it absorbs terrestrial radiation over most of the infrared spectrum. Even though humans are not altering the average amount of water vapor in the atmosphere very much by direct injections of this gas, increases in other greenhouse gases that warm the surface cause an increase in evaporation, which increases atmospheric water vapor concentrations, leading to an amplifying, or "positive" feedback, process known as the "water vapor-surface temperature-greenhouse feedback." The latter is believed responsible for the bulk of the climate sensitivity (Ramanathan, 1981). Carbon dioxide is another major greenhouse gas. Although it absorbs and re-emits considerably less infrared radiation than does water vapor, CO_2 is of intense interest because its concentration is increasing because of human activities, creating what is known as "anthropogenic radiative forcing." Ozone, nitrogen oxides, some hydrocarbons, and even some artificial compounds such as chlorofluorocarbons are also greenhouse gases. The extent to which these gases are important to climate depends on their atmospheric concentrations, the rates of change of those concentrations, and their effects on depletion of stratospheric ozone. In turn, lower levels of stratospheric ozone can indirectly modify the radiative forcing of the atmosphere below the ozone layer, thus offsetting a considerable fraction of the otherwise expected greenhouse warming signal.

The Earth's temperature, then, is determined primarily by the planetary radiation balance, through which the absorbed portion of the incoming solar radiation is nearly exactly balanced over a year's time by the outgoing terrestrial infrared radiation emitted by the climatic system to Earth. Because both of these quantities are determined by the properties of the atmosphere and the Earth's surface, major climate theories that address changes in those properties have been constructed. Many of these theories remain plausible hypotheses of climatic change. Certainly *the natural greenhouse effect is established beyond a reasonable scientific doubt*, accounting for natural warming that has allowed the coevolution of climate and life to proceed to this point (e.g., see Schneider and Londer, 1984). The extent to which human augmentation of the natural greenhouse effect (i.e., global warming) will prove serious is, of course, the current debate, along with discussions over the potentially offsetting cooling effects of "anthropogenic aerosols" (particles created in the atmosphere primarily from emissions of sulfur dioxide, primarily from burning of high-sulfur coal and oil: Schneider, 1994; IPCC, 1996a).

6.2.4 Model Validation

There are many types of parameterizations of processes that occur at a smaller scale than our models can resolve, and scientists debate which type is best. In effect, the question is, are these parameterizations an accurate representation of the large-scale consequences of processes that occur on smaller scales than we can explicitly treat? These processes include cloudiness, radiative energy transport, turbulent convection, evapotranspiration, oceanic mixing processes, chemical processes, ecosystem processes, sea ice dynamics, precipitation, mountain effects, and surface winds.

In forecasting climatic change, then, validation of the model becomes important. In fact, we cannot easily know in principle whether these parameterizations are "good enough." We must test them in a laboratory. That's where the study of paleoclimates has proved valuable (e.g., Hoffert and Covey, 1992). We also can test parameterizations by undertaking detailed small-scale field or modeling studies aimed at understanding the high-resolution details of some parameterized process that the large-scale model has told us is important. In the Second Assessment Report of IPCC (IPCC, 1996a) Working Group, I devoted more than one chapter to the issue of validation of climatic models, concluding that

> the most powerful tools available with which to assess future climate are coupled climate models, which include three-dimensional representations of the atmosphere, ocean, cryosphere and land surface. Coupled climate modeling has developed rapidly since 1990, and current models are now able to simulate many aspects of the observed climate with a useful level of skill. [For example, as noted earlier, good skill is found in simulating the very large annual cycle of surface temperatures in the Northern and Southern Hemispheres or the cooling of the lower atmosphere following the injection of massive amounts of dust into the stratosphere after explosive volcanic eruptions such as Mt. Pinatubo in the Philippines in 1991.] Coupled model simulations are most accurate at large spatial scales (e.g., hemispheric or continental); at regional scales skill is lower.

One difficulty with coupled models is known as *flux adjustment*, a technique for accounting for local oceanic heat transport processes that are not well simulated in some models. Adding this element of empirical-statistical "tuning" to models that strive to be based as much as possible on first principles has been controversial (see Shackley et al., 1999). However, not all models use flux adjustments (e.g., Barthelet et al., 1998; Bryan, 1998; Gregory and Mitchell, 1997; see also Rahmstorf and Ganopolski, 1999). Nearly all models, with or without this technique, produce climate sensitivities within or near to the standard IPCC range of 1.5 °C to 4.5 °C. Flux adjustments do, however, have a large influence on regional climatic projections, even though they do not seem to have a major impact on globally averaged climate sensitivity. Improving coupled models is thus a high priority for climate researchers because it is precisely such regional projections that are critical to the assessment of climatic impacts on environment and society (e.g., IPCC, 1996b, 1997, 1998).

6.2.5 Transient Versus Equilibrium Simulations

One final issue must be addressed in the context of coupled climate simulations. Until the past few years, climate modeling groups did not have access to sufficient computing power to routinely calculate time-evolving runs of climatic change given several alternative future histories of greenhouse gases and aerosol concentrations. That is, they did not perform so-called transient climate change scenarios. (Of course, the real Earth is undergoing a transient experiment – e.g., Schneider, 1994.) Rather, the models typically were asked to estimate how the Earth's climate would eventually be altered (i.e., in equilibrium) after CO_2 was artificially doubled and held fixed indefinitely rather than increased incrementally over time as it has in reality or in more-realistic transient model scenarios. The equilibrium climate sensitivity range has remained fairly constant for more than 20 years of assessments by various national and international groups, with the assessment teams repeatedly suggesting that, were CO_2 to double, climate would eventually warm at the surface somewhere between 1.5 °C and 4.5 °C. (Later we address the issue of the probability that warming above or below this range might occur and discuss how probabilities can even be assigned to this sensitivity.)

Transient model simulations exhibit less immediate warming than do equilibrium simulations because of the high heat-holding capacity of the thermally massive oceans. However, that unrealized warming eventually expresses itself decades to centuries later. We can account for this thermal delay, which retards the climate signal and can lull us into underestimating the long-term amount of climate change, by coupling models of the atmosphere to models of the oceans, ice, soils, and an interactive biosphere (so-called Earth System Models, or ESMs). Early generations of such transient calculations with ESMs give much better agreement with observed climate changes on Earth than previous calculations in which equilibrium responses to CO_2 doubling were the prime simulations available. When the transient models at the Hadley Center in the United Kingdom and the Max Planck Institute in Hamburg, Germany, were also driven both by greenhouse gases (which heat) and by sulfate aerosols (which cool), these time-evolving simulations yielded much more realistic "fingerprints" of human effects on climate (e.g., Chapter 8 of IPCC, 1996a). More such computer simulations are needed if we are to achieve greater confidence levels in the models. Nevertheless, scientists using coupled, transient simulations are now beginning to express growing confidence that current projections are plausible. (A number of still unpublished or in press papers have been performed very recently at a number of centers showing that the best explanation of the 20th-century climate trend is a combination of natural factors, such as solar variations, and human forcings, such as greenhouse gases, aerosols, and ozone changes – e.g., Wigley et al., 1998.) In my opinion, the "discernible" human impact is becoming clearer.

6.2.6 Transients and Surprises

However, a very complicated, coupled system such as an ESM is likely to have unanticipated results when forced to change very rapidly by external disturbances such as CO_2 and aerosols. Indeed, some of the transient models that are run out for hundreds of years exhibit dramatic change to the basic climate state (e.g., radical change in global ocean currents). Thompson and Schneider (1982) were the first to investigate the question of whether the time-evolving patterns of climate change might depend on the *rate* at which CO_2 concentrations increased. They used very simplified transient energy balance models to illustrate the importance of rates of forcing on regional climate responses. For slowly increasing CO_2 build-up scenarios, their model predicted the standard outcome: The temperature at the poles warmed more than that at the Tropics.

Any changes in equator-to-pole temperature difference help to create altered regional climates, because temperature differences over space influence large-scale atmospheric wind patterns. However, for very rapid increases in CO_2 concentrations a reversal of the equator-to-pole difference occurred in the Southern Hemisphere. If sustained over time, this would imply difficult-to-forecast transient climatic conditions during the century or so that the climate adjusts toward its new equilibrium state. In other words, the harder and faster the enormously complex climate system is forced to change, the higher is the likelihood for unanticipated responses. Or, in a phrase, *the faster and harder we push on nature, the greater the chance for surprises* – some of which are likely to be nasty.

Noting this possibility, the Summary for Policymakers of IPCC Working Group I concluded with the following paragraph (IPCC, 1996a, p.7):

> Future unexpected, large and rapid climate system changes (as have occurred in the past) are, by their nature, difficult to predict. This implies that future climate changes may also involve "surprises." In particular these arise from the non-linear nature of the climate system. When rapidly forced, non-linear systems are especially subject to unexpected behavior. Progress can be made by investigating non-linear processes and sub-components of the climatic system. Examples of such non-linear behavior include rapid circulation changes in the North Atlantic and feedbacks associated with terrestrial ecosystem changes.

Of course, if the climate system were somehow less "rapidly forced" by virtue of policies designed to slow the rate at which human activities modify the land surfaces and atmospheric composition, this would lower the likelihood of nonlinear surprises. Whether the risks of such surprises justify investments in abatement activities is the question that integrated assessment (IA) activities (see next section) are designed to inform (IPCC, 1996c). The likelihood of various climatic changes and estimates of the probabilities of such potential changes are the kinds of information IA modelers need from climate scientists in order to perform IA simulations (Schneider, 1997b). We turn next, therefore, to a discussion of methods to evaluate the subjective probability distributions of scientists on one important climate change issue, climate sensitivity.

Table 6.1. Experts Interviewed in the Study. Expert Numbers Used in Reporting Results are Randomized. They do not Correspond with Either Alphabetical Order or the Order in which the Interviews were Performed. From Morgan and Keith, 1995

James Anderson, Harvard University	Ronald Prinn, Massachusetts Institute of Technology
Robert Cess, State University of New York at Stony Brook	Stephen Schneider, Stanford University
Robert Dickinson, University of Arizona	Peter Stone, Massachusetts Institute of Technology
Lawrence Gates, Lawrence Livermore National Laboratories	Starley Thompson, National Center for Atmospheric Research
William Holland, National Center for Atmospheric Research	Warren Washington, National Center for Atmospheric Research
Thomas Karl, National Climatic Data Center	Tom Wigley, University Center for Atmospheric Research/National Center for Atmospheric Research
Richard Lindzen, Massachusetts Institute of Technology	Carl Wunsch, Massachusetts Institute of Technology
Syukuro Manabe, Geophysical Fluid Dynamics Laboratory	
Michael MacCracken, U.S. Global Change Research Program	

6.2.7   Subjective Probability Estimation

What defines a scientific consensus? Morgan and Keith (1995) and Nordhaus (1994) are two attempts by non-climate scientists who are interested in the policy implications of climate science to tap the knowledgeable opinions of what they believe to be representative groups of scientists from physical, biological, and social sciences on two separate questions: first, the climate science itself, and, second, policy-relevant impact assessment. The Morgan and Keith surveys show that although there is a wide divergence of opinion, nearly all scientists (e.g., Table 6.1) assign some probability of negligible outcomes and some probability of very highly serious outcomes, with few exceptions, such as Richard Lindzen at MIT (who is scientist number 5 in Figure 6.1, taken from Morgan and Keith).

In the Morgan and Keith study, each of the 16 scientists listed in Table 6.1 was put through a formal decision-analytic elicitation of their subjective probability estimates for a number of factors. Each survey took several hours. Figure 6.1 shows the elicitation results for the important climate sensitivity factor. Note that 15 of 16 scientists surveyed (including several 1995 IPCC Working Group I Lead Authors; I am scientist 9) assigned something like a 10% subjective likelihood of small (less than 1 °C) surface warming from doubling of CO_2. These scientists also typically assigned a 10% probability for extremely large climatic changes – greater than 5 °C, roughly equivalent to the temperature difference experienced between a glacial and interglacial age, but occurring some 100 times more rapidly. In addition to the lower probabilities assigned to the mild and catastrophic outcomes, the bulk of the scientists interviewed (with the one exception) assigned the bulk of their subjective cumulative probability distributions in the center of the IPCC range for climate sensitivity. What is most striking about the exception,

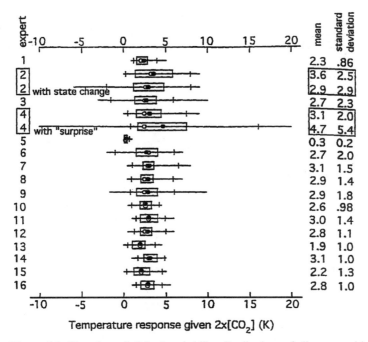

Figure 6.1. Box plots of elicited probability distributions of climate sensitivity, the change in globally averaged surface temperature for a doubling of CO_2 ($2x$ [CO_2] forcing). Horizontal line denotes range from minimum (1%) to maximum (99%) assessed possible values. Vertical tick marks indicate locations of lower (5) and upper (95) percentiles. Box indicates interval spanned by 50% confidence interval. Solid dot is the mean, and open dot is the median. The two columns of numbers on right-hand side of the figure report values of mean and standard deviation of the distributions. From Morgan and Keith, 1995.

scientist 5, is the lack of variance in his estimates, suggesting a very high confidence level in this scientist's mind that he understands how all the complex interactions within the Earth System described above will work. None of the other scientists displayed that confidence, nor did the Lead Authors of IPCC. However, several scientists interviewed by Morgan and Keith expressed concern for "surprise" scenarios; for example, scientists 2 and 4 explicitly display this possibility in Figure 6.1, whereas several other scientists – myself among them – implicitly allow for both positive and negative surprises inasmuch as they assigned a considerable amount of their cumulative subjective probabilities for climate sensitivity outside the standard 1.5 to 4.5 range. This concern for surprises is consistent with the concluding paragraph of the IPCC Working Group I Summary for Policymakers quoted above and the studies of Rahmstorf (1997), Broecker (1997), and Stocker and Schmittner (1997).

IPCC Lead Authors, who wrote the Working Group I Second Assessment Report, were fully aware both of the wide range of possible outcomes and of the broad distributions of attendant subjective probabilities. After a number of sentences highlighting such uncertainties, the report concluded, "Nevertheless, the balance of evidence suggests that there is a discernible human influence on global climate" (IPPC, 1996a, p. 5). The reasons for this now-famous subjective judgment were many, such as the kinds of factors listed above. These include a well-validated theoretical case for the natural greenhouse effect, validation tests of both model parameterizations and performance against present and paleoclimatic data, and the growing "fingerprint" evidence that suggests that horizontal and vertical patterns of climate change predicted to occur in coupled atmosphere-ocean models have been increasingly evident in observations over the past several decades. Clearly, more research is needed, but enough is already known to warrant assessments of the possible impacts of such projected climatic changes and the relative merits of alternative actions to mitigate emissions and/or to make adaptations less costly. That is the ongoing task of integrated assessment analysts, a task that will become increasingly critical in the next century. To accomplish this task, it is important to recognize what is well established in climate data and modeling and to separate this from aspects that are more speculative. That is precisely what IPCC (1996a) attempted to accomplish.

6.3   Assessing the Environmental and Societal Impacts of Climatic Change Projections

One of the principal tools used in the integrated assessment of climate change is integrated assessment models (IAMs). These models often comprise many submodels adopted from a wide range of disciplines. In IAMs, modelers "combine scientific and economic aspects of climate change in order to assess policy options for climate change" control (Kelly and Kolstad, in press).

One of the most dramatic of the standard "impacts" of climatic warming projections is the increase in sea level typically associated with warmer climatic conditions. A U.S. Environmental Protection Agency (EPA) study used an unusual approach: combining climatic models with the subjective opinions of many scientists on the values of uncertain elements in the models to help bracket the uncertainties inherent in this issue. Titus and Narayanan (1996) – including teams of experts of all persuasions on the issue – calculated the final product of their impact assessment as a statistical distribution of future sea level rise, ranging from negligible change as a low probability outcome to a meter or more rise, also with a low probability (see Figure 6.2). The midpoint of the probability distribution is something like a half-meter sea level rise by the end of the next century.

Because the EPA analysis stopped there, this is by no means a complete assessment. To take integrated assessment to its logical conclusion, we must ask what the economic costs of various control strategies might be and how the costs of abatement compare to the economic or environmental losses (i.e., impacts, or *damages*, as they are called) from sea level rises. That means putting a value – a dollar value typically – on climate change, coastal wetlands, fisheries, environmental refugees, and so on. Hadi Dowlatabadi at Carnegie Mellon University leads a team of integrated assessors who, like Titus, combined a wide range of scenarios of climatic changes and impacts but, unlike the EPA studies, added a wide range of abatement cost estimates into the mix. The group's integrated assessment was presented in statistical form as a probability that investments in CO_2 emissions controls would either cost more than the losses from averted climate change or the reverse (Morgan and Dowlatabadi, 1996). Because its results do not include estimates for all conceivable costs (e.g., the human or political consequences

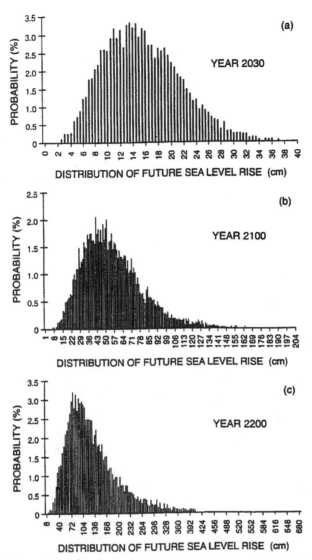

Figure 6.2. Plots showing the probability of various rises of sea level in the years 2030, 2100, and 2200, calculated on the basis of the Monte Carlo estimation technique, combining experts' probability distributions for model parameters. From Titus and Narayanan, 1994.

of persons displaced from coastal flooding), the Carnegie Mellon group offered its results only as illustrative of the capability of IA techniques. Its numerical results have meaning only after the range of physical, biological, and social outcomes and their costs and benefits have been quantified – a Herculean task. Similar studies have been made in Holland by a Dutch effort to produce integrated assessments for policy makers. Jan Rotmans, who headed one of the efforts, likes to point out that such modeling of complex physical, biological, and social factors cannot produce credible "answers" to current policy dilemmas but can provide "insights" to policy makers that will put decision making on a stronger analytical basis (Rotmans and van Asselt, 1996). Understanding the strengths and weaknesses of any complex analytic tool is essential to rational policy making, even if quantifying the costs and benefits of specific activities is controversial (e.g., Schneider, 1997b).

William Nordhaus, an economist from Yale University, has taken heroic steps to put the climatic change policy debate into an optimizing framework. He is an economist who has long acknowledged that an efficient economy must internalize externalities (in other words, find the full social costs of our activities and not only the direct cost reflected in conventional "free market" prices to private firms or individuals). He tried to quantify this external damage from climate change and then tried to balance it against the costs to the global economy of policies designed to reduce CO_2 emissions. His "optimized" solution was a carbon tax, designed to internalize the externality of damage to the climate by increasing the price of fuels in proportion to how much carbon they emit, thereby providing an incentive for society to use less of these fuels – in essence, a "pollutor pays" principle.

Nordhaus (1992) imposed carbon tax scenarios ranging from a few dollars per ton to hundreds of dollars per ton of carbon emitted; the latter would effectively limit coal use in the world economy. He showed that, in the context of his model and its assumptions, these carbon emission fees would cost the world economy anywhere from less than 1% annual loss in gross national product (GNP) to a several percent loss by the year 2100. The efficient, optimized solution from classical economic cost-benefit analysis is that carbon taxes should be levied sufficient to reduce the GNP as much as it is worth to avert climate change (e.g., the damage to GNP from climate change). He assumed that the impacts of climate change were equivalent to a loss of about 1% of GNP. This led to an "optimized" initial carbon tax of about five dollars or so per ton of carbon dioxide emitted, rising by severalfold to A.D. 2100. In the context of his modeling exercise, this would avert only a few tenths of a degree of global warming to the year 2100, a very small fraction of the 4 °C warming his model projected.

How did Nordhaus arrive at climate damage being about 1% of GNP? He assumed that agriculture was the economic market sector most vulnerable to climate change. For decades, agronomists had calculated potential changes to crop yields from various climate change scenarios, suggesting some regions, now too hot, would sustain heavy losses from warming, whereas others, now too cold, could gain. Noting that the United States lost about one-third of its agricultural economy in the heat waves of 1988, and that agriculture then represented about 3% of the U.S. GNP, Nordhaus thought the typically projected climatic changes might thus cost the U.S. economy something like 1% annually in the 21st century. This figure was severely criticized because it neglected damages from health impacts (e.g., expanded areas of tropical diseases, heat-stress deaths, etc.), losses from coastal flooding or severe storms, security risks from the presence of "boat people" as the result of coastal disruptions in South Asia, or any damages to wildlife (e.g., Sorenson et al., 1998), fisheries, or ecosystems (e.g., IPCC, 1996b) that would almost surely accompany temperature rises at rates of degrees per century as are typically projected. It also was criticized because Nordhaus's estimate neglected potential increases in crop or forestry yields from the direct effects of increased CO_2 in the air on the photosynthetic response of these marketable plants. Nordhaus responded to his critics by conducting a survey, similar to that undertaken by Morgan

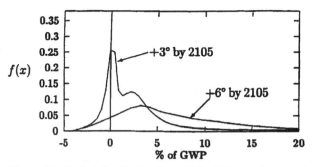

Figure 6.3. Probability distributions (f(x)) of climate damages (market and nonmarket components combined) from an expert survey in which respondents were asked to estimate 10th, 50th, and 90th percentiles for the two climate change scenarios shown. From Roughgarden and Schneider, 1999. Data from Nordhaus, 1994.

and Keith, but this time focused on the impacts of several scenarios of climatic change on world economic product, including standard market sector categories (e.g., forestry, agriculture, heating and cooling demands) as well as so-called nonmarket sectors such as biological conservation, international equity, and national security.

When Nordhaus surveyed the opinions of mainstream economists, environmental economists, and natural scientists (I am respondent 10 in Nordhaus, 1994), he found that the former expressed a factor of 20 less anxiety about the economic or environmental consequences of climate change than the latter (see Figure 6.3). However, the bulk of even the conservative group of economists Nordhaus surveyed considered there to be at least a 10% probability that typically projected climate changes could still cause economic damages worth several percent of gross world product (the current U.S. GNP is around five trillion dollars, about 20% of the global figure). And some of these economists didn't include estimates for possible costs of nonmarket damages (e.g., harm to nature). One ecologist who did explicitly factor in nonmarket values for natural systems went so far as to assign (for 6 °C warming) a 10% chance of a 100% loss of GNP – the virtual end of civilization! Although Nordhaus observed that those who know the most about the economy are "comparatively unconcerned," I countered with the obvious observation that those who know the most about nature are comparatively very concerned (e.g., see Roughgarden and Schneider, 1999).

We will not easily resolve the paradigm gulf between the relatively optimistic and pessimistic views of these specialists, who have different training, traditions, and world-views. But the one thing that is clear from both the Morgan and Keith and the Nordhaus studies is that the vast majority of knowledgeable experts from a variety of fields admits to a wide range of plausible outcomes in the area of climate change – including both mild and catastrophic eventualities – under the broad umbrella of possibilities. This is a condition ripe for misinterpretation by those who are unfamiliar with the wide range of probabilities most scientists attach to climate change issues. The wide range of probabilities follows from recognition of the many uncertainties in data and assumptions still inherent in climate models, climatic impact models, and economic models, or their synthesis via integrated assessment models (see Schneider, 1997a,b). In a highly interdisciplinary enterprise such as the integrated assessment of climate change problems, it is necessary to include a wide range of possible outcomes, along with a representative sample of the subjective probabilities that knowledgeable assessment groups such as the IPCC believe accompany each of those possible outcomes. In essence, the "bottom line" of estimating climatic impacts is that both "the end of the world" and "it is good for business" are the two lowest-probability outcomes, and the vast majority of knowledgeable scientists and economists considers there to be a significant chance of climatic damage to both natural and social systems. Under these conditions – and the unlikelihood that research will soon eliminate the large uncertainties that still persist – it is not surprising that most formal climatic impact assessments have called for cautious but positive steps, both to slow the rate at which humans modify the climatic system and to make natural and social systems more resilient to whatever changes eventually materialize (e.g., National Academy of Sciences, 1991).

6.4   Policy Implications

6.4.1   What Are Some Actions to Consider?

What are the kinds of actions that can be considered to deal with global problems such as climate change? The following list is a consensus from a multidisciplinary, business, university, and government assessment conducted by the U.S. National Research Council in 1991. It is encouraging that this multidiscipline, ideologically diverse group could agree that the United States, for example, could reduce or offset its greenhouse gas emissions by between 10% and 40% of 1990 levels at low cost, or even at some net savings, if proper policies are implemented. Here is the Council's (National Academy of Sciences, 1991) entire suggested list (pp. 72–82):

1. Continue the aggressive phaseout of CFC and other halocarbon emissions and the development of substitutes that minimize or eliminate greenhouse gas emissions.
2. Study in detail the "full social cost pricing" of energy, with a goal of gradually introducing such a system. On the basis of the principle that the polluter should pay, pricing of energy production and use should reflect the full costs of the associated environmental problems.
3. Reduce the emissions of greenhouse gases during energy use and consumption by enhancing conservation and efficiency.
4. Make greenhouse warming a key factor in planning for our future energy supply mix. The United States should adopt a systems approach that considers the interactions among supply, conversion, end use, and external effects in improving the economics and performance of the overall energy system.
5. Reduce global deforestation.
6. Explore a moderate domestic reforestation program and support international reforestation efforts.
7. Maintain basic, applied, and experimental agricultural research to help farmers and commerce adapt to climate change and thus ensure ample food.

8. Make water supply more robust by coping with present variability by increasing efficiency of use through water markets and by better management of present systems of supply.

9. Plan margins of safety for long-lived structures to take into consideration possible climate change.

10. Move to slow present losses in biodiversity.

11. Undertake research and development projects to improve our understanding of both the potential of geoengineering options to offset global warming and their possible side effects. This is not a recommendation that geoengineering options be undertaken at this time, but rather that we learn more about their likely advantages and disadvantages.

12. Control of population growth has the potential to make a major contribution to raising living standards and to easing environmental problems like greenhouse warming. The United States should resume full participation in international programs to slow population growth and should contribute its share to their financial and other support.

13. The United States should participate fully with officials at an appropriate level in international agreements and in programs to address greenhouse warming, including diplomatic conventions and research and development efforts.

This National Academy of Sciences (1991) assessment produced a remarkable list, considering the diversity of the participants' backgrounds and their varying ideological perspectives. It even recommends "international agreements . . . including diplomatic conventions," something that indeed occurred at Kyoto in 1997. In the crucible of open debate that permeated this scientific and policy assessment activity, the self-interest polemics and media grandstanding typical of the Kyoto event (e.g., see Schneider, 1998a for an account) did not occur. This NAS group didn't assert that catastrophe was inevitable, nor that it was improbable. The National Research Council assessors simply believed that "despite the great uncertainties, greenhouse warming is a potential threat sufficient to justify action now." Integrated assessments of the policy options offered by the National Research Council report are actively being pursued with a variety of models.

This comprehensive list of recommendations from the National Research Council report still overlooks two fundamental policy options: the desperate need for (1) an intelligent, nonpolemical public debate about climate change and (2) interdisciplinary public education that also teaches about whole systems and long-term risk management and not only traditional areas of isolated, disciplinary specialization.

6.4.2   Environment and (or Versus) Development?

Although the NRC report acknowledged the importance of international dimensions of climate change policy making, it was still largely a developed-country perspective. Less developed countries (LDCs) often have very different perspectives and priorities. Many more pressing concerns critical to human health and well-being are competing for priority attention. First, LDCs are struggling to raise literacy rates, lower death rates, increase life expectancy, provide employment for burgeoning populations, and reduce local air and water pollution that poses imminent health hazards to their citizens and environments. Protecting species or slowing climate change are simply low on their priority lists as compared with those of more mature economic powers such as the Organization for Economic Cooperation and Development (OECD) nations. It is unfortunate that LDCs place a low priority on the abatement of global climate change because nearly all impact assessments suggest that it is these very countries that are most vulnerable to climatic change (e.g., Rosenzweig and Parry, 1994).

There is a phrase in economics known as "the marginal dollar." In our context it means that given all the complexity of interconnected physical, biological, and social systems, climate abatement may not be perceived as the best place to invest the next available dollar so as to bring the maximum social benefit to poor countries. I have heard many representatives of LDCs exclaim that until poverty is corrected, preventable disease stamped out, injustice redressed, and economic equity achieved, they will invest their precious resources on these priorities. My response has been that climatic changes can exacerbate all those problems they rightly wish to address, and thus we should seek to make investments that both reduce the risks of climate change and help with economic development (transfer of efficient technologies being a prime example). It is a great mistake, I believe, to get trapped in the false logic of the mythical "marginal dollar," for it is not necessary that every penny of the next available dollar go exclusively to the highest-priority problem whereas all the rest (particularly problems with surprise potential and the possibility of irreversible damages) must wait until priority 1 is fully achieved. To me, the first step is to get that "marginal dollar" cashed into "small change" so that many interlinked priority problems can be at least partially addressed. Given the large state of uncertainty surrounding both the costs and the benefits of many human and natural events, it seems most prudent to address many issues simultaneously and to constantly reassess which investments are working and which problems – including climate change – are growing more or less serious.

It takes resources to invest, of course, and because the bulk of available capital is in developed countries, it will require international negotiations – "planetary bargaining" it has been called – to balance issues of economic parity and social justice with environmental protection. Such negotiations are under way under UN auspices as the Conference of the Parties (COP), and it will likely take many years to work out protocols that weigh the diverse interests and perceptions of the world's nations (e.g., Schneider, 1998b).

There is a lively debate among economists, technologists, industrialists, and environmentalists about what are the most cost-effective strategies for abating carbon emissions that also can reduce potential impacts of climatic changes to below the undefined "dangerous" levels referred to in the language of the UN Framework Convention on Climate Change. Most economists argue that some policy to "internalize the externality" of potential climate damage is already appropriate, reflecting the recommendations already published by the U.S. National Research Council in 1991. Environmentalists usually argue that major efforts to spur immediate abatement of carbon emissions are necessary if climatic changes less than one more degree Celsius (which they typically define as "dangerous") are to, be avoided. Most economists, on the other hand, often argue that new technologies will be able to accomplish carbon abatement more cheaply in the future as such technologies are developed and deployed (Wigley et al., 1996). Thus, their logic suggests that a cost-effective time profile of abatement would be to postpone most carbon reductions until later in the 21st century. This seemingly implacable debate will echo well past the Kyoto meeting.

My colleague, Stanford University economist Lawrence Goulder, and I have used state-of-the-art economic modeling tools to study this debate, and we conclude that both the stereotypical environmentalist position (to abate now) and economist position (abate later) are actually not incompatible; rather, they are complementary. We show (Schneider and Goulder, 1997) that although the economist view that future abatement is likely to be cheaper is probably correct, so too is the environmentalist argument that current actions are urgently needed, because such technologies referred to in economic cost-effectiveness studies won't simply invent themselves. In other words, policy actions to help induce technological changes are needed now in order to bring about a profile of cost-effective abatement in the decades ahead (see also Hoffert et al., 1998). Schneider and Goulder also address the relative economic efficiency of alternative policy instruments: contrasting carbon taxes versus research and development (R&D) subsidies. Although we recognize the political reluctance of many people to embrace any new taxes, in truth, most economic analyses show that a fee for the use of the atmosphere (currently a "free sewer") will reduce incentives to pollute, will increase incentives to develop and deploy less-polluting technologies, and can be more economically efficient than other policies, particularly if some of the revenues generated by a carbon tax were recycled back into the economy (e.g., Hamond et al., 1997). R&D subsidies can be economically efficient, our conventional economic analyses suggest, to the extent that current R&D markets are already subsidized or otherwise not optimally efficient – a very real likelihood (e.g., Goulder and Schneider, 1999).

Therefore, it is my personal view that all parties to the climate negotiations should recognize that potential damages to a global commons like the Earth's climate are not mere ideological rhetoric, nor are solutions necessarily unaffordable. Moreover, "win-win" solutions, in which economic efficiency, cost-effectiveness, and environmental protection can happily coexist, are possible – if only we put aside hardened ideological positions (e.g., Schneider, 1998b).

6.5   Concluding Remarks

A condensed summary of my principal conclusions is as follows, beginning with the more narrowly technical issues and proceeding to broader generalizations about human impacts, uncertainties, and policy choices.

6.5.1 Hierarchy of Models

A hierarchy of models – ranging from simple zero- or one-dimensional, highly parameterized models up to coupled three-dimensional models that simulate the dynamics and thermodynamics of connected physical and biological subsystems of the

Earth System – are needed for climatic effects assessment. The simpler models are more transparent – allowing cause-and-effect processes to be more easily traced – and are much more tractable to construct, run and diagnose, whereas multidimensional, dynamic models can provide geographic and temporal resolution needed for regional impact assessments and, hopefully, provide more-realistic and detailed simulations, even if at much higher costs for construction, computation, diagnosis, and interpretability. Because the real climate system is undergoing a transient response to regionally hetero-geneous (patchy) forcings (e.g., aerosols and greenhouse gases combined, which both vary over time and space), eventually it will be necessary to run fully coupled three-dimensional Earth Systems models in order to "hand off" their results to a variety of regional impact assessment models. In the interim, lower-resolution, "simple" climate models can be hybridized into more-comprehensive models to produce hybrid estimates of time-evolving regional patterns of climatic changes from a variety of emissions and land use change scenarios. Such estimates may be instructive to policy makers interested in the differential climatic impacts of various climate forcing scenarios and/or various assumptions about the internal dynamics of both climate and impact models.

6.5.2   Sensitivity Studies Are Essential

It is unlikely that all important uncertainties in either climatic or social and environmental impact models will be resolved to the satisfaction of the bulk of the sci-entific community in the near future. However, this does not imply that model results are uninformative. On the contrary, sensitivity analyses in which various policy-driven alternative radiative forcing assumptions are made can offer insights into the potential effectiveness of such policies in terms of their differential climatic effects and impacts. Even though absolute accuracy is not likely to be claimed for the foreseeable future, considerable precision concerning the sensitivity of the physical and biological subsys-tems of the Earth can be studied via carefully planned and executed sensitivity studies across a hierarchy of models.

6.5.3   Validation and Testing Are Required

Although it may be impractical, if not theoretically impossible, to validate the precise future course of climate given the uncertainties that remain in forcings, internal dynamics, and unpredictable surprise events, many of the basic features of the coupled physical and biological subsystems of the Earth can already be simulated to a consider-able degree. Testing models against each other when driven by the same sets of forcing scenarios, testing the overall simulation skill of models against empirical observations, testing model parameterizations against high-resolution process models or data sets, testing models against proxy data of paleoclimatic changes, and testing the sensitivity of models to radiative forcings of anthropogenic origin by computing their sensitivity to natural radiative forcings (e.g., seasonal radiative forcing, volcanic dust forcing, orbital element variation forcings, meltwater-induced rapid ocean current changes, etc.) con-stitute a necessary set of validation-oriented exercises that all modelers should agree to perform. Similarly, impacts models should also be subjected to an analogous set of validation protocols if their insights are to gain a high degree of credibility.

6.5.4 Subjective Probability Assessment

In addition to standard simulation modeling exercises in which various parameters are specified or varied over an uncertainty range, formal decision-analytic techniques can be used to provide a more consistent set of values for uncertain model parameters or functional relationships (e.g., Moss and Schneider, 1997). The embedding of subjective probability distributions into climatic models is only beginning (e.g., Titus and Narayanan, 1996), but it may become an important element of IA modeling in future generations of model building (e.g., see the discussion of the hierarchy of IAMs in Schneider, 1997b).

6.5.5 Rolling Reassessment

It is obvious that the projection of climatic effects and related impacts will continue to change as the state of the art in both kinds of models improves over the next few decades. Therefore, the most flexible management possible of a global commons like the Earth's climate seems a virtual necessity; the potential seriousness of the problem – or even the perception of that seriousness – is virtually certain to change with new discoveries and actual climatic and other environmental or societal events. Therefore, a series of assessments of climatic effects, related impacts, and policy options to prevent potentially dangerous impacts will be needed periodically – perhaps every five years, as IPCC has chosen for the repeat period of its major Assessment Reports, which treat climatic effects, impacts, and policy issues as separable assessments. It seems important that whatever policy instruments are employed (either to mitigate anthropogenic forcings or to help reduce damage from projected climatic effects) be flexible enough to respond quickly and cost-effectively to the evolving science that will emerge from this rolling reassessment process.

6.5.6 Consider Surprises and Irreversibility

Given the many uncertainties that still attend most aspects of the climatic change and impacts debate, priority should be considered for those aspects that could exhibit irreversible damages (e.g., extinction of species whose already-shrinking habitat is further stressed by rapid climatic changes) or for which imaginable "surprises" (e.g., Schneider et al., 1998) have been identified (e.g., alterations to oceanic currents from rapid increases in greenhouse gases; see Broecker, 1997, or Rahmstorf, 1997). For these reasons, management of climatic risks must to be considered well in advance of more certain knowledge of climatic effects and impacts.

6.5.7 Win-Win Strategies

Economically efficient, cost-effective, and environmentally sustainable policies have been identified, and others can be found, to help induce the kinds of technological innovations needed to reduce atmospheric emissions in the decades ahead. Some mix of emissions "cap and trade," carbon taxes with revenue recycling, or technology development incentives can provide win-win solutions if all parties to the environment-development debate would lower the intensity of their ideological preconceptions (e.g., Schneider, 1998b) and work together for cost-effective and equitable measures to protect the global commons.

Controversy will still remain, of course, because total emissions are the product, of world population size, per capita economic product, and the activities that produce that economic activity (e.g., Yang and Schneider, 1998). Technological innovations to reduce emissions are less controversial than are social policies, which affect affluence and population growth. As a result, incentives for technology development and deployment are likely to be the focus of climate policy for the immediate future. But the social factors will eventually need to be considered if very large human impacts on the environment are to be averted.

REFERENCES

Barthelet, P., Terry, L., and Velcke, S. (1998). Transient CO_2: Experiment using the ARPEGE/OPAICE non flue corrected couple model. Geophysical Research Letters, 25, 2277–2280.

Broecker, W. S. (1997). Thermohaline circulation, the Achilles heel of our climate system: will man-made CO_2 upset the current balance? Science, 278, 1582–1588.

Bryan, F. O. (1998). Climate drift in a multicentury integration of the NCAR climate system model. Journal of Climate, 11, 1455–1471.

Crowley, T. J., and North, G. R. (1991). Paleoclimatology. New York: Oxford University Press.

Eddy, J. A., and Oeschger, H., eds. (1993). Global Changes in the Perspective of the Past. New York: John Wiley and Sons.

Edwards, P. N., and Schneider, S. H. (1997). The 1995 IPCC report: Broad consensus or "scientific cleansing"? Ecofables/Ecoscience. Stanford University: Center for Conservation Biology.

Goulder, L. H., and Schneider, S. H. (1999). Induced technological change and the attractiveness of CO_2 abatement policies. Resource and Energy Economics, 21, 211–253.

Gregory, J. M., and Mitchell, J. F. B. (1997). The climate response to CO_2 of the Handley centre coupled AOGCM with and without flux adjustment. Geophysical Research Letters, 24, 1943–1946.

Hamond, M. J., DeCanio, S., Duxbury, P., Sanstad, A. H., and Stinson, C. H. (1997). Tax Waste, Not Work: How Changing What We Tax Can Lead to a Stronger Economy and a Cleaner Environment. San Francisco: Redefining Progress.

Hoffert, M. I., and Covey, C. (1992). Deriving global climate sensitivity from paleoclimate reconstructions. Nature, 360, 573–576.

Hoffert, M. I., Caldeira, K., Jain, A. K., Harvey, L. D. D., Haites, E. F., Potter, S. D., Schlesinger, M. E., Schneider, S. H., Watts, R. G., Wigley, T. M. L., and Wuebbles, D. J. (1998). Energy implications of future stabilization of atmospheric CO_2 content. Nature, 395, 881–884.

Intergovernmental Panel on Climatic Change (IPCC) (1996a). Climate Change 1995. The Science of Climate Change: Contribution of Working Group I to the Second Assessment Report of the Intergovernmental Panel on Climate Change, eds. J. T. Houghton, L. G. Meira Filho, B. A. Callander, N. Harris, A. Kattenberg, and K. Maskell. Cambridge: Cambridge University Press.

Intergovernmental Panel on Climatic Change (IPCC) (1996b). Climate Change 1995. Impacts, Adaptations and Mitigation of Climate Change: Scientific-Technical Analyses. Contribution of Working Group II to the Second Assessment Report of the Intergovernmental Panel on Climate Change, eds. R. T. Watson, M. C. Zinyowera, and R. H. Moss. Cambridge: Cambridge University Press.

Intergovernmental Panel on Climatic Change (IPCC) (1996c). Climate Change 1995. Economic and Social Dimensions of Climate Change. Contribution of Working Group III to the Second Assessment Report of the Intergovernmental Panel on Climate Change, eds. J. P. Bruce, H. Lee, and E. F. Haites. Cambridge: Cambridge University Press.

Intergovernmental Panel on Climatic Change (IPCC) (1997). Workshop on Regional Climate Change Projections for Impact Assessment, Imperial College, London, 24–26 September 1996.

Intergovernmental Panel on Climatic Change (IPCC) (1998). The Regional Impacts of Climate Change. An Assessment of Vulnerability. A Special Report of IPCC Working Group II, eds. R. T. Watson, M. C. Zinyowera, and R. H. Moss. Cambridge: Cambridge University Press.

Kelly, D. L., and Kolstad, C. D., in press. Integrated assessment models for climate change control. International Yearbook of Environmental and Resource Economics 1999/2000: A Survey of Current Issues. Cheltenham, UK: Edward Elgar.

Mann, M. E., Bradley, R. S., and Hughes, M. K. (1999). Northern hemisphere temperatures during the past millennium: Inferences, uncertainties, and limitations. Geophysical Research Letter, 26, 759.

Mass, C., and Schneider, S. H. (1977). Influence of sunspots and volcanic dust on long-term temperature records inferred by statistical investigations. Journal of Atmospheric Science, 34, 1995–2004.

Morgan, M. G., and Dowlatabadi, H. (1996). Learning from integrated assessment of climate change. Climatic Change, 34, 337–368.

Morgan, M. G., and Keith, D. W. (1995). Subjective judgments by climate experts. Environmental Science and Technology, 29, 468A–476A.

Moss, R., and Schneider, S. H. (1997). Characterizing and communicating scientific uncertainty: Building on the IPCC second assessment. In: Elements of Change, 1996, eds. S. J. Hassol and J. Katzenberger. Colorado: Aspen Global Change Institute.

National Academy of Sciences (1991). Policy Implications of Greenhouse Warming. Washington, D.C.: National Academy Press.

Nordhaus, W. D. (1992). An optimal transition path for controlling greenhouse gases. Science, 258, 1315–1319.

Nordhaus, W. D. (1994). Expert opinion on climate change. American Scientist, 82, 45–51.

Overpeck, J. T., Webb, R. S., and Webb III, T. (1992). Mapping eastern North American vegetation change over the past 18,000 years: No analogs and the future. Geology, 20, 1071–1074.

Rahmstorf, S. (1997). Risk of sea-change in the Atlantic. Nature, 388, 825–826.

Rahmstorf, S., and Ganopolski, A. (1999). Long-term global warming scenarios computed with an efficient coupled climate model. Climatic Change, 43, 353–67.

Ramanathan, V. (1981). The role of ocean-atmospheric interactions in the CO_2 climate problem. Journal of Atmospheric Science, 38, 918–930.

Raval, A., and Ramanathan, V. (1989). Observational determination of the greenhouse effect. Nature, 342, 758.

Robinson, A. B., and Robinson, Z. W. (1997). Science has spoken: Global warming is a myth. Wall Street Journal, December 4.

Root, T. L., and Schneider, S. H. (1995). Ecology and climate: research strategies and implications. Science, 269, 331–341.

Rosenzweig, C., and Parry, M. (1994). Potential impact of climate change on world food supply. Nature, 367, 133–138.

Rotmans, J., and van Asselt, M. (1996). Integrated assessment: a growing child on its way to maturity – an editorial. Climatic Change, 34, 327–336.

Roughgarden, T., and Schneider, S. H. (1999). Climate change policy: Quantifying uncertainties for damages and optimal carbon taxes. Energy Policy 27(7), 415–429.

Santer, B. D., Taylor, K. E., Wigley, T. M. L., Johns, T. C. , Jones, P. D., Karoly, D. J. Mitchell, J. F. B., Oort, A. H., Penner, J. E., Ramaswamy, V., Schwarzkopf, M. D., Stouffer, R. J., and Tett, S. (1996). A search for human influences on the thermal structure of the atmosphere. Nature, 382, 39–46.

Schneider, S. H. (1993). Can paleoclimatic and paleoecological analyses validate future global climate and ecological change projections? In: Global Changes in the Perspective of the Past, eds. J. A. Eddy and H. Oeschger. New York: John Wiley and Sons.

Schneider, S. H. (1994). Detecting climatic change signals: Are there any "fingerprints"? Science, 263, 341–347.

Schneider, S. H. (1997a). Laboratory Earth: The Planetary Gamble We Can't Afford to Lose. New York: Basic Books.

Schneider, S. H. (1997b). Integrated assessment modeling of global climate change: Transparent rational tool for policy making or opaque screen hiding value-laden assumptions? Environmental Modeling and Assessment, 2, 229–249.

Schneider, S. H. (1998a). Kyoto Protocol: The Unfinished agenda. An editorial essay. Climatic Change, 39, 1–21.

Schneider, S. H. (1998b). The climate for greenhouse policy in the U.S. and the incorporation of uncertainties into integrated assessments. Energy and Environment, 9, 425–440.

Schneider, S. H., and Goulder, L. (1997). Achieving carbon dioxide concentration targets. What needs to be done now? Nature, 389, 13–14.

Schneider, S. H., and Londer, R. (1984). The Coevolution of Climate and Life. San Francisco: Sierra Club Books.

Schneider, S. H., and Mesirow, L. E. (1976). The Genesis Strategy: Climate and Global Survival. New York: Plenum.

Schneider, S. H., Turner II, B. L., and Morehouse Garriga, H. (1998). Imaginable surprise in global change science. Journal of Risk Research, 1, 165–185.

Shackley, S., Risbey, J., Stone, P., and Wynne, B. (1999). Adjusting to policy expectations in climate change modeling: An interdisciplinary study of flux adjustments in coupled atmosphere-ocean general circulation models. Climatic Change, 43, 413–454.

Sorenson, L. G., Goldberg, R., Root, T. L., and Anderson, M. G. (1998). Potential effects of global warming on waterfowl populations breeding in the northern Great Plains. Climatic Change, 40, 343–369.

Stocker, T. F., and Schmittner, A. (1997). Influence of CO_2 emission rates on the stability of the thermohaline circulation. Nature, 388, 862–865.

Thompson, S. L., and Schneider, S. H. (1982). CO_2 and climate: The importance of realistic geography in estimating the transient response. Science, 217, 1031–1033.

Titus, J., and Narayanan, V. (1996). The risk of sea level rise: A delphic monte carlo analysis in which twenty researchers specify subjective probability distributions for model coefficients within their respective areas of expertise. Climatic Change, 33, 151–212.

Trenberth, K. E. (ed.). (1992). Climate System Modeling. Cambridge: Cambridge University Press.

Washington, W. M., and Parkinson, C. L. (1986). An Introduction to Three-Dimensional Climate Modeling. New York: Oxford University Press.

Wigley, T. M. L., Richels, R., and Edmonds, J. A. (1996). Economic and environmental choices in the stabilizations of atmospheric CO_2 concentrations. Nature, 379, 240–243.

Wigley, T. M. L, Smith, R. L., and Santer, B. D. (1998). Anthropogenic influence on the auto correlation structure of hemispheric-mean temperatures. Science, 282, 1676–1679.

Wright, H. E., Kutzbach, J. E., Webb III, T., Ruddiman, W. F., Street-Perrott, F. A., and Bartlein, P. J., eds. (1993). Global Climates Since the Last Glacial Maximum. Minneapolis: University of Minnesota Press.

Yang, C., and Schneider, S. H. (1998). Global carbon dioxide emissions scenarios: Sensitivity to social and technological factors in three regions. Mitigation and Adaptation Strategies for Global Change, 2, 373–404.

PART THREE

MODELING THE EARTH'S SYSTEM

7 Earth System Models and the Global Biogeochemical Cycles

DAVID SCHIMEL

ABSTRACT

Earth System Models have become a holy grail of the earth sciences. Earth System Models are a class of simulation that model a significant number of interactions between the atmosphere, the oceans, the land, the cryosphere, and the biogeochemical cycles. Such models are an evolution of climate and physical ocean models developed for disciplinary purposes and of the land surface models that have developed from ecology and hydrology. Increasingly they also include some components of the carbon cycle, ecosystems, and atmospheric photochemistry. The development of a new class of models successfully capturing the behavior of a system is itself a demonstration of a certain level of scientific knowledge. However, the development of Earth System Models has been strongly forced by a series of important scientific questions. Of special interest are questions coupling forcing (atmospheric greenhouse gases) and response (climate and ocean circulation). The carbon cycle is the best understood of the major global biogeochemical cycles, and so I focus on issues linked to carbon; however, the next generation of challenges will grow and include the nitrogen, ozone, sulfur, and iron cycles. In this chapter, I discuss some emerging questions and identify some key research areas associated with these new areas of inquiry.

7.1 Scientific Challenges

7.1.1 Where Does the Carbon Go?

For the past decades, as research has focused on the carbon cycle, there has been keen interest in the sinks of anthropogenic CO_2 and especially the so-called missing sink. This is an Earth System modeling problem for two reasons. First, measurements over the period 1958–present and over the extended record (preindustrial–present) are too sparse in space and time to give more than hints as to the detailed fate of CO_2 and of the mechanisms that have controlled carbon dynamics. Second, simulation of observed changes is a key credibility test for models that will be used in a prognostic mode. This question has become more interesting as the data records have become richer. Consider the following:

- Atmospheric data show substantial interannual variability of the growth rate of CO_2. This variability is not caused by variations in fossil fuel use, and thus it demonstrates dynamic behavior of the carbon-climate system.
- Multiple lines of evidence suggest that there exists significant interannual variability of land and ocean CO_2 exchange. Terrestrial CO_2 exchange appears to vary with climate, responding to large annual temperature anomalies, probably dominated by the mid- and northern latitudes. Both natural net ecosystem productions (NEP) and anthropogenic biomass burning in the Tropics may vary with precipitation and the El Niño southern oscillation (ENSO) cycle.
- Intriguing evidence of trends in the seasonal cycle of CO_2, satellite "greenness" and in situ measurements of plant phenology (timing of leaf growth and death) suggests a terrestrial biotic response to climate trends and changes in the growing season.

For some time there has been energetic discussion of feedbacks to atmospheric CO_2 from changing ecosystem and oceanic CO_2. Many mechanisms have been advanced as dominating the responses of the oceans and ecosystems. Although trends over the next century could be large enough to trigger responses beyond those functioning over the past decade, the atmosphere tells us that ecosystems and the oceans have responded to climate during our period of observation. This provides an opportunity to evaluate which mechanisms on land (climate, carbon dioxide fertilization, nitrogen deposition, anthropogenic disturbance, etc.) may have contributed to observed changes in CO_2 sources and sinks, and it offers a focus for observations over the coming years. Similarly, observations of climate, circulation, and CO_2 in the oceans provide information on the effects on CO_2 exchange of sea surface temperatures and circulation anomalies. The challenges of the recent observational record offer a great opportunity for Earth System modeling experiments and for both the identification and validation of key mechanisms.

7.1.2   What Will Happen to Future Terrestrial Carbon Storage?

Early models assumed that terrestrial carbon storage would track atmospheric CO_2 via a β factor. In this paradigm, while atmospheric CO_2 was increasing (at least up to some asymptotic level) photosynthesis would exceed respiration, and carbon storage on land would increase. Early inverse results suggested a large Northern Hemisphere sink, which suggested a more complex picture inasmuch as a CO_2-driven sink should have more even global distribution. Based on climate records, atmospheric CO_2 data, satellite data, and models, workers such as Dai and Fung (1993) and Myneni et al. (1997) suggested that climate change and variability over the past decades could have contributed to land uptake. Work on the nitrogen cycle by Bruce Peterson and Jerry Melillo (1985), Dave Schindler and Suzanne Bailey (1993), and Beth Holland et al. (1999) suggested that changes to the global nitrogen cycle could be contributing to carbon uptake in nitrogen-limited ecosystems (linking the carbon cycle to the atmospheric chemistry and transport of oxidized and reduced N compounds). The IPCC (1995) suggested that recovery from historical land use could be a cause of large land uptake in the northern mid-latitudes based on forest inventory data. Most recently, work using atmospheric inversion techniques by Song-miao Fan and co-workers (1998) and by Peter Rayner and others (1999) has suggested a large (even a very large) sink in regions affected by intense land use (North America, Eurasia). This has led to a much more complex picture of controls over terrestrial carbon storage and the potential for complex Earth System interactions.

Most models of vegetation and biogeochemistry assume a procession of states ("biome" types based on a mix of plant functional types) based on a modified "optimal" response of plants' use of water and energy to capture carbon in a given climate regime. Given the central role water and carbon play in plant metabolism, this is a reasonable point of departure and may apply in the long term. However, if we consider the trajectory τ of some ecosystem response variable (such as carbon storage), the response of a region to an altered environment is conditioned on (at least) three factors. These are, first, the physiological plasticity of the organisms present in the site, which can adjust rapidly (e.g., by changing leaf area). Second is the genotypic variability present in the populations, which may add significant additional flexibility in environmental responses. Third is the rate at which species and even life forms (trees vs. grasses) can change on a site. The time scales for this change can be fast, and the time scale for an optimal response is unknown for most ecosystems, especially in the presence of intense human activity. The extent to which the response of vegetation will be optimal with respect to climate change is a great unknown in projections of the future carbon cycle.

In addition, the genomes of plants correspond to a large covariance matrix of plant properties. The effects of climate changes on ecosystems will depend on how a host of associated properties will change as systems are forced by changes to water, energy, and CO_2. For example, changes to growth rates are often accompanied by changes to plant carbon chemistry and C:N ratio. We don't clearly understand how plant chemistry will change with phenotypic, genotypic, or successional change, and thus changes to decomposition, soil processes, and nutrient cycling are uncertain. Accompanying plant chemical changes, effects on higher trophic levels are likely. Other, more complex changes may also occur. Changes to photosynthesis and plant carbon metabolism may also be accompanied by changes to volatile organic carbon (VOC) emissions. Isoprene, in particular, serves to protect plant photosynthetic enzymes from high levels of light and heat, and its specific emission varies considerably between species. Thus, changes to the main carbon/water pathways could also be accompanied by changes to emissions affecting regional-global air quality. Finally, changes to the water/energy exchange characteristics of vegetation themselves feed back directly to climate via surface temperature, Bowen ratio, and albedo, and thus the trajectory of the coupled system may be a function of complex interactions.

As we consider the development of coupled models, we commit ourselves to bringing increased biological realism to ecosystem and carbon cycle modeling, and also to considering the biological covariance of changes to carbon, water, and energy processing with other aspects of ecosystem function and of ecosystem-atmosphere interaction.

7.1.3   How Will Carbon Respond As the Oceans Change?

Consideration of ocean carbon changes to climate change aren't new in global change research. The effects of sea surface temperature on CO_2 exchange have been explored in many ways. The impacts of ocean circulation changes have, likewise, received considerable attention in the context of paleo-oceanography and in view of potential climate change effects on the thermo-haline circulation. In the oceans, circulation is driven by heat and freshwater inputs and is thus tightly coupled to the atmosphere. Circulation affects carbon by the subduction of carbon-rich surface water and by the sustaining of biological productivity through the upwelling of nutrients. Changes to ocean circulation could have large effects on ocean carbon and apparently have had large effects observable in the paleorecord.

Several new questions have arisen with respect to the future of the ocean carbon cycle. First, marine biological productivity appears to be linked to the bioavailability of iron. Iron, in the surface ocean, is derived largely or mainly from terrestrial dust. In the present day, much of this dust is derived from erosion on managed or degraded lands. Over time, dust entrainment into the atmosphere will vary with the state of climate and vegetation on land. Thus a key marine control over carbon is linked to the Earth System, that is, to climate-and-vegetation-defined source regions in the land biosphere, and through atmospheric transport to deposition in the oceans. Thus, understanding controls over marine carbon and marine ecosystems requires a global perspective.

Most calculations of the "biological pump" of the biological component of the marine carbon cycle rely on Redfield ratios: ratios of organic carbon to limiting nutrients. Although Redfield ratios in the ocean are astonishingly constant, they do vary. Just as in terrestrial systems, Redfield ratios may vary within a marine taxon as environmental conditions change, and they can vary as the dominant marine taxa change. These changes could occur if environmental change affects which phytoplankton taxa are best adapted to the new circumstances, possibly modulated by effects of higher trophic levels. The controls over broad biome distributions in the relatively stable environment of the land have long been a prime focus of ecological research, and a first generation of "dynamic global vegetation models" is now in use. Little similar work has been done with marine ecosystems, and global marine ecosystem models simplify biology substantially more than terrestrial models do. Clearly, linking marine community ecological models to atmosphere-ocean physical models is a much-needed next step.

7.1.4   Chemistry, Biogeochemistry, and Climate

A number of the principal greenhouse gases are, unlike CO_2, reactive in the atmosphere or in fact the product (ozone) of reactive species. Great progress has been made in the understanding of chemical reactions controlling the greenhouse (and "anti-greenhouse," i.e., aerosols) constituents. The sources of the principal greenhouse gases methane and nitrous oxide are approximately quantified, as are, although less exactly, the sources of many ozone precursors. We know that the atmospheric concentrations of many key atmospheric species have varied in the past, in some cases strikingly coherently with climate (as is the case for methane and N_2O). This implies links between climate and the sources and transformations of these atmospheric species. Species related to the ozone cycle (oxidized nitrogen, some reactive organics) are also preserved in ice cores and also show both oscillations and secular changes in the past, and so the controls over atmospheric ozone precursors may well vary with climate.

This illuminates a dangerous uncertainty. Greenhouse gases other than CO_2 are a substantial fraction of total radiative forcing and could become more important if CO_2 emissions are controlled. In addition, we do not understand the contemporary trends in methane and nitrous oxide quantitatively. The growth rates of these two gases have been dynamic over the historic record, and we understand neither the long-term reduction in the methane growth rate nor its rapid change in the cool year following the eruption of Mt. Pinatubo. Much of today's nitrogen trace gas emission is caused by fertilizer use in agriculture and the mobilization of nitrogen in disturbed soils undergoing land use. Human land use, in combination with climate changes, could affect future nitrogen trace gas emissions significantly, as well as affect the mix of direct greenhouse gases (N_2O), ozone precursors (NO), and aerosol constituents (NH_3). The nitrogen cycle could be a source of surprises in future radiative forcing, thus behooving us to better understand the relationship between environmental change and future trace gas emissions. A similar argument could be made for methane, whose future emissions from northern wetlands could either increase or decrease, depending on the interactions of climate change with permafrost and high-latitude hydrology.

7.2   Next Steps for Earth System Models

A first generation of models coupling physical, chemical, and biological components of the Earth System now exists, and progress is occurring at a rapid pace in incorporating new processes into models. Significant focused effort must be placed on exploring the coupled behavior of such systems. Although much work must be done to improve existing coupled model components, a parallel effort to evaluate coupled behavior involving the biogeochemical cycles is crucial. There are three obvious lines of attack. The first of these is to evaluate models against major coupled climate-biogeochemical transients in the paleorecord to ascertain whether we can reproduce the magnitude and phasing of paleo-climate and trace gas changes. Such an effort has been initiated within the IGBP as the paleo-trace gas initiative. Second, the rich record of variability in climate and biogeochemistry over the recent past should be explored. In general, this will explore the responses of the biogeochemical systems to climate forcing, rather than the effects of biogeochemistry on climate, which is key in the past and future. However, it allows the climate forcing of biogeochemistry to be both modeled and observed fairly directly. Finally, there is substantial public and scientific uncertainty about how feedbacks between radiatively active trace gases and aerosols and climate will operate in the future. Climate affects sources and hence concentrations, and concentrations, in turn, affect climate. There is a need to begin a systematic assessment of the sensitivity of at least those biogeochemical mechanisms we understand. Such an activity has been proposed by the IGBP-Global Analysis and Modeling Program in partnership with the WCRP (the "Great Leap," denoting the ambitious first coupled calculation proposed).

In parallel with excercising Earth System Models that embody the mechanisms we understand (however incompletely), we must also begin to evaluate how new mechanisms may operate. Examples of such relatively new ideas include the coupling of land and ocean via iron aerosols, the constraints on terrestrial ecosystem adaptation that arise from population processes, and the potential consequences of large ecosystem changes in the oceans. Such mechanisms can be only crudely quantified in today's global models but may have influenced past dynamics and could control future processes. In addition, coupled climate-biogeochemistry models have at best crudely captured the synergistic effects of humans via direct impacts (not through the climate system). This must be a first-order effect in land ecosystems, in which disturbance responses may dominate the current magnitude of land sinks and may contribute substantially to sources. It may be of growing significance in marine ecosystems, where human effects on higher trophic levels are today large, and in marine coastal zones, where direct human impacts may be overwhelming.

In short, there is enormous scope for scientific accomplishment in Earth System science and Earth System modeling. The challenges facing us are of great scientific difficulty, but the answers will be of great value to society.

REFERENCES

Dai, A. G., and Fung, I. Y. (1993). Can climate variability contribute to the missing CO_2 sink? Global Biogeochemical Cycles, 7(3), 599–609.

Fan, S., Gloor, M., Mahlman, J., Pacala, S., Sarmiento, J., Takahashi, T., and Tans, P. (1998). A large terrestrial carbon sink in North America implied by atmospheric and oceanic carbon dioxide data and models. Science, 282(5388), 442–446.

Holland, E. A., Dentener, F. J., Braswell, B. H., and Sulzman, J. M. Contemporary and pre-industrial global reactive nitrogen budgets. Biogeochemistry, 46(1), 7–43.

Myneni, R. B., Keeling, C. D., Tucker, C. J., Asrar G., and Nemani, R. R. Increased plant growth in the northern high latitudes from 1981 to 1991. Nature, 386(6626), 698–702.

Peterson, B. J., and Melillo, J. M. The potential storage of carbon caused by eutrophication of the biosphere. Tellus Series B – Chemical and Physical Meteorology, 37(3), 117–127.

Rayner, P. J., Enting, I. G., Francey, R. J., and Langenfelds, R. Reconstructing the recent carbon cycle from atmospheric CO_2, delta C-13 and O-2/N-2 observations. Tellus Series B – Chemical and Physical Meteorology, 51(2), 213–232.

Schindler, D. W., and Bayley, S. E. The biosphere as an increasing sink for atmospheric carbon: Estimates from increased nitrogen deposition. Global Biogeochemical Cycles, 7(4), 717–733.

8 The Role of CO_2, Sea Level, and Vegetation During the Milankovitch-Forced Glacial-Interglacial Cycles

ANDRÉ BERGER

ABSTRACT

Sensitivity experiments have been made over the last glacial-interglacial climatic cycle using the Louvain-la-Neuve two-dimensional Northern and Southern Hemispheres climate model. The continental ice volume was simulated for the past 122 kyr in response to changes in both the insolation and the CO_2 atmospheric concentration. The sensitivity of such a response to sea level changes and to the vegetation-snow albedo feedback indicates that the 100-kyr cycle cannot be sustained if these processes are not taken into account. The adoption of the factor separation method by Stein and Alpert allows the identification of the contribution of the processes involved in a climate model as well as their synergistic effects. Here, this technique was restricted to two variables – the sea level and the albedo of vegetation when covered by snow – to quantify their individual impacts and mutual contributions to the global ice volume variations over the last glacial-interglacial cycle. The simulated sea level drop of about 100 m at the Last Glacial Maximum leads to an increase of emerged continental surfaces over the present-day value by about 13%. As a consequence, the growth of the ice sheets up to the shoreline leads to an ice volume change 20% larger than if the sea level would have been kept constant. On the other hand, if the vegetation-snow albedo feedback is not included in the simulation, the Northern Hemisphere ice volume is overestimated most of the time, especially in the experiment where the sea level is allowed to vary. Over the past 10 kyr, the difference between the control case (all processes are included) and the reference case (sea level fixed and no vegetation effect) is explained mainly by the vegetation-snow albedo feedback, the pure positive contribution of sea level being almost canceled by the negative contribution of the synergism between the sea level and vegetation impacts.

8.1 The Astronomical Theory of Paleoclimates

The astronomical theory of paleoclimates aims to explain the climatic oscillations between glacial and relatively ice-free environments over the past million years. Milutin Milankovitch (1941) is given much credit for explaining the quasi-periodic nature of these Pleistocene ice ages. He hypothesized that the periodic variations in solar irradiance received in the high latitudes of the Northern Hemisphere during the summer season were critical to the waxing and waning of continental ice sheets. But he also recognized the importance of feedback mechanisms (mainly the albedo–temperature feedback) to amplify the response to the astronomical forcing.

Changes in the geometry of the Earth's orbit about the Sun and in the inclination of its axis of rotation cause the amount of solar energy received on the Earth surface to vary by season and latitude. Three astronomical parameters have therefore received great attention in recent climatological studies: the eccentricity of the Earth's orbit, which varies with a mean periodicity of ~100,000 years; the tilt of the Earth's rotational axis, which varies from about 21.5° to 24.5° over an average period of 41,000 years; and the climatic precession of the equinoxes, which causes the Earth-Sun distance on each equinox and solstice to change with an average period of 21,000 years. Actually, the full spectra of variations of these astronomical parameters are more complex (Berger, 1978) and unstable with time (Berger et al., 1998a). It is possible to demonstrate, for example, that there is a relationship between the amplitude and the frequency modulations of these astronomical parameters. Moreover, all these properties also characterize the behavior of insolation, in which harmonics can be significantly present, such as the 10.5 kyr related to precession. In the tropics, the Sun indeed passes over each latitude twice a year, leading to a double insolation maximum. Assuming that the climate will respond systematically to the largest of them creates such a quasi-cycle of 10.5 kyr. The spectra of insolation therefore depend on which insolation parameter is considered. In particular, the daily amount of energy received from the Sun is essentially a function of precession. But as soon as this insolation is integrated over a season it becomes exclusively a function of obliquity, according to Kepler's second law.

At the present, the value of e is 0.0167 (Table 8.1) and is decreasing (we have just passed a maximum 0.0197, 14 kyr ago). Obliquity (currently 23.45°) is decreasing from a maximum of 24.23° reached 9 kyr ago, and climatic precession is decreasing both because of e and because the longitude of the perihelion is decreasing (Figure 8.1). This means that the present-day summer solstice, which occurs at the aphelion, will be at the perihelion within 9 kyr. But more importantly, we are at the end of a 400-kyr cycle for eccentricity. One consequence is that the 100-kyr cycle we entered ~50 kyr ago will be shorter and will last only 30 kyr more, leading to an absolute minimum of e at 27 kyr AP, when it will reach almost 0 (0.0027). This implies that the climatic precession parameter ($e \sin \tilde{\omega}$) will be very small over the next 50 kyr. At the same

Table 8.1. Astronomical Parameters and Insolation
(Berger, 1978; Berger and Loutre, 1994)

	Present	6 kyr BP	21 kyr BP	115 kyr BP
e	0.017	0.019	0.019	0.041
ε (°)	23.4	24.1	22.9	22.4
Perihelion	3 Jan	20 Sept.	16 Jan	15 Jan
Summer (length in days)	93.6	89.2	94.2	97.5
Seasonal contrast 50°N	395	425	386	353
June–Dec. (Wm^{-2})		$\Delta = +30$	$\Delta = -9$ tt	$\Delta = -42$
Latitud. gradient in June	107	126	96	80
70°N–Equator (Wm^{-2})		$\Delta = +19$	$\Delta = -11$	$\Delta = -27$

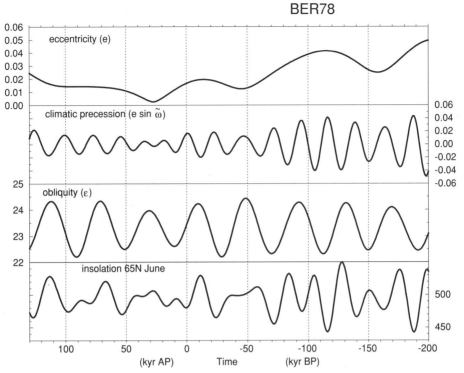

Figure 8.1. Long-term variations of eccentricity, climatic precession, obliquity, and insolation at 65°N in June (Wm⁻²) from 200 kyr BP to 130 kyr AP (Berger, 1978).

time the amplitude of obliquity is also reduced (obliquity going from 22.6° at 10 kyr AP to slightly less than 24° at 31 kyr AP). Because the daily insolation is mainly a function of precession (Berger et al., 1993a), all these particular characteristics lead to an insolation that will vary little over the next 50 kyr, a feature that is strikingly different from the last interglacial (with an amplitude of 25 Wm⁻² against 105 for 65°N in June). This particular feature is unique over the past 3 Myr, and the closest analog occurred 400 kyr ago, at stage 11 (Berger and Loutre, 1996). This might have an influence on the sensitivity of the climate system to other forcings such as the greenhouse gases because, among the driving forces, one (the insolation) will remain almost constant, giving the others a possibility of playing a more important role.

The discovery in 1976 (Hays et al.) that the main periods of these three elements were preserved in the ice volume signal of ocean cores was a major step toward understanding the glacial–interglacial cycles. With time, the astronomical theory of paleoclimates has been extended and is now related mainly to the response of the climate system to the astronomically driven changes of the latitudinal and seasonal distributions of the energy received from the Sun. In particular, it is worth pointing out that the hypothesis by Milankovitch, requiring a low summer insolation in high northern latitudes to trigger a glacial, is actually coherent with the occurrence of a reduced seasonal contrast and of an enhanced insolation in winter, essentially in lower latitudes. Indeed, according to the astronomical theory, it is possible to demonstrate that the long-term variations of the daily insolation depend mainly on precession (except for the latitudes and days close to the polar night, Berger et al., 1993a). As a consequence, for a given latitude, there is a phase lag of about 2 kyr (one-twelfth of a precessional cycle) between the insolation of two consecutive months (Loutre and Berger, 1995). Therefore, when the high latitudes receive less insolation in summer, it is also the case for the mid- and low latitudes, which, in turn, receive more insolation in winter. This has a direct impact on climate, and mild winters in regions where the evaporation is the largest (the tropical oceans) allow more water vapor to be available at the global scale. At the same time, a larger winter latitudinal gradient strengthens the atmospheric circulation and the water vapor transport to the northern high latitudes, where it falls as snow.

To investigate the complexity of such a response, a series of numerical experiments have been performed with climate models of different complexity (Berger, 1995), in addition to the detailed analysis of many proxy records. Among them, the Louvain-la-Neuve (LLN) 2.5-D climate model (Gallée et al., 1991) has been used to reconstruct the long-term climatic variations over the Quaternary Ice Age. Sensitivity analyses to the insolation changes calculated by Berger (1978) and to the CO_2 atmospheric concentration reconstructed by Jouzel et al. (1993) have been performed over the last 200 kyr (Gallée et al., 1992; Berger et al., 1998b). Because the LLN model does not yet have a carbon cycle component, the atmospheric CO_2 concentration is used as a forcing, although it acts as a feedback in the real world. On long time scales (tens and hundreds of thousands of years), CO_2 covaries with the Vostok isotopic temperatures (Yiou et al., 1991) and other climatic variables (Yiou et al., 1994). Records show minimum glacial CO_2 concentrations around 190 ppmv, glacial-interglacial transitions accompanied by a rapid increase in CO_2 concentrations to a maximum of about 290 ppmv, and a gradual return to low CO_2 values during glaciation. These variations have been attributed to climate-induced changes in the carbon cycle, but they also amplify climate variation by the accompanying greenhouse effect. However, the Vostok CO_2 spectrum is far from reproducing all the astronomical frequencies.

Significant peaks appear near 143, 59, 31, and 22 kyr and at shorter periods (Yiou et al., 1991), although peaks near the orbital frequencies are more clearly visible in the Vostok temperature record at 41.7, 24.1, and 18.3 kyr. The peaks near 100 kyr are strongly shifted to the left in both Vostok records (133.3 kyr for temperature and 143 kyr for CO_2), and the peak near 41 kyr has very little significance in the CO_2 record. Periods close to 60 and 30 kyr (also present in the temperature record) do not appear to be related to the orbital frequencies themselves nor to their combination tones but rather seem to be caused by ice sheet/bedrock interactions. On shorter time scales, the picture is even more complex (Fischer et al., 1999). During all the last three terminations (between marine isotopic stages [MIS] 8 to 7, 6 to 5, and 2 to 1), the rise in CO_2 concentration lags behind temperature change by 400 to 1000 years. Between 14 and 13 kyr a brief small decline in CO_2 lags the Antarctic cold reversal (ACR) in the Antarctic isotope temperatures by $\sim$400 years, but it occurs $\sim$1 kyr before the Younger Dryas cooling event (Blunier et al., 1997). As a consequence, the ACR leads the Younger Dryas by at least 1.8 kyr, and CO_2 rises steadily during the Younger Dryas. During the Holocene and Eemian interglacials, atmospheric CO_2 concentrations drop by $\sim$10 ppmv after an initial maximum reached, respectively, at 10 kyr BP and 128 kyr BP. This drop might be attributed to a substantial increase in the terrestrial biospheric carbon storage extracting CO_2 from the atmosphere. In the case of the Eem, CO_2 concentration does not show a substantial change in the following 15 kyr, despite a distinct cooling over the Antarctic ice sheet. Not until 6 kyr after the major cooling of MIS 5.4 does a substantial decline in CO_2 occur, another 5 kyr being required to return to an approximate phase relationship of CO_2 with the temperature variations. During the Holocene, atmospheric CO_2 concentrations even increase during the last 8 kyr. In contrast, high CO_2 concentrations are not sustained during MIS 7, but rather CO_2 follows the rapid temperature drop into MIS 7.4, reaching a minimum 1 to 2 kyr after the minimum in the isotope temperature.

In the experiments made with the LLN climate model, the insolation changes alone act as a pacemaker for the glacial-interglacial cycles, but CO_2 changes help to better reproduce past climatic changes and, in particular, the air temperature and the southern extent of the ice sheets. Actually we have shown that the albedo and water vapor–temperature feedbacks play a fundamental role in amplifying the astronomical perturbations. Forced by the insolation, albedo, and CO_2 of the Last Glacial Maximum (LGM), the model by Berger et al. (1993b) leads to a 4.5 °C cooling in the Northern Hemisphere (recall that the equilibrium response of the LLN model for a doubling of the CO_2 concentration is a warming of 2 °C). Two-thirds of the LGM cooling (3 °C) is explained by the astronomical and albedo forcing when allowing for the water vapor feedback, which, by itself, accounts for a cooling of 1.2 °C. In a CO_2-alone forcing experiment the cooling amounts to 1.5 °C, 40% of it (0.6 °C) being due to the water vapor feedback. These results therefore stress the fundamental role of the albedo and water vapor feedbacks. They also show that the combined effect of a change in the CO_2 and in the insolation plus albedo forcings, all of which generate the water vapor feedback, is about equal to the sum of the responses of the climate system when one or the other is kept constant. This type of linear response might be due to the fact that the perturbation in all cases remains small (a few degrees) as compared with the basic state of reference ($\sim$280 K), although the system itself is described with a set of highly nonlinear equations. This kind of behavior has also been found recently in more-complex climate models (Ramaswamy and Chen, 1997).

Leads and lags have also been analyzed in the response of the LLN climate model to the insolation and CO_2 forcings during the Eemian interglacial and over the whole last glacial-interglacial cycle (Berger et al., 1996). If we take the June insolation at 65°N as a guide for the time scale, the maxima of the simulated Northern Hemisphere ice volume – reached at 134 and 109 kyr BP, respectively, for isotopic stages 6 and 5d – lags behind this insolation by 6 kyr. The minimum ice volume is reached at 126 kyr BP and lasts 10 kyr, covering the whole period during which insolation is decreasing by 20% from a maximum of about 550 Wm^{-2} to a minimum of 440 Wm^{-2}. Summer temperature of seawater at the surface (SST), at middle to high latitudes, lags behind insolation by a few thousand years. As a consequence, SST starts to decrease well before (actually 11 kyr before) the ice sheets start to grow on the continents (as, for example, at 116 kyr BP). In the 50–55°N band, the simulated SST then starts to rise again at that time and is therefore in antiphase with the continental ice volume. Cooling rises at the end of the melting phase of the ice sheets, a result in agreement with the reconstruction by Cortijo et al. (1994), who showed an abrupt cooling in the high latitude ocean only a few thousand years after the start of the last interglacial; and warming occurs at the beginning of their growing phase, in agreement with the hypothesis of Ruddiman and McIntyre (1979), who claimed that the ice sheets were growing under warm SST conditions. At higher latitudes (70–75°N), the behavior of the zonal mean temperature at the surface is more complex, reflecting not only the direct influence of insolation but also the influence of sea ice, seawater, and the ice sheets.

To further investigate the importance of CO_2 changes, in addition to the insolation, an atmospheric CO_2 concentration decreasing linearly from 320 ppmv at 3 Myr BP (late Pliocene) to 200 ppmv at the LGM was also used to force the model. Under such conditions, the model simulates the entrance into glaciation around 2.75 Myr BP (Li et al., 1998a), the late Pliocene–early Pleistocene 41 kyr cycle, the emergence of the 100 kyr cycle around 900 kyr BP (Berger et al., 1999), and the glacial-interglacial cycles of the last 600 kyr (Li et al., 1998b). The hypothesis was put forward that during the Late Pliocene (in an ice-free, warm world) ice sheets can develop only during times of sufficiently low summer insolation. This occurs during large eccentricity times when climatic precession and obliquity combine to obtain such low values, leading to the 41 kyr period between 3 and 1 Myr. Sensitivity analyses have also demonstrated that to prevent ice sheets from developing, CO_2 must be relatively large between 4 and 3 Myr BP (more than 450 ppmv in our model), in particular around 3.8 Myr BP when all the astronomical conditions are most favorable. But, at the same time, CO_2 concentration must decrease (less than 370 ppmv in our model) after 3 Myr BP to allow the ice sheets, triggered by the insolation forcing, to grow sufficiently at around 2.7 Myr BP. On the other hand, it is interesting to point out that the 100-kyr signal appears in the climatic record at around 1 Myr BP, when the 100-kyr component of eccentricity starts to fade. Very strong between 2 and 1 Myr BP, it weakens progressively between 1 Myr BP and now, and it finally disappears over the next 500 kyr for the benefit of the 400-kyr cycle, in relation to the ∼2-Myr period of term number 6 in the expansion of eccentricity.

In a glacial world, ice sheets persist most of the time except when insolation is very high in polar latitudes, again requiring large eccentricity but leading this time to an interglacial and finally to the 100-kyr period of the last 1 Myr. Using a CO_2 concentration reconstructed over the last 600 kyr from a regression based on SPECMAP (Li et al., 1998b), it has been shown that stage 11 and stage 1 require a high CO_2 to reach the interglacial level. The insolation profile and modeling results at both stages tend to show that stage 11 might be a better analog for our future climate than the last Eemian interglacial. Such a CO_2 reconstruction was used because no other CO_2 record (Vostok in particular, Petit et al., 1999) was yet available over such a long period at the time the experiment was performed. It must be understood, however, that this is only a temporary surrogate because the SPECMAP $\delta^{18}O$ record is not very well correlated with the CO_2 record over the period for which both data were existing (Li et al., 1998b). Using the calculated insolation and a few scenarios for CO_2, the climate of the next 130 kyr has been simulated (Loutre and Berger, 2000), showing that our interglacial will most probably last particularly long (50 kyr). This conclusion is reinforced if we take into account the possible intensification of the greenhouse effect as a result of human activities over the coming centuries.

Given the complexity of the interactions between the different components of the climate system, a major challenge remains: to understand how the astronomical forcing is amplified by different processes that control the response of the climate system. In this chapter, we focus on the role of the atmospheric CO_2 concentration changes, sea level changes, and the vegetation-climate feedback over the last glacial-interglacial cycle.

8.2   CO₂ and Insolation Thresholds

Experiments made with the LLN NH climate model sought to explain its response to the astronomical and CO_2 forcings. These experiments show, in particular, that the sensitivity of the Northern Hemisphere ice volume to CO_2 is not constant through time.

To test the Hays et al. (1976) hypothesis that the orbital forcing acts as a pacemaker of the ice ages, experiments were made in which insolation was allowed to change, but CO_2 was kept constant to either 210, 250, or 290 ppmv (Berger et al., 1998a). As CO_2 varied around 230 ppmv most of the time during the last 200 kyr, these concentrations correspond to an average for, respectively, glacial, intermediate, and interglacial times (the present-day CO_2 concentration is already 60% above the average value of the last glacial-interglacial cycle). These simulations show (Figure 8.2) that the

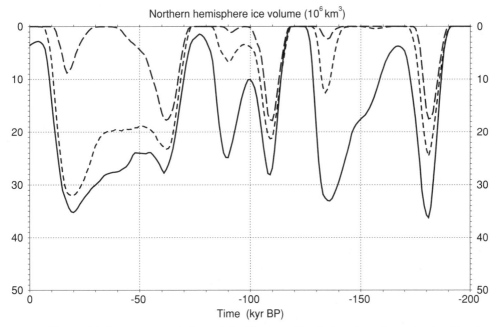

Figure 8.2. Long-term variations of the Northern Hemisphere ice volume simulated by the Northern Hemisphere 2-D LLN model (Gallée et al., 1991) in response to the astronomical forcing (Berger, 1978) and a constant CO_2 (210 ppmv solid line, 250 ppmv small dashed line, 290 ppmv dashed line) (Berger et al., 1998b).

modeled ice volume variations are comparable to the geological reconstructions only when the CO_2 is low (210 ppmv), but, more importantly, that the response of the climate system is far from being linear in CO_2. For a CO_2 of 210 ppmv, the glacial maxima are reached at 181, 136, and 20 kyr BP, with about 35×10^6 km^3 of continental ice in the Northern Hemisphere. Secondary maxima of about 25×10^6 km^3 occur at 109, 90, and 61 kyr BP.

Analysis of the range of ice volumes simulated under these three constant CO_2 concentrations show that the sensitivity of the simulated Northern Hemisphere ice sheets volume to CO_2 is far from being constant in time. It is larger at 136, 90, and 20 kyr BP, when the range of ice volume is much broader. These times correspond to secondary minima in the insolation curve, with a moderate value situated between 462 and 472 Wm^{-2} at 65°N in June. For the other three ice maxima (181, 109, and 61 kyr BP), the insolation reaches its deepest minimum ($\sim$440 Wm^{-2}) 7 to 11 kyr before ice sheets form under the three CO_2 concentrations. At these times eccentricity is large, solstice occurs close to aphelion, and obliquity is low. A tentative conclusion may therefore be that the sensitivity of the climate system to CO_2 is larger (i.e., a broader range of ice volume occurs in response to a given range of CO_2) for intermediate values of the insolation minima. For the other ice maxima, the insolation minima are deep enough (lower than 460 Wm^{-2}) to drive the system into a glacial stage for the three CO_2 values.

The amplitude of the ice volume change for the three experiments is very different, but the timing of the build-up of the ice sheets is very similar. For the 181, 109, and 61 kyr BP ice maxima, the 250 ppmv curve is situated halfway between those of 210 and 290 ppmv. This is not the case for the other maxima. At 136 and 90 kyr BP, the 250 ppmv simulation is definitely closer to the 290 ppmv one. In contrast, at 20 kyr BP the 250 ppmv simulated ice maximum is tied very closely to the 210 ppmv one. This different sensitivity at the Last Glacial Maximum compared with what happens around 90 and 136 kyr BP is actually related to the state of the climate system before these ice maxima: the Earth is in a glacial mode during isotopic stages 4 to 2, whereas the climate is interstadial before the 90 and 136 kyr BP maxima. The very large difference in the response of the model for the 250 ppmv CO_2 level also seems to indicate the existence of critical CO_2 concentrations (which are time-dependent) around which the climate system may be responding either as a high or a low CO_2 level. At the Last Glacial Maximum, even a rather high CO_2 does not prevent the Earth from being largely glaciated. This indicates that the threshold value may be as large as 270 ppmv, with only a CO_2 concentration lower than this value leading to a glacial. On the contrary, the situations at 136 and 90 kyr BP require a rather small CO_2 for the Earth to enter into glaciation, and the threshold might be as low as 230 ppmv. The similar behavior of the response of the model under a 250 and a 290 ppmv forcing during all interglacials and interstadials (except isotopic stage 3) confirms that its sensitivity to CO_2 is different in a cold-glaciated Earth than in a warm, ice-free Earth. This implies that during the Holocene, an ice age can be initiated, in our model, only with a rather low CO_2 concentration (below 250 ppmv at least).

8.3   Sea Level and Vegetation Changes

At the geological time scale, the amplitude of the mean sea level variations might be relatively large. This has a direct impact on the size of the continents by allowing or not allowing the continental platforms to emerge. In particular, the reconstruction of the configuration of the Earth's surface at the Last Glacial Maximum by CLIMAP (1976) shows the emergence of the Sunda Isles and of large areas from the China Sea, a bridge between Australia and New Guinea and between Siberia and Alaska, and the disappearance of the North Adriatica, of the Channels, and of the North Sea.

Vegetation and land surface cover have also changed significantly over time. At the LGM (Prentice et al., 2000), the most obvious features are the equatorward regression of forest types in North America and Eurasia and a compression and fragmentation of the forest zones in these regions. These vegetation changes provide evidence for drier conditions than present across large areas of the mid-latitudes. The boreal evergreen forest (taïga) occupied a far smaller area than today, and the temperate deciduous forest was apparently almost nonexistent. Refugia for this temperate forest and the tropical rain forest biomes may have existed offshore at LGM, but their characteristic taxa also persisted as components of other biomes. Tropical moist forests in Africa were reduced.

At the peak of the Holocene (6 kyr ago), in the northern circumpolar region, taïga extended poleward at the expense of tundra, indicating greater than present growing-season warmth (Prentice, 1998). This shift appears relatively slight (200–300 km) when viewed on global maps, and furthermore it is not symmetrical around the pole. In the northern mid-latitudes of Eurasia, the forest belts shifted poleward. In many cases, these shifts implicate warmer winter conditions even though the orbital forcing alone would tend to produce colder winters (the biome paradox). In the circum-Mediterranean region, temperate deciduous forests encroached southward and xerophytic woodland/scrub were absent, suggesting a moister climate than today. But the largest changes are seen in the monsoon regions, mainly in northern Africa, where the Sahara desert was drastically reduced and the Sahelian vegetation belts shifted systematically northward. The basic mechanisms of this monsoon amplification are caused by the early to mid-Holocene astronomical forcing, but further positive feedback mechanisms, involving changes in land surface and/or sea surface, must be invoked to account for the magnitude of the biome shifts in Africa.

These vegetation changes are important for climate because the surface albedo is related not only to the snow and ice fields but also to the nature of vegetation that covers the land surfaces. It can indeed be anticipated that a vegetation-albedo-climate coupling should exist, because changing global climate produces modifications in large-scale precipitation and temperature patterns, which, in turn, modify the vegetation of land surfaces and hence change the surface albedo, thus producing further climatic change. Charney et al. (1977) were among the first to investigate such an effect of albedo change on precipitation in semiarid regions. Otterman et al. (1984) then showed the importance to climate sensitivity of a lower effective snow albedo on forests and shifts in the tree line, using a mean annual model. Indeed, snow on forests has a considerably smaller effective snow-cover fraction than snow of the same depth on tundra, leading to a smaller albedo for the snow-covered part of the zone when forest is present in place of tundra.

Later, Harvey (1988) analyzed the potential role of the shifts in the tundra-forest ecotone using a seasonal energy balance climate model. Sensitivity experiments were performed in which the fraction of tundra was parameterized in terms of July land air temperature. In the snow-covered regions, the albedo was given by the vegetation and snow albedos, weighted by an effective snow-cover fraction that depends on snow depth and vegetation roughness. In an experiment in which the orbital parameters were changed from those of 125 to 114 kyr BP, Harvey (1989) showed that changes in tundra fraction enhance the zonally averaged July temperature response over land by as much as 3 °C, and they change the global mean temperature response from slight warming to slight cooling. In addition, allowing the low-latitude vegetation fraction to vary with summer land-sea temperature difference tends to cool the NH and warm the SH. But the NH effect dominates, causing an increase in the global mean cooling. Because these experiments confirmed that vegetation feedbacks at high and low latitudes contribute significantly to the temperature response to the orbital change, the same parameterization was introduced in the LLN 2.5-D model by Gallée et al. (1992). This led, during the entrance into glaciation, to significant changes in the latitude of the taïga-tundra border and in the removal of high-latitude forests, which, in turn, play a key role in enhancing the surface albebo, in increasing the persistence of the snow field, in changing the sea-ice cover and the oceanic transport, and, therefore, in lowering the surface temperature over the continents (Berger et al., 1992, 1993c).

Over the past five years, many climate models, mostly of general circulation types coupled to land-surface schemes, have attempted to simulate this kind of impact of vegetation changes on climate, mainly at isotopic stage 5d, at the Last Glacial Maximum, and at the peak of the Holocene (Tables 8.2a and 8.2b). The results obtained from a climate model of intermediate complexity by Ganopolski et al. (1998) for 6 kyr BP help to synthesize this vegetation-climate interaction. These modeling efforts converge to show that changes in vegetation cover during the mid-Holocene modify and amplify the response of the climate system to the astronomical forcing, both directly (primarily through the changes in surface albedo) and indirectly (through changes in oceanic temperature, sea-ice cover, and oceanic circulation). But such changes are not the only ones to play a key role in the climate system (see more comments below).

In response to mid-Holocene insolation changes, the Ganopolski et al. atmosphere-only model shows an important warming over the continents of the Northern Hemisphere in summer and a global annual increase of precipitation due to the intensification of the summer monsoon in North Africa and Asia. Similar results are obtained by their atmosphere-ocean model. However, the ocean dampens the signal owing to its heat capacity. At high latitudes, the reduction of sea ice amplifies the Arctic warming throughout most of the year. In the Tropics, a northward shift of the intertropical convergence zone in autumn is simulated, but no increase occurs in precipitation in North Africa nor in Southeast Asia (as compared with the atmospheric experiment). Simulation with their atmosphere-vegetation model shows a pronounced expansion of the forests to the north in the northern high latitudes and in the subtropics.

Table 8.2a. Vegetation-Climate Feedbacks During the Last Glacial-Interglacial and in Glacial Times.

Authors	Model	Results
	Last Glacial-Interglacial Cycle	
Gallée et al., 1992 Berger et al., 1992, 1993c	• LLN NH 2.5-D climate atmosphere, ocean, ice sheets, sea-ice, land surface	Taïga-tundra-snow albedo feedback enhances significantly the response to the orbital forcing.
	115 kyr BP	
Harvey, 1988, 1989	• Energy balance model	Vegetation feedbacks at high and low latitudes contribute significantly to the temperature response to the orbital change.
Harrison et al., 1995	• CCM1 P+T+, P−T− • BIOME	P+T+ → warm summer → less sea ice → warm winter → taïga northward P−T−cold → biome southward
Gallimore-Kutzbach, 1996	• CCM1 ML ocean Sea ice • biome	tundra increase → snow albedo feedback → glaciation
de Noblet et al., 1996	• LMD • +BIOM1 iterative	taïga + cold deciduous forests → tundra → snow feedback → ice sheet
	LGM	
Friedingstein et al., 1992	• EBM Sellers • bioclimatic scheme	• Forests decrease (−30%) • Desert and tundra increase (+25%)
Crowley and Baum, 1997	• GENESIS • +land surface vegetation	• exp. dryland veget. → add. cooling Africa and Australia • conifer → tundra → add. cooling in W. Europe
Kubatzki and Claussen, 1998	• ECHAM-3 • +BIOME-1 (asynchronous)	• vegetation-snow albedo → improved high lat. climate • 2 stable sol.: bright and dark desert
Prentice et al., 1998	Reconstruction	• Greatly reduced forests at all latitudes • Extensive tundra and steppes in mid-latitudes • Expanded deserts

Tundra and polar deserts decrease by 3×10^6 km^2 as compared with the present day, and the vegetation snow-albedo feedback amplifies the high northern latitude warming. In the subtropics, precipitation increases with a reduction of the desert areas (between 10 and 30°N, the desert fraction becomes 14% instead of 71% as at the present). The results of their coupled atmosphere-ocean-vegetation model show strong interactions between the sea surface temperature, sea ice, and vegetation cover at boreal latitudes, but in the subtropics the atmosphere-vegetation feedback is the most important.

As already cautioned, it must be stressed that such results cannot be considered definitive, and more research must be done before robust conclusions can be drawn. Indeed, Kutzbach and Liu (1997), for example, showed that, if we allow a dynamical

Table 8.2b. Vegetation-Climate Feedbacks at 6 kyr BP.

Authors	Model	Results
	6 kyr BP	
Prentice et al., 2000	Reconstruction	• Northern temperate and boreal forests farther north than present • Greening of Sahara
Guiot et al., 1996	Reconstruction	• Cool conifer and cold deciduous forests replace taïga • Temperate deciduous forests in Mediterranean region
Foley et al., 1994	GENESIS 1.02 mixed layer ocean	prescribed boreal forest extended → add. warm.

TEMPO, 1996	• GENESIS 1
	  GENESIS 2
	  CCM1
	  BIOME
	• GENESIS 1 + BIOME iter.
		• tundra → taïga + cold deciduous forests
		
		• Forests increased by 23%

Kutzbach et al., 1996	• NCAR CCM 2 • land surface model	orbit + prescribed 15–30 N desert converted into grass $\Delta R = 28\% = +12$ (orbit) $+ 16$ (vegetation) Δ desert $= -20\% = -11$ (orbit) $- 9$ (vegetation)
Texier et al., 1997	• LMD • BIOME 1 iterative	• high lat. warming amplified + monsoon • Summer rain in West Africa double; tundra: -25%; Sahara: -20% and Southern limit → 18°N • More feedbacks necessary
Claussen and Gayler, 1997	• ECHAM • BIOME 1 asynchronous	• Same as Texier et al. but Southern limit Sahara → 20°N
Harrison et al., 1998	• 10 AGCM (PMIP) • BIOME	• Northward shift in tundra–forest boundary • Increase in warm grass/shrub in NH continental interiors • Expansion of Afro-Asian monsoon • Reduced tropical rain forest

Ganopolski et al., 1998	CLIMBER
	  atmosphere
	  ocean
	  land surface
		• Vegetation feedback:
		  direct = albedo
		  indirect = SST, sea-ice, ocean circulation
		• NH high lat.: summer warming → boreal forests exp. north → winter warming through sea-ice feedback
		• Subtropics: veget-precip. feedback → Greening Sahara
		• North Alt.: strong warming + fresh water → weak THC → SH warm

Claussen et al., 1999	CLIMBER 2	• Abrupt start of desertification Sahara ~5.5 kyr BP due to vegetation change and Charney feedback

ocean to interact with the atmosphere, the increased amplitude of the seasonal cycle of insolation in the Northern Hemisphere 6 kyr ago could have increased tropical Atlantic sea surface temperature in late summer and, in turn, could have enhanced the summer monsoon precipitation of northern Africa. The processes mainly responsible for this precipitation increase in north Africa are therefore different from those of Ganopolski et al. (1998), involving aspects of changes in sea surface temperature and in atmospheric circulation that cannot be simulated in models other than the GCMs. The explanation of a given climate change is therefore still model-dependent, and the effort must be sustained to build climate models in which the most important processes acting in the real world would be treated equally well.

8.4   Impact of Sea Level Change in the LLN Model

To investigate the sea level–ice volume feedback, the NH climate model of LLN has been extended to the whole Earth, and sea level has been allowed to vary. As a consequence, during glacials the ice sheets can grow up to the shoreline, leading to a larger ice volume (Figure 8.3). At the Last Glacial Maximum, for example (Table 8.3), the North American ice sheet increases by about 7%, the Eurasian ice sheet by about 25%, and the Greenland ice sheet by 20% when compared with the experiment using a fixed sea level. Consequently, when the sea level is allowed to change, the simulated total ice volume amounts to 71×10^6 km³, with 17 over North America, 14 over Eurasia,

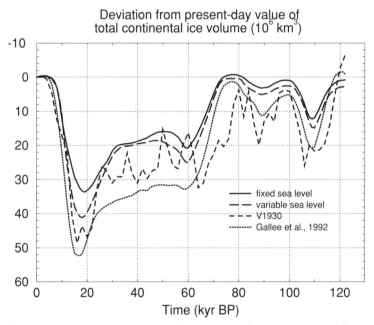

Figure 8.3. Long-term variations of the deviation from present day of the total continental ice volume over the last 122 kyr. The experiments with fixed and variable sea level are from Dutrieux (1997), and the V1930 reconstruction is from Duplessy et al. (1988) and Labeyrie et al. (1987). The dotted line represents the simulation made with the LLN 2-D Northern Hemisphere climate model by Gallée et al. (1992).

Table 8.3. Ice Volume of the Individual Ice Sheets, and Total Ice Volume Over the Northern Hemisphere and the Whole Earth for the Present and for the Last Glacial Maximum, According to Different Authors.

| | Gallée et al. (1992) | | Dutrieux (1997) Fixed SL | | Dutrieux (1997) Variable SL | | Huybrechts (1990a,b, 1996) Huybrechts and T'Siobbel (1995, 1997) | | Peltier (1994; 1996; 1998 a,b) | Lambeck (1993, 1995a, 1995b), (Lambeck et al. (1998) |
	PRS	LGM	PRS	LGM	PRS	LGM	PRS	LGM	LGM-PRS[a]	LGM-PRS[b]
Greenland	3	7.1	1.1	6.5	~0	7.8	2.9	3.6	2.3	2
Eurasia	0.2	15.3	0	11.6	0	14.4	[c]	[c]	11.9	8
N. America	0.1	24.9	0	16.3	0	17.4	[c]	[c]	24.2	28
Antarctica	27.9	37.7[d]	30.3	30.6	29.5	31.1	25.0	32.4	8.7	6
NH	3.3	47.3	1.2	34.4	~0	39.6	4.3	58.1	38.4	40[e]
NH + SH	31.2	85.0	31.4	65.0	29.5	70.7	29.3	90.5	47.1	50[f]

[a] Estimates made by Berger from the eustatic sea-level changes given by Peltier for each ice sheet (with a total of 118 m at the LGM; 1998a,b) by multiplying his values by 0.4×10^6 km³ to compare with (LGM-PRS) of Dutrieux (1997). See text.

[b] Same as in note (a) but for Lambeck maximum reconstruction of eustatic sea-level changes, with a total of 125 m at LGM.

[c] Contribution of the individual Eurasia and N. America ice sheet is not calculated.

[d] Estimated from data, not calculated by the model.

[e] Northern hemisphere value contains an additional eustatic sea level of 5 m for mid-latitudes mountain glaciers.

[f] NH + SH value contains an imbalance of 10 m in sea-level change.

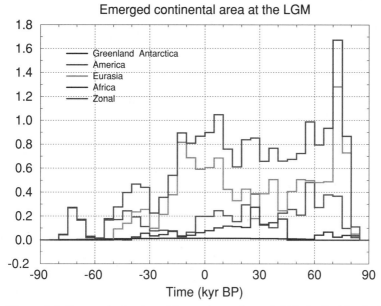

Figure 8.4. Increase in the size of the continents for each 5° latitudinal bands, from the South Pole to the North Pole, at the Last Glacial Maximum due to the lowering of the sea level by about 100 m, as simulated by the NSH model of Louvain-la-Neuve (Dutrieux, 1997). See also color plate section.

8 over Greenland, and 32 over Antarctica, which is 5.7×10^6 km^3 of ice more than if the sea level is kept fixed.

The accompanying sea level drop increases the size of the continents; to test the reliability of our simulation, we compared our results to those used in the Paleoclimate Model Intercomparison Programme (Joussaume and Taylor, 1995) and calculated by Peltier (1994, 1996). Most of our simulated values for the LGM are smaller than the Peltier ones. However, the difference between our results and the Peltier results, averaged over each continent, decreases when the sea level is allowed to change: for Greenland, it drops from 15 to 6%, for North America from 15 to 4%, for Eurasia from 21 to 4%, and for Antarctica from 11 to 8%. For the whole globe, our simulated sea level drop of about 100 m at the LGM leads to an increase of emerged continental surfaces of about 20×10^6 km^2 (13.7 in the Northern Hemisphere and 5.9 in the Southern Hemisphere). This is a 13% increase over the present-day value of 149×10^6 km^2.

Figure 8.4 shows, for each 5° latitudinal belt, this variation of the land surfaces over each continent. The largest increase is situated around 72.5°N and corresponds to the Bering Strait and the North Sea closings. In the Tropics, the platforms associated with Indonesian coral reefs, the southern China Sea, and Southeast Asia are responsible for a secondary maximum just south of the equator.

8.5 Sea Level and Vegetation-Snow Feedbacks

Given the importance of the vegetation-snow albedo feedback (VSAF) and of the sea level change in the Northern Hemisphere high latitudes, a detailed analysis of their contribution to the Northern Hemisphere ice volume variations (V) and of their synergism has been made for the whole last glacial–interglacial cycle. To isolate the respective contribution of the vegetation–snow albedo and of the sea level–ice volume feedbacks, the factor separation technique of Stein and Alpert (1993) was used. This method allows the identification of the contribution of the processes involved in a climate model as well as their synergistic effects. When applied to two processes, as in this chapter, this method requires four experiments to be made. In the first simulation, V_{11}, called the *control* experiment, both the sea level and the albedo of the vegetated area covered by snow are varied. In the *reference* experiment, V_{00}, they are kept fixed. In the two others, only one factor is allowed to vary at a time (the VSAF in V_{10} and the sea level in V_{01}). From the Taylor expansion of any function V of two variables, the factor separation technique gives

$$V_{11} = \hat{V}_{00} + \hat{V}_{10} + \hat{V}_{01} + \hat{V}_{11} \tag{1}$$

where $\hat{V}_{00} = V_{00}$ is the value of V when none of the factors is active, $\hat{V}_{10}$ and $\hat{V}_{01}$ represent the pure contribution of, respectively, the first and the second factor, and $\hat{V}_{11}$ gives the pure contribution of the interactions (synergism) between the two factors. When identified by the Taylor series expansion, this last term sums the contribution of all the partial derivatives of V with respect to both variables (second order and higher; Dutrieux et al., 1996).

These pure contributions are calculated according to the following relationships:

$$\hat{V}_{10} = V_{10} - V_{00} \tag{2}$$
$$\hat{V}_{01} = V_{01} - V_{00} \tag{3}$$
$$\hat{V}_{11} = V_{11} - (V_{10} + V_{01}) + V_{00} \tag{4}$$

Two remarks must be made here:

1. In a highly nonlinear system in which most of the variables are interacting with each other, determining the pure contribution of each factor would require all the variables to be taken into account. For n variables, this means that 2^n experiments must be performed. Here, we assume that all the variables other than the albedo of the snow-covered vegetation and the sea level do not significantly affect their synergism. However, the influence of other processes is clearly visible in the defined pure contribution of the two processes analyzed here.
2. The classical analysis of the sensitivity of the climate system to one of the factors leads to a different (sometimes spurious) interpretation. To perform a classical analysis of the sensitivity of the model to the taïga-tundra effect, for example, we would compare the results of the experiment in which this effect is not active – that is, V_{01} – to those of the control experiment V_{11}. This can be written, according to (2), (3) and (4), as follows:

$$V_{11} - V_{01} = \hat{V}_{10} + \hat{V}_{11}$$

This clearly introduces, in addition to $\hat{V}_{10}$, the synergism, $\hat{V}_{11}$, between vegetation and sea level. Therefore, the classical method does not provide the expected

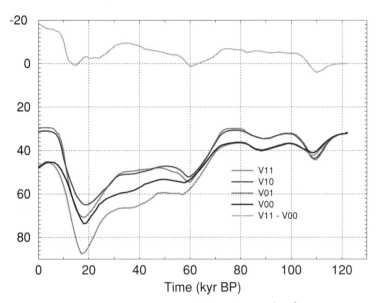

Figure 8.5. Long-term variations of the ice volume (in 10^6 km^3) simulated by the NSH model for the four experiments where only the vegetation-snow albedo feedback is considered, V_{10}, where only the sea level–ice volume feedback is considered, V_{01}, where the two effects are included, V_{11}, and where none of them is taken into account, V_{00}. The difference, $V_{11} - V_{00}$, is also reproduced (Dutrieux, 1997). See also color plate section.

contribution of the process analyzed. Instead, it shows this contribution plus its interactions with the other.

Figure 8.5 gives the long-term variations of the ice volume simulated in each of the experiments V_{11}, V_{10}, V_{01}, and V_{00} (Dutrieux, 1997). In the experiments in which VSAF is not included (V_{01} and V_{00}), the Northern Hemisphere ice volume is overestimated most of the time during the last glacial-interglacial cycle. This leads to an unrealistic amount of ice for the present day, in both cases $\sim 18 \times 10^6$ km^3 above the current value of 30×10^6 km^3. The critical point is that the large volume of ice produced at the LGM melts only partly between 18 kyr BP and now, the insolation and CO_2 forcings being insufficient to melt the excessive amount of ice and to counterbalance the related albedo and water vapor–temperature feedbacks. This is true mainly in the V_{01} experiment, where the absence of the VSAF when the sea level is allowed to change helps to produce about 87.5×10^6 km^3 of ice at the LGM, well above the 70×10^6 km^3 simulated in V_{11} (Table 8.3). Actually the ice volume simulated during glacial climates is larger in V_{01} (only sea level is allowed to vary) than in V_{00} (both factors are kept fixed). In particular, the ice volume at the LGM amounts to 73.6×10^6 km^3 in the V_{00} experiment, 14×10^6 km^3 less than in V_{01}. The rest of the time, both ice volumes are almost equal, with $V_{00} \leq V_{01}$. These results were expected because the difference between V_{01} and V_{00} gives exactly the pure contribution of sea level, $\hat{V}_{01}$. In V_{01}, during glacials, the ice volume grows in response to larger available continental surfaces without any negative counterpart.

The reverse is true during interglacials, but with a much smaller magnitude of the difference $V_{01} - V_{00}$. That is because the sea level is only slightly above the present-day level during these warm climates, whereas it is much lower during cold conditions (in V_{10} and V_{00}, sea level is fixed to its present-day value).

On the other hand, the two experiments that include the vegetation–snow albedo feedback (V_{11} and V_{10}) seem to be more realistic and lead to very similar ice volumes, except at the Last Glacial Maximum. In a classical sense, one might conclude that the sea level change plays no role most of the time during the last glacial-interglacial cycle, something that is contradicted by the importance of $\hat{V}_{01}$ described above. In these V_{11} and V_{10} experiments, taking into account the sea level change (V_{11}) as compared with keeping it fixed (V_{10}) increases the ice volume during the glacials ($V_{11} > V_{10}$) at isotopic stages 5d, 4, and 2 (~110, 60, and 20 kyr BP, respectively) and decreases it ($V_{11} < V_{10}$) during interglacials, the difference being smaller during the interglacials than during the glacials. Actually, $V_{11} - V_{10} = \hat{V}_{01} + \hat{V}_{11}$, and this difference therefore represents the sum of the pure contribution of sea level (strictly positive only during glacial times) and the synergism, $\hat{V}_{11}$, between sea level and VSAF.

If we compare the experiments in which the vegetation–snow albedo feedback is considered (V_{11} and V_{10}) to the experiments in which it is not, all the rest being kept the same (respectively, V_{01} and V_{00}), we can see that the vegetation effect helps to moderate the ice volume most of the time (with $V_{11} < V_{01}$ and $V_{10} < V_{00}$), reinforcing it only at stage 5d (where $V_{11} > V_{01}$ and $V_{10} > V_{00}$). At this isotopic stage 5d, the VSAF can play its normal role, amplifying the initial cooling through an increase of the tundra fraction and of the albedo in winter. This is because before 5d – that is at isotopic stage 5e – the Northern Hemisphere ice volume was indeed minimum, and the pure VSAF could not be masked by the albedo-temperature feedback. This is not the case at the other glacial stages. After 5d, in the experiments with no VSAF (V_{01} and V_{00}), ice is much less removed than in V_{11} and V_{10} and the ice-albedo feedback rapidly helps to produce more ice in V_{01} and V_{00} than in V_{11} and V_{10}, hiding the importance of the VSAF. This screening effect of the ice-albedo feedback is actually responsible for the trend visible in $\hat{V}_{10}$ and $\hat{V}_{01}$.

Actually, the difference $V_{10} - V_{00}$ is exactly equal to the pure contribution of the VSAF feedback in a two-factor problem. Here, it is negative everywhere except at stage 5d (for the reason explained above) and is larger at the present day than at the LGM (Figure 8.5): in V_{10}, there are 16×10^6 km^3 of ice less than in V_{00} now, whereas there are only 9×10^6 km^3 less at the LGM. This means that this feedback produces in V_{10}, when compared with V_{00}, less ice at LGM but much less during the Holocene, reinforcing the amplitude of the variation between these two climatic states.

The difference between the ice volumes of V_{11} and V_{01} is more difficult to explain because it reflects both the pure contribution of the VSAF, $\hat{V}_{10}$, and the synergism, $\hat{V}_{11}$. As seen in Figure 8.5, when the sea level varies, the total vegetation-snow albedo feedback, $\hat{V}_{10} + \hat{V}_{11}$, decreases the ice volume significantly, whether the climate is glacial or interglacial. Both at the LGM and at the peak of the Holocene, there are 18×10^6 km^3 of ice less in V_{11} than in V_{01}.

The difference between the ice volumes simulated in the control, V_{11}, and in the reference, V_{00}, experiments is actually the sum of the individual contributions of the vegetation and the sea level feedbacks and of their synergism (see Equation 1). Most of the time, V_{11} is smaller than V_{00}, the combined effect of both the vegetation–snow albedo and the sea level feedbacks reducing progressively the amount of ice stored in the ice sheets of V_{00}. After one full glacial–interglacial cycle, this reduction amounts to 20×10^6 km^3 of ice. It is only at stage 5d that the difference $V_{11} - V_{00}$ is slightly positive, the vegetation and sea level mechanisms contributing to increase the amount of ice during this abortive glaciation. At stage 4 and a little after the LGM of stage 2, their global contribution is negligible and $V_{11} \cong V_{00}$. But the detailed explanation is not as simple nor as straightforward, and the calculation of $\hat{V}_{10}$, $\hat{V}_{01}$, and $\hat{V}_{11}$ is necessary (Figure 8.6).

During the whole stage 5, from 122 to 70 kyr BP, it is essentially the vegetation–snow albedo feedback, $\hat{V}_{10}$, that explains the difference $V_{11} - V_{00}$, first increasing the ice volume from 120 to 106 kyr BP, and then decreasing it from 106 to 70 kyr BP. (During this whole time interval, both $\hat{V}_{01}$ and $\hat{V}_{11}$ are negligible except slightly after 110 kyr BP where $\hat{V}_{01}$ is positive and counterbalanced by $\hat{V}_{11}$.) After 70 kyr BP, the situation becomes more complex. The pure contribution of sea level change, $\hat{V}_{01}$, increases continuously up to the LGM, reaching a maximum of $\sim 16 \times 10^6$ km^3 at 17 kyr BP. This increase is slowed by the negative effect of both the VSAF and the synergism between this effect and the sea level. From 65 to slightly after 30 kyr BP, $V_{11} - V_{00} \cong \hat{V}_{10}$, the other

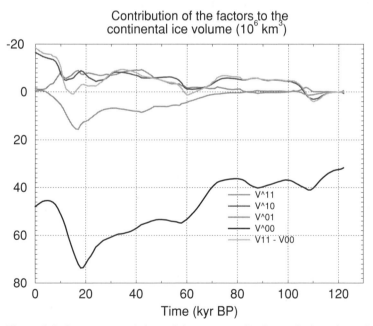

Figure 8.6. Long-term variations of the pure contribution to the ice volume simulated with the NSH model in experiment V_{11}, of the vegetation–snow albedo feedback, $\hat{V}_{10}$, of the sea level, $\hat{V}_{01}$, and of the synergism between the two, $\hat{V}_{11}$. The ice volume of V_{00} and the difference $V_{11} - V_{00}$ are also reproduced (Dutrieux, 1997). See also color plate section.

two factors being significant but almost counterbalancing each other, and $\hat{V}_{01} > 0$ and $\hat{V}_{11} < 0$ being of the same order of magnitude. Between 30 and about 10 kyr BP (i.e., over the whole isotopic stage 2 and slightly more), the sea level factor becomes increasingly important and controls the increase of the ice volume up to its maximum at around 19 kyr BP where $\hat{V}_{01} \cong |\hat{V}_{10} + \hat{V}_{11}|$.

Over the last 10 kyr the entire difference ($V_{11} - V_{00}$) can be explained mainly by the vegetation-snow albedo feedback, as was also the case during isotopic stage 5. However, there is a difference between the Holocene and stage 5: over the last 10 kyr, the synergism between the vegetation and the sea level mechanisms, $\hat{V}_{11}$, is significantly negative, almost canceling the pure positive contribution of sea level, $\hat{V}_{01}$, whereas both $\hat{V}_{11}$ and $\hat{V}_{01}$ are close to zero during stage 5.

To illustrate the efficiency of this factor separation method, it can be applied to the results otained by Ganopolski et al. (1998) and discussed in section 8.3. Their simulation using an atmosphere-ocean-vegetation model demonstrates a strong synergism between the climate components. Tables 8.4a and 8.4b show that these interactions are larger than the pure contribution of any of the components, the importance of the ocean and vegetation subsystems depending on the climatic parameter under consideration. In addition to the synergism, the warming of the boreal winter, for example, is caused mainly by the ocean, whereas it is caused by the vegetation for the summer and annual means. The warming (up to 3 to 4 °C in the northern continental areas in winter and

Table 8.4a. Pure Contribution and Synergism of the Different Components of the Climate System (Atmosphere-Ocean-Vegetation) to Temperature Change at 6 kyr BP Simulated by Ganopolski et al. (1998). This Table has been Constructed from their Table 1, Using the Stein-Alpert Method (1993) Explained in this Chapter.

		Temperature (°C)				
		$A\hat{O}$	$A\hat{V}$	$A\hat{O}V$	AOV	$AOV-AO$
Boreal Summer	NH$_L$	−0.5	0.5	0.8	2.5	1.3
	NH	−0.4	0.3	0.7	1.6	1.0
	SH	0	0.1	0.5	0.8	0.6
Boreal Winter	NH$_L$	0.3	0.1	0.8	0.4	0.9
	NH	0.2	0.1	0.8	0.6	0.9
	SH	0.1	0	0.7	0.5	0.7
Annual Average	NH$_L$	−0.1	0.3	0.8	1.2	1.1
	NH	−0.1	0.2	0.8	1.0	1.0
	SH	−0.1	0.1	0.7	0.7	0.8

$A\hat{O}$ represents the pure contribution of the ocean
$A\hat{V}$ represents the pure contribution of the vegetation
$A\hat{O}V$ represents the synergism between the ocean and vegetation
AOV gives the sum of all contributions: $A + A\hat{O} + A\hat{V} + A\hat{O}V$
$AOV - AO = A\hat{V} + A\hat{O}V$ gives what we would conclude from a classical sensitivity experiment for the contribution of vegetation changes (control case: fixed vegetation)
NH Northern Hemisphere; NH$_L$ Northern Hemisphere lands; NAfr$_L$ North African lands
SH Southern Hemisphere

Table 8.4b. Same as Table 8.4a but for Precipitation.

		Precipitation (mm/day)				
		$A\hat{O}$	$A\hat{V}$	$A\hat{O}V$	AOV	$AOV-AO$
Boreal Summer	NH$_L$	−0.06	0.24	0.11	0.54	0.35
	SH$_L$	−0.03	0.02	0.04	0.07	0.06
	NAfr$_L$	−0.06	1.09	0.24	1.85	1.33
Boreal Winter	NH$_L$	0	0.01	0.05	0.05	0.06
	SH$_L$	0.04	0	0.12	−0.01	0.12
	NAfr$_L$	0.01	0.03	0.02	0.05	0.05
Annual Average	NH$_L$	−0.02	0.11	0.09	0.26	0.20
	SH$_L$	0	0	0.09	0.06	0.09
	NAfr$_L$	−0.02	0.45	0.11	0.72	0.56

summer, respectively) is caused by a decrease of the planetary albedo related to an extension of the boreal forests and a decrease of the subtropical deserts. In high northern latitudes, a northward expansion of the boreal forest due to summer warming leads to an annual warming via the vegetation-snow-albedo feedback, strongly amplified by the sea-ice feedback. In the subtropics, a strong positive feedback between vegetation and precipitation leads to a greening of the Sahara. Contrary to the temperature, precipitation changes in summer over the NH continents, North Africa in particular, are related mainly to the pure contribution of vegetation changes, as shown in Table 8.4b. Finally, a strong warming of the North Atlantic, together with an increased freshwater flux into the Atlantic basin, leads to a weakening of the thermo-haline circulation, which in turn results in a warming of the Southern Hemisphere (up to 2 °C near the Antarctic) and in a negative feedback for the Northern Hemisphere high latitudes.

8.6   Conclusions

During the whole glacial-interglacial cycle, the synergism, $\hat{V}_{11}$, between the sea level and taïga-tundra mechanisms is negatively correlated with the contribution of sea level alone, $\hat{V}_{01}$. This synergism tends therefore to definitely damp the pure positive contribution of sea level to an increase in the ice volume. This leads us to assume that $\hat{V}_{11}$ corresponds, at least in part, to the afforestation/deforestation of the continental emerged/submerged platforms following a decrease/increase in sea level ($\hat{V}_{11}$ also contains many indirect interactions between sea level and vegetation).

Although these results are only from a model that does not take all processes into account, they show clearly (i) the difficulty of explaining the ice volume changes if the number of experiments performed is too limited (e.g., if we perform only $V_{11} - V_{01}$ or $V_{11} - V_{10}$) and (ii) the important role played by the dynamics of the forests and, in particular, in high latitudes their associated albedo feedback when they are covered by snow, $\hat{V}_{10}$. However, as soon as the ice sheets becomes sufficiently large to induce a significant sea level drop and a subsequent significant emergence of the continental platforms, two effects appear and oppose each other. The first one, the sea level-ice volume feedback, $\hat{V}_{01}$, contributes to an increase in the ice volume by allowing more space for the ice sheets to grow; the second one, the vegetation-snow albedo feedback, $\hat{V}_{10}$ (and the synergism between the two, $\hat{V}_{11}$), contributes to slow this increase in ice volume through the reduction of the global albedo. It remains true, however, that these conclusions can be only partial because only two factors have been considered. One next step would be to reconsider the main role played by the ice albedo and the water vapor–temperature feedbacks (Berger et al., 1993b), but this time in conjunction with the sea level and vegetation-snow albedo factors. Additional factors might also be included, such as the isostatic rebound of the bedrock beneath the ice sheets, continentality, and snow aging factors. However, it must be remembered that four factors require 16 experiments, which involves not only a considerable amount of computer time but also a great deal of human effort to analyze the results.

All the sensitivity experiments made with the LLN 2.5-D models, including those discussed here, show that the response – both in amplitude and in phase – of the climate system to a given forcing is clearly a function of the actual climatic state, the latitude, and the season. This seriously complicates the elaboration of any simple universal rule to describe the causes-to-effects relationship in a glacial-interglacial cycle. According to the simulations done with the LLN models, the following factors play a key role in shaping the 100 kyr cycle: insolation, albedo, water vapor, and vegetation-temperature feedbacks, and the isostatic response of the lithosphere to ice sheet loading. This means that when one of these parameters is kept fixed, the model produces or melts too much ice, preventing it from sustaining the 100 kyr cycle. It is therefore almost impossible to claim which process is the most important for explaining an entrance into or termination of glaciation. It is much more a series of events, to be seen in a dynamical and synergistic point of view, that explains the waxing and waning of the ice sheets at the astronomical time scale. Figure 8.7 gives such a flow chart, describing the response of the climate system between isotopic stages 5e and 5d. The ice volume maximum occurs at 109 kyr BP in our simulation, 7 kyr later than the minimum of the June 65°N insolation (Berger et al., 1996). The time lag between the ice volume and the insolation forcing depends on the month that is chosen as a reference (Loutre and Berger, 1995), but this does not affect the overall conclusions. Moreover, a similar flow chart can be made for any other climatic variable, such as the air surface temperature at a given latitude for a whole zonal belt, over the oceans, or over the continents, but the timing of the variations will not be the same. For example, the surface temperature of the 50–55°N belt from July to September is almost in phase with the 65°N June insolation (lagging behind by only ~1 to 2 kyr), a much faster response than at higher latitudes, where a much more profound influence of the ice sheets is discernable.

8.7   Recommendations

With a global average surface temperature of 0.58 °C warmer than the 1961–1990 baseline, 1998 was the hottest year on record since 1860 (WMO, 1998) and maybe the warmest over the past 1200 years (Overpeck, 1998). More analysis of temperature

ENTERING INTO GLACIATION

1. INSOLATION
high latitude summer $\Big\}$ ↓ 127 – 116 kyr BP

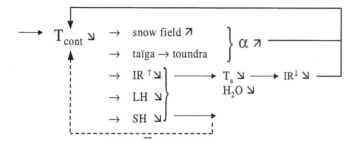

$\longrightarrow \overline{T}_s$ ↘ $\longrightarrow$ $\overline{T}_a$ ↘ → snow fall ↗
  → radiative budget < 0
  ⇔ melting = 0 $\longrightarrow$ **net ice accumulation > 0**

2. ICE VOLUME INCREASE : 116 – 109 kyr BP

(although insolation increases from 116 to 105 ky BP at 65°N in June, for example)

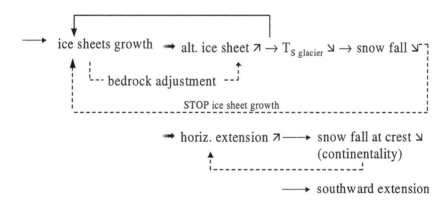

GLACIATION : MAX ICE VOLUME   109 kyr BP

Figure 8.7. Flow charts giving a series of processes that (1) result from the insolation decrease (at 65°N in June, for example) between 127 and 116 kyr BP, and (2) finally lead to the ice volume maximum at 109 kyr BP, although the insolation begins increasing at 116 kyr BP. ↘ stands for decrease, ↗ for increase; → for implies; IR$^↑$ upward and IR$^↓$ downward infrared radiation; LH for latent heat; SH for sensible heat; α for albedo, T$_a$ for air temperature; T$_s$ for surface temperature; full horizontal line for positive feedback; dashed horizontal line for negative feedback; alt for altitude.

data recorded over the past century provides additional evidence (Wigley et al., 1998) that anthropogenic greenhouse gas emissions are at least partially responsible for global temperature increase this century (Houghton et al., 1996). Global CO_2 emissions could grow 75% by 2020 (IEA, 1998). This would result from a world energy demand that could increase by 65% between 1995 and 2020, all scenarios suggesting that the world will continue to rely heavily on fossil fuels until at least 2020. New models (Fells, 1998) even predict that the energy consumption will increase two to five times by the end of the 21st century. Achieving stabilization of atmospheric CO_2 levels, anywhere up to 750 ppmv, will therefore require an enormous increase in carbon-free energy production (Hoffert et al., 1998) and greater improvements in energy efficiency, two things that it seems almost impossible to sustain over the next century. In addition, model results project that full compliance with the Kyoto agreement will reduce the projected warming only 0.05 °C by 2050, and even a 20% reduction in emissions by industrialized nations would decrease projected warming only another tenth of a degree or so. This means that regardless of how effectively society will be at limiting its greenhouse gas emissions, global concentrations of greenhouse gases are certain to increase over the coming decades, making some change inevitable (Mitchell et al., 1998). It is therefore urgent that we better understand the behavior of the climate system by using as many climatic situations as possible to test the reliability of the models used to predict the future climate. This must be done for all time scales because high-frequency changes, including abrupt changes (characteristic of human life), develop along the longer-term trends and develop differently in different climatic states. The factor separation techniques can help to quantify the individual effect of the processes and feedbacks involved and their synergism along the lines developed in this chapter. Other lessons from the LLN experiments are that (i) terrestrial biospheric changes (whether climatically induced or brought about by land use) can have major implications for the climate and therefore must be carefully taken into account in climate models, (ii) greenhouse gas emissions have already altered the climate, and (iii) they have the potential to influence the course of the glacial-interglacial cycles over the next 50,000 years. As a consequence, in addition to improving our fundamental knowledge of how the climate system works, we must start to investigate climates of the past that were warmer than today by a few degrees and during which the CO_2 concentration was much higher than now (this is especially important if adaptation becomes the main response to the threat of climate change). The late Pliocene might be a good candidate, inasmuch as most of the other boundary conditions were not too far from those of the present day. Finally, all this scientific effort must allow us to quantify the impacts of global warming at the regional scale and must give policy makers the necessary tools to help them to take the right decisions.

ACKNOWLEDGMENTS

Many thanks to Dr. M. F. Loutre for her sustained collaboration in the astronomical theory of paleoclimates. The work of Dr. A. Dutrieux and the help of Michel Crucifix are also sincerely acknowledged. I also thank N. Depoorter for a careful typing of the manuscript.

REFERENCES

Berger, A. (1978). Long-term variations of daily insolation and Quaternary climatic changes, *J. Atmos. Sci.*, **35**, 2362–2367.

Berger, A., Fichefet, T., Gallée, H., Tricot, C., and van Ypersele, J. P. (1992). Entering the glaciation with a 2-D coupled climate model, *Quaternary Science Reviews*, **11(4)**, 481–493.

Berger, A., Loutre, M. F., and Tricot, Ch. (1993a). Insolation and Earth's orbital periods. *J. Geophys. Research*, **98** n° **D6**, 10,341–10,362.

Berger, A., Tricot, C., Gallée, H., and Loutre, M. F. (1993b). Water vapour, CO₂ and insolation over the last glacial-interglacial cycles, *Philosophical Transactions of the Royal Society, London*, **B, 341**, 253–261.

Berger, A., Gallée, H., and Tricot, Ch. (1993c). Glaciation and deglaciation mechanisms in a coupled two-dimensional climate-ice-sheet model, *J. of Glaciology*, **39(131)**, 45–49.

Berger, A., and Loutre, M. F. (1994). Long term variations of the astronomical seasons, In: *Topics in Atmospheric and Interstellar Physics and Chemistry*, Cl. Boutron (ed.), pp. 33–61, Les Editions de Physique, Les Ulis, France.

Berger, A. (1995). Modeling the response of the climate system to the astronomical forcing. In: *Future Climates of the World, A Modelling Perspective*, A. Henderson-Sellers (ed.), World Survey of Climatology (H. E. Landsberg, ed), vol. **16**, pp. 21–69, Elsevier Science Publishers B.V., Amsterdam.

Berger, A., and Loutre, M. F. (1996). Modeling the climate response to the astronomical and CO₂ forcings, *Comptes Rendus de l'Académie des Sciences de Paris*, t. **323**, série II a, 1–16.

Berger, A., Gallée, H., Li, X. S., Dutrieux, A., and Loutre, M. F. (1996). Ice-sheet growth and high-latitudes sea-surface temperature, *Climate Dynamics*, **12(7)**, 441–448.

Berger, A., Loutre, M. F., and Mélice, J. L. (1998a). Instability of the astronomical periods from 1.5 Myr BP to 0.5 Myr AP. *Palaeoclimate Data and Modeling*, **2(4)**, 239–280.

Berger, A., Loutre, M. F., and Gallée, H. (1998b). Sensitivity of the LLN climate model to the astronomical and CO₂ forcings over the last 200 kyr. *Climate Dynamics*, **14**, 615–629.

Berger, A., Li, X. S., and Loutre, M. F. (1999). Modelling northern hemisphere ice volume over the last 3 Ma, *Quaternary Science Reviews*, **18(1)**, 1–11.

Blunier, T., Schwander, J., Stauffer, B., Stocker, T., Dällenbach, A., Indermühle, A., Tschumi, J., Chappellaz, J., Raynaud, D., and Barnola, J. M. (1997). Timing of the Antarctic cold reversal and the atmospheric CO₂ increase with respect to the Younger Dryas event, *Geophysical Research Letter*, **245(21)**, 2683–2686.

Charney, J., Quirk, W. J., Chow, S. H., and Kornfield, J. (1977). A comparative study of the effects of albedo change on drought in semi-arid regions, *J. Atmos. Sc.*, **34(9)**, 1366–1401.

Claussen, M., and Gayler, V. (1997). The greening of Sahara during the mid-Holocene results of an interactive atmosphere-bionic model, *Global Ecol. Biogeogr. Lett.*, **6**, 369.

Claussen, M., Kubatzki, C., Brovkin, V., and Ganopolski, A. (1999). Simulation of an abrupt change in Saharan vegetation at the end of the mid-Holocene, *Geophysical Research Letters*, **26(16)**, 2037–2040.

CLIMAP Project Members (1976). The surface of the ice-age Earth, *Science*, **191**, 1131–1137.

Cortijo, E., Duplessy, J. C., Labeyrie, L., Leclaire, H., Duprat, J., and van Weering, T. C. E. (1994). Eemian cooling in the Norwegian Sea and North Atlantic ocean preceding continental ice-sheet growth, *Nature*, **372(6505)**, 446–449.

Crowley, T. J., and Baum, S. L. (1997). Effect of vegetation on an ice-age climate model simulation, *Journal of Geophysical Research*, **102**, n° **D14**, 16,463–16,480.

de Noblet, N. I., Prentice, I. C., Joussaume, S., Texier, D., Botta, A., and Haxeltine, A. (1996). Possible role of atmosphere-biosphere interactions in triggering the last glaciation, *Geophysical Research Letters*, **23**, n° **22**, 3191–3194.

Duplessy, J. C., Shackleton, N. J., Fairbanks, R. G., Labeyrie, L., Oppo, D., and Kallel, N. (1988). Deepwater source variations during the last climatic cycle and their impact on the global deepwater circulation, *Paleoceanography*, **3(3)**, 343–360.

Dutrieux, A., Berger, A., Loutre, M. F., and Tricot, Ch. (1996). Classical Feedback Method and Separation Factor Method: Comparison and application to 0-D Energy Balance Model. *Progress Report 1996/3*, *Institut d'Astronomie et de Géophysique G. Lemaitre (ASTR), Université Catholique de Louvain (UCL), Louvain-la-Neuve, Belgique*.

Dutrieux, A. (1997). Etude des variations à long terme du climat à l'aide d'un modèle global à deux dimensions du système climatique. Doctor of Sciences dissertation, Institut d'Astronomie et de Géphysique G. Lemaitre, Université catholique de Louvain, Louvain-la-Neuve. Unpublished manuscripts, 274 pages (+73 pp. figures).

Fells, I. (1998). The need for energy. *Europhysics News*, **29**(6), 193–195.

Fischer, H., Wahlen, M., Smith, J., Mastroianni, D., and Deck, B. (1999). Ice core records of atmospheric CO_2 around the last three glacial terminations, *Science*, **283**(5408), 1712–1714.

Foley, J. A., Kutzbach, J. E., Coe, M. T., and Levis, S. (1994). Feedbacks between climate and boreal forests during the Holocene epoch, *Nature*, **371**, n° **6492**, 52–54.

Friedlingstein, P., Delire, C., Müller, J. F., and Gérard, J. C. (1992). The climate induced variation of the continental biosphere: a model simulation of the last glacial maximum, *Geophysical Research Letters*, **19**, n°**9**, 897–900.

Gallée, H., van Ypersele, J. P., Fichefet, T., Tricot, Ch., and Berger, A. (1991). Simulation of the last glacial cycle by a coupled, sectorially averaged climate-ice sheet model. Part 1. The climate model, *J. Geophys. Res.*, **96**, 7, 13,139–13,161.

Gallée, H., van Ypersele, J. P., Fichefet, T., Marsiat, I., Tricot, Ch., and Berger, A. (1992). Simulation of the last glacial cycle by a coupled, sectorially averaged climate-ice sheet model. Part 2. Response to Insolation and CO_2 variations, *J. Geophys. Res.*, **97**, 14, 15,713–15,740.

Gallimore, R. G., and Kutzbach, J. E. (1996). Role of orbitally induced changes in tundra area in the onset of glaciation, *Nature*, **381**, n° **6582**, 503–505.

Ganopolski, A., Kubatzki, C., Claussen, M., Brovkin, V., and Petoukhov, V. (1998). The influence of vegetation-atmosphere-ocean interaction on climate during the Mid-Holocene, *Science*, **280**, n°**5371**, 1916–1919.

Guiot, J., Cheddadi, R., Prentice, I. C., and Jolly, D. (1996). A method of biome and land surface mapping from pollen data: application to Europe 6000 years ago, *Palaeoclimates, Data and Modelling*, **1**, n°**4**, 311–324.

Harrison, S. P., Kutzbach, J. E., Prentice, I. C., Behling, P. J., and Sykes, M. T. (1995). The response of Northern Hemisphere extratropical climate and vegetation to orbitally induced changes in insolation during the last interglacial, *Quaternary Research*, **43**, 174–184.

Harrison, S. P., Jolly, D., Laarif, F., Abe-Ochi, A., Dong, B., Herterich, K., Hewitt, C., Joussaume, S., Kutzbach, J. E., Mitchell, J., De Noblet, N., and Valdes, P. (1998). Intercomparison of simulated global vegetation distribution in response to 6 kyr BP orbital forcing, *Journal of Climate*, **11**(11), 2721–2742.

Harvey, L. D. D. (1988). On the role of high latitude ice, snow, and vegetation feedbacks in the climatic response to external forcing changes, *Climatic Change*, **13**(2), 191–224.

Harvey, L. D. D. (1989). Milankovitch forcing, vegetation feedback, and North Atlantic deep-water formation, *Journal of Climate*, **8**(2), 800–815.

Hays, J. D., Imbrie, J., and Shackleton, N. J. (1976). Variations in the Earth's orbit: Pacemaker of the Ice Ages, *Science*, **194**, 1121–1132.

Hoffert, M. I., Caaldeira, K., Jain, A. K., Haites, E. F., Harvey, L. D., Potter, S. D., Schlesinger, M. E., Schneider, S. H., Watts, R. G., Wigley, T. M. L., and Wuebbles, D. J. (1998). Energy implications of future stabilization of atmospheric CO_2 content, *Nature*, **395**(6705), 881–884.

Houghton, J. T., Meira Filho, L. G., Callander, B. A., Harris, N., Kattenberg, A., and Maskell, K. (1996). Climate Change 1995: The Science of Climate Change, *Intergovernmental Panel on Climate Change*, Cambridge University Press, 572 pp.

Huybrechts, Ph. (1990a). A 3-D model for the Antarctic ice sheet: a sensitivity study on the glacial-interglacial contrast, *Climate Dynamics*, **5**, 79–92.

Huybrechts, Ph. (1990b). The Antarctic ice sheet during the last glacial-interglacial cycle: a three-dimensional experiment, *Annals of Glaciology*, **14**, 115–119.

Huybrechts, Ph., and T'Siobbel, S. (1995). Thermomechanical modelling of northern hemisphere ice sheets with a two-level mass-balance parameterization, *Annals of Glaciology*, **21**, 111–116.

Huybrechts, Ph. (1996). Basal temperature conditions of the Greenland ice sheet during the glacial cycles, *Annals of Glaciology*, **23**, 226–236.

Huybrechts, Ph., and T'Siobbel, S. (1997). A three-dimensional climate/ice-sheet model applied to the Last Glacial Maximum, *Annals of Glaciology*, **25**, 333–339.

International Energy Agency (1998). World Energy Outlook, *Organization for Economic Cooperation and Development*, Paris.

Joussaume, S., and Taylor, K. E. (1995). Status of the paleoclimate Modelling Intercomparison Project. *Proceedings of the First International AMIP Scientific Conference, Monterey, CA, USA*, **WCRP-92**, 425–430.

Jouzel J., Barkov, N. I. Barnola, J. M., Bender, M., Chappelaz, J., Genthon, C., Kotlyakov, V. M., Lipenkov, V., Lorius, C., Petit, J. R., Raynaud, D., Raisbeck, G., Ritz, C., Sowers, T., Stievenard, M., Yiou F., and Yiou, P. (1993). Extending the Vostok ice-core record of palaeoclimate to the penultimate glacial period. *Nature*, **364(29 July 1993)**, 407–412.

Kubatzki, C., and Claussen, M. (1998). Simulation of the global-geophysical interactions during the last Glacial Maximum, *Climate Dynamics*, **14**, 461–471.

Kutzbach, J., Bonan, G., Foley, J., and Harrison, S. P. (1996). Vegetation and soil feedback on the response of the African monsoon to orbital forcing in the early to middle Holocene, *Nature*, **384**, 623–626.

Kutzbach, J. E., and Liu, Z. (1997). Response of the African monsoon to orbital forcing and ocean feedbacks in the Middle Holocene, *Science*, **278(5337)**, 440–443.

Labeyrie, L. D., Duplessy, J. C., and Blanc, P. L. (1987). Variations in mode of formation and temperature of oceanic deep waters over the past 125,000 years. *Nature*, **327**, 477–482.

Lambeck, K. (1993). Glacial rebound of the British Isles – II. A high-resolution, high-precision model, *Geophys. J. Int.*, **115**, 960–990.

Lambeck, K. (1995a). Constraints on the Late Weichselian ice sheet over the Barents Sea from observations of raised shorelines. *Quaternary Science Reviews*, **14**, 1–16.

Lambeck, K. (1995b). Late Pleistocene and Holocene sea-level change in Greece and south-western Turkey: a separation of eustatic, isostatic and tectonic contributions. *Geophys. J. Int.*, **122**, 1022–1044.

Lambeck, K., Smither, C., and Johnston, P. (1998). Sea-level change, glacial rebound and mantle viscosity for northern Europe, *Geophys. J. Int.*, **134**, 102–144.

Li, X. S., Berger, A., Loutre, M. F., Maslin, M. A., Haug, G. H., and Tiedemann, R. (1998a). Simulating late Pliocene Northern Hemisphere climate with the LLN 2-D model. *Geophysical Research Letters*, **25(6)**, 915–918.

Li, X. S., Berger, A., and Loutre, M. F. (1998b). CO₂ and Northern Hemisphere ice volume variations over the middle and late Quaternary. *Climate Dynamics*, **14**, 537–544.

Loutre, M. F., and Berger, A. (1995). Solar radiation. In: *Paleoclimates of the Northern and Southern Hemispheres*, The PANASH Project, The Pole-Equator-Pole Transects, **PAGES series 95-1**, IGBP-PAGES, Bern, 7–11.

Loutre, M. F., and Berger, A. (2000). Future climatic changes: are we entering an exceptionally long interglacial? *Climatic Change*, **46**, 61–90.

Milankovitch, M. (1941). Kanon der Erdbestrahlung und seine Anwendung auf das Eiszeiten-problem. *Royal Serbian Sciences, Spec. pub.* **132**, *section of Mathematical and Natural Sciences*, vol. **33**, Belgrade, 633 pp. ("Canon of Insolation and the Ice Age problem," English Translation by Israël Program for Scientific Translation and published for the U. S. Department of Commerce and the National Science Foundation, Washington, D. C., 1969).

Mitchell, J. F. B., Johns, T. C., and Senior, C. A. (1998). Transient response to increasing greenhouse gases using models with and without flux adjustment, *Hadley Centre Technical Note*, **2**, 1–15.

Overpeck, J. (1998). Warmest year in 1200 years, *Global Environmental Change Report*, **X(24)**, 2.

Otterman, J., Chou, M. D., and Arking, A. (1984). Effects of nontropical forest cover on climate, *J. Appl. Meteorol.*, **23**, 762–767.

Peltier, W. R. (1994). Ice age paleotopography, *Science*, **265**, 195–201.

Peltier, W. R. (1996). Mantle viscosity and ice-age ice sheet topography, *Science*, **273**, 1359–1364.

Peltier, W. R. (1998a). "Implicit Ice" in the global theory of glacial isostatic adjustment, *Geophysical Research Letters*, **25(21)**, 3955–3958.

Peltier, W. R. (1998b). Postglacial variations in the level of the sea: implications for climate dynamics and solid Earth geophysics, *Rev. Geophys.*, **36(4)**, 603–689.

Petit, J. R., Jouzel, J., Raynaud, D., Barkov, N. I., Barnola, J. M., Basile, I., Bender, M., Chappellaz, J., Davis, M., Delaygue, G., Delmotte, M., Kotlyakov, V. M., Legrand, M., Lipenkov, V. Y., Lorius, C., Pépin, L., Ritz, C., Saltzman, E., and Stievenard, M. (1999). Climate and atmospheric history of the past 420,000 years from the Vostok ice core, Antarctica, *Nature*, **399**, 429–436.

Prentice, C., Jolly, D., and Biome 6000 participants (2000). Mid-Holocene and glacial maximum geography of the Northern continents and Africa, *J. Biogeography*, **27**, in press.

Prentice, C. (1998). Interactions of climate change and the terrestrial biosphere, Pontifical Academy, Roma, 9–13 November 1998. Chap. 11, this volume.

Ramaswamy, V., and Chen, C. T. (1997). Linear additivity of climate response for combined albedo and greenhouse perturbations, *Geophysical Research Letters*, **24(5)**, 567–570.

Ruddiman, W. F., and McIntyre, A. (1979). Warmth of the subpolar North Atlantic Ocean during Northern Hemisphere ice-sheet growth, *Science*, **204(4389)**, 173–175.

Stein, U., and Alpert, P. (1993). Factor separation in numerical simulations. *J. Atmos. Sci.*, **50**, 2107–2115.

TEMPO – Kutzbach, J. E., Bartlein, P. J., Foley, J. A., Harrison, S. P., Hostetler, S. W., Liu, Z., Prentice, I. C., and Webb, T. III (1996). Potential role of vegetation feedback in the climate sensitivity of high-latitude regions: a case study at 6000 years BP, *Global Biogeochemical Cycles*, **10, n°4**, 727–736.

Texier, D., de Noblet, N., Harrison, S. P., Haxeltine, A., Jolly, D., Joussaume, S., Laarif, F., Prentice, I. C., and Tarasov, P. (1997). Quantifying the role of biosphere-atmosphere feedbacks in climate change: coupled model simulations for 6000 years BP and comparison with palaeodata for northern Eurasia and northern Africa, *Climate Dynamics*, **13, n°12**, 865–882.

Wigley, T. M. L., Smith, R. L., and Santer, B. D. (1998). Anthropogenic influence on the auto-correlation structure of hemispheric-mean temperatures, *Science*, **282(5394)**, 1676.

World Meteorological Organization (WMO) (1998). Hottest year this century, *Global Environmental Change Report*, **X(24)**, 2.

Yiou, P., Ghil, M., Le Treut, H., Genthon, C., Jouzel J., Barnola, J. M., Lorius, C., and Korotkevitch, Y. N. (1991). High-frequency paleovariability in climate and in CO_2 levels from Vostok ice-core records, *Journal of Geophysical Research*, **96(B12)**, 20365–20378.

Yiou, P., Ghil, M., Jouzel, J., Paillard, D., and Vautard, R. (1994). Nonlinear variability of the climatic system, from singular and power spectra of Late Quaternary records, *Climate Dynamics*, **9(8)**, 371–389.

9 Nonlinearities in the Earth System: The Ocean's Role

THOMAS F. STOCKER

ABSTRACT

The climate system contains, even in its simplest possible representation, nonlinearities that can give rise to multiple equilibria. The role of the ocean in this context is discussed, and recent progress is reviewed. The paleoclimatic record indicates that such different equilibria are relevant to our understanding of past changes and likely are fundamental for a correct assessment of future changes.

9.1 Introduction

During the past two decades, paleoclimatic research has been the key to a quantitative understanding and appreciation of the full dynamics of the climate system. It has long been thought that the major climatic shifts have been caused by changes in the Earth's orbit, evidenced most dramatically by the sequence of ice ages during the Pleistocene (Imbrie et al., 1992). The advent of high-resolution archives such as ice cores from the polar regions modified this view considerably. Measurements of the stable isotopes of polar ice from Greenland indicated a succession of abrupt events during the last glacial (Dansgaard et al., 1984). The late Hans Oeschger, one of the pioneers in ice core research, demonstrated that the changes seen during the last deglaciation about 15,000–11,000 years ago are coeval with those registered in Swiss Gerzensee (Oeschger et al., 1984). This historical and bold hypothesis is reproduced here (Figure 9.1). Oeschger and his colleagues showed a strong correlation between these two records for the Younger Dryas (YD), the last of the series of abrupt events found by Dansgaard et al. (1984). This was surprising because these two paleoclimatic archives have very different characteristics and are located far apart. In honor of this finding, these events have become known as Dansgaard/Oeschger events, of which 24 have been identified during approximately the last 100,000 years (Dansgaard et al., 1993).

Earlier paleoceanographic reconstructions by Ruddiman and McIntyre (1981) showed that the polar front in the Atlantic moved farther south during the YD. This suggested that the common cause of these events was the ocean, in particular changes in the sea surface temperature distribution and hence circulation patterns. The hypothesis of an oceanic flip-flop was formulated (Oeschger et al., 1984; Broecker et al., 1985).

147

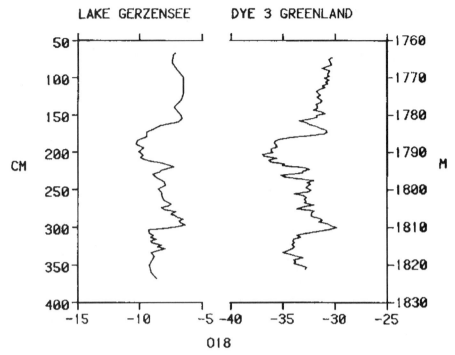

Figure 9.1. Reproduction of the original figure of Oeschger et al. (1984) demonstrating that changes in the isotopic composition of the ice in Greenland is an over-regional signal that is registered also in paleoclimatic archives several thousand kilometers farther south. The sequence of abrupt events during the last termination – the abrupt warming into Bolling, the Allerod cooling, the Younger Dryas, and its termination – are all visible in the stable isotopes of ice from Greenland (right) and carbonates from Lake Gerzensee in Switzerland (from Oeschger et al., 1984).

Since then, evidence for abrupt climate change in various paleoclimatic archives from different locations has been rapidly accumulating (e.g., Bond et al., 1992; Lehman and Keigwin, 1992; Bond et al., 1993; Grimm et al., 1993; Behl and Kennet, 1996). Thanks to various high-resolution paleoclimatic records, the term "rapid change" can now be quantified. Dansgaard et al. (1989) and Taylor et al. (1993) have demonstrated, by analyzing the ice cores from Summit (Greenland), that the termination of YD occurred within a few decades (for the stable isotope of precipitation, a proxy for temperature) and as rapidly as a few years (for dust, an indicator for atmospheric circulation). These signals are not manifesting only regional variability in Greenland, caused, for instance, by small shifts in frontal locations; instead, they characterize large-scale climatic change of at least hemispheric extent (for comprehensive reviews see Broecker and Denton, 1989; Broecker, 1997; and Stocker and Marchal, 2000).

The purpose of this chapter is to summarize briefly the current knowledge of the mechanisms leading to abrupt climate changes. They are a consequence of nonlinearities in the coupled ocean-atmosphere system and give rise to three structurally different responses to perturbations. Irreversibility can occur in the course of such changes. Its relevance for the future evolution of the climate system is discussed.

9.2   Multiple Equilibria in the Earth System

Multiple equilibria in the Earth System have intrigued climate scientists for decades. In the late 1960s and early 1970s, simple energy balance models of the atmosphere were developed in order to estimate the influence of changes in the solar activity (Sellers, 1969) or the consequence of human climate modification (Budyko, 1969). Assuming a particular temperature dependence of the planetary albedo and arguing that seasonal snow cover decreases with increasing temperature, these models indicated that energy balance could be achieved in more than one way for a given value of insolation. The solution exhibited multiple equilibria. Perturbations in the radiative balance of the planet could induce a shift to a fully glaciated world. In this cold state most of the incident solar energy would be reflected back to space. When the perturbations were again removed, the system remained in the cold state. This is a typical hysteresis behavior of a nonlinear physical system (Sellers, 1969).

Likewise, multiple equilibria were also found in simple ocean models (Stommel, 1961; Welander, 1982). Stommel's simple two-box model simulated the basic aspect of the density-driven or thermo-haline circulation of the ocean (THC). The exchange flow between the two boxes, representing low and high latitudes, was assumed to be proportional to the density contrast between these boxes and to be maintained by atmosphere-ocean heat and freshwater fluxes. The use of multiple equilibria reflects the different roles of salt and heat in determining the density of seawater and their different response time scales to atmosphere-ocean flux anomalies. Possible implications of these results for the climate system have been recognized by Stommel (1961), but these attempts were long considered academic games with little reality. Then the advent of time series of climatic change with high temporal resolution from the ocean and the polar ice sheets, along with a fortunate coincidence in ocean modeling progress (Bryan, 1986), convinced the climate community that the existence of multiple equilibria of the ocean circulation was more than a result of theoretical value. It was recognized that these mechanisms have actually played a major role in the history of past climates.

Such equilibria were subsequently found in an entire hierachy of ocean models (Manabe and Stouffer, 1988; Marotzke et al., 1988; Stocker and Wright, 1991; Hughes and Weaver, 1994; Mikolajewicz and Maier-Reimer, 1994; Rahmstorf, 1995). Common to all these results is that a climatic variable (e.g., the meridional oceanic heat flux in the Atlantic) exhibits a hysteresis behavior as a function of a control variable (Figure 9.2). This was first demonstrated with a zonally averaged ocean circulation model, and it was shown that the surface freshwater balance is an important controlling variable for the Atlantic THC (Stocker and Wright, 1991). More-complex models exhibit a richer hysteresis structure (Mikolajewicz and Maier-Reimer, 1994; Rahmstorf, 1995), but the essence is unchanged. The existence of hysteresis has major implications for the reaction of the climate system to perturbations. The location of the initial state on the hysteresis branch is crucial in determining the response. Small perturbations will cause only linear changes of the system. However, if threshold values are crossed, a transition to a different hysteresis branch can occur. The system thus moves to a second equilibrium state that is fundamentally different from the preceding one. This

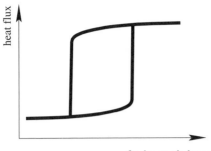

Figure 9.2. The ocean-atmosphere system is a nonlinear physical system that can exhibit hysteresis behavior of the deep circulation in the ocean (Stocker and Wright, 1991). Depending on the surface freshwater balance of the Atlantic Ocean, the meridional heat flux in the Atlantic is not unique and multiple equilibria exist. Changes are linear as long as they remain on the same branch of the hysteresis loop. If certain threshold values in the atmosphere-ocean system are passed, the climate state can change abruptly by switching from one branch to the other. This is a robust feature of the climate system, as demonstrated by the entire hierarchy of climate models.

manifests itself usually as an abrupt change in the structure and strength of the THC. The possibility of irreversibility also exists: although a perturbation is applied for only a finite time, changes in the circulation or, more generally, in the state of the climate system can be permanent.

The upper branch of the hysteresis curve (Figure 9.2) represents the current state of the Atlantic THC. Warm water is flowing from low latitudes as a western boundary current northward (Gulf Stream); then it crosses the Atlantic and reaches the Greenland-Iceland-Norwegian Sea, where much of the heat is given off to the atmosphere. This circulation is most important for the climate of the North Atlantic region in that mean annual temperatures are about 7 °C warmer than at corresponding latitudes in the Pacific. The THC therefore operates like a giant heat pump for this region (Model, 1950). Upon losing its heat, the water becomes denser, sinks, and flows southward at 2 to 3 km depth along the continental margins of the Americas into the Southern Ocean (Stommel, 1958). From the Southern Ocean, deep water then flows into the Indian and Pacific oceans, where it is broadly upwelling and mixing back into the near-surface waters (Gordon, 1986). This global THC has become known as the "great conveyor belt" (Broecker, 1987), and it consists of several paths of water masses with different characteristics (Warren, 1981; Schmitz, 1995). All paleoclimatic indicators known today suggest that the ocean circulation, the conveyor belt, has resided on the upper branch of the hysteresis during the past 10,000 years.

The lower branch of the hysteresis, in its original form, is characterized by a reversed circulation in the Atlantic, that is, the Atlantic has become rather like the Pacific today. This implies that surface air temperatures around the North Atlantic would fall considerably (Manabe and Stouffer, 1988; Mikolajewicz et al., 1997). However, it depends on the model, on parameterizations of mixing and boundary conditions, on what exactly other equilibrium states of the THC look like. Some models exhibit a fully collapsed THC in the Atlantic (Manabe and Stouffer, 1988; Stocker et al., 1992; Mikolajewicz and Maier-Reimer, 1994; Rahmstorf, 1995; Seidov et al., 1996; Schiller et al., 1997; Marchal et al., 1998a), whereas others simulate a reduced overturning (Manabe and Stouffer, 1997) or even a multiplicity of states with different convection sites in the North Atlantic (Rahmstorf, 1995). This suggests that the real ocean-atmosphere system has quite a complex array of different equilibrium states and that Figure 9.2 merely explains the organizing principle.

One of the major challenges of climate modeling is to find out where on the hysteresis branch our current climate state is, and how it will evolve in future times. This then tells us something about the distance of the state to the next threshold value, that is, about the stability of the atmosphere-ocean system.

9.3   The North-South Seesaw

Today, the meridional heat flux in the South Atlantic is actually northward because of the conveyor belt circulation (Rintoul, 1991; Macdonald and Wunsch, 1996). An active THC in the Atlantic draws heat from the Southern Ocean (Crowley, 1992). This is a consequence of the net input of warm surface waters from the Southern Ocean into the south Atlantic and a corresponding flux of colder water exiting the Atlantic at depth as north Atlantic deep water (NADW). Although NADW is warmer than both Antarctic intermediate water and Antarctic bottom water, both flowing into the Atlantic basin, the zonally and depth-averaged heat flux is northward. If the THC is reduced or absent, the Southern Ocean is no longer cooled and can actually warm up. However, because of the huge thermal inertia involved, temperature changes will be small and time scales comparatively long. A very crude estimate of the thermal inertia of the Southern Ocean ($55\,°S-70\,°S$, 4 km depth) is 5×10^{23} J/K. The meridional heat transport into the Atlantic is estimated at 0.25×10^{15} W (Rintoul, 1991). If that heat export stopped, all else unchanged, the Southern Ocean would warm by about $1.6\,°C$ per century (Stocker, 1998). Clearly, the assumption that all other fluxes, in particular the air-sea heat exchange, remain unchanged implies that the above estimate is an upper bound. Furthermore, regional differences in the Southern Ocean are to be expected if, for instance, sea-ice coverage and deepwater formation areas change.

Therefore, if the Atlantic THC switches off, the south should react with a warming, while the north exhibits cooling, particularly strong around the north Atlantic. North and south are coupled and operate as an interhemispheric seesaw (Broecker, 1998). This antiphase behavior has been simulated in a simplified climate model (Stocker et al., 1992). It also shows up in three-dimensional, coupled atmosphere-ocean general circulation models provided that the THC is sufficiently reduced (Manabe and Stouffer, 1988; Schiller et al., 1997). Other models do not show a clear signal (Manabe and Stouffer, 1995).

However, this antiphase behavior was only recently detected in the paleoclimatic archives. A strong and very well-documented cooling event was Younger Dryas, which terminated about 11,500 years ago. One therefore expects that this distinct north-south behavior should then be present if YD was indeed caused by a collapse of the Atlantic

THC. It has long been debated whether YD was actually a global cooling event (Peteet, 1995). Clearly, YD is not restricted to the north Atlantic region; distinct changes during that time can be detected also in the Pacific (Santa Barbara Basin, Kennet and Ingram, 1995) and in the Tropics (Cariaco Basin, Hughen et al., 1996). The finding that a glacier in New Zealand advanced around the time of YD suggested that cooling also occurred in the Southern Hemisphere (Denton and Hendy, 1994), but pollen analyses do not support this (Singer et al., 1998).

Recent results from two ice cores from Antarctica, which were synchronized with an ice core from Greenland, indicate that the north-south antiphase behavior was operating for three distinct events during the last deglaciation. At 14,500 yr BP, an abrupt warming is registered in the Greenland ice core, at which time a moderate cooling event, the Antarctic cold reversal (ACR), starts in the south (Jouzel et al., 1995; Blunier et al., 1997). When the YD cold event begins in the north, ACR comes to a close, and continuous warming all through YD is registered in the Antarctic ice cores. One more time, the antiphase behavior is seen at the termination YD, when the north warms abruptly, at which time the warming in the south stops. There are two earlier climatic events during the glacial that suggest a similar phase relation between north and south (Blunier et al., 1998). At the time of the abrupt warming during the two major Dansgaard/Oeschger events 8 and 12, the south reacts with a cooling. Warming events, coeval with Heinrich cooling events in the north, are registered in marine sediments from the south Atlantic (Vidal et al., 1999). This is also consistent with the north-south seesaw.

However, the picture has recently become more puzzling again. Another ice core from Antarctica (Taylor Dome) exhibits one abrupt event of warming around the time of the abrupt warming transition to the Bolling in the north (Steig et al., 1998). Although this transition appears as abrupt as in the Greenland records and therefore casts doubt on the above considerations, it is also remarkable that a clear Younger Dryas is not registered in this ice core. A tentative conclusion is that the regionality of climatic responses in Antarctica must be studied more thoroughly. In particular, the Southern Ocean with its marginal seas has the potential to strongly influence regional climate through changes in the sea-ice cover and ocean-to-atmosphere heat fluxes.

A few of the abrupt events of the north have a corresponding signal in the south, but others evolve seemingly independently. The simple interhemispheric seesaw as suggested by Broecker (1998) therefore describes an important aspect of the dynamics of abrupt climate changes. But not surprisingly, the climate system is more complicated and many open questions still remain. A hierarchy of climate models will contribute to hypothesis building and will provide an indication of which scenarios are physically possible.

9.4   How Well Are Rapid Transitions Modeled?

Abrupt climate change has been modeled with comprehensive three-dimensional, coupled atmosphere-ocean general circulation models (Manabe and Stouffer, 1995; Fanning and Weaver, 1997; Manabe and Stouffer, 1997; Mikolajewicz et al., 1997; Schiller et al., 1997), with ocean models (Maier-Reimer and Mikolajewicz, 1989;

Mikolajewicz and Maier-Reimer, 1994; Weaver and Hughes, 1994; Rahmstorf, 1995), and with low-order, coupled climate models (Stocker et al., 1992; Wright and Stocker, 1993; Stocker and Wright, 1996; Marchal et al., 1998a). All these models react to perturbation freshwater fluxes applied at high latitudes in the Atlantic Ocean. The response of the THC depends on the location and on the amplitude of the perturbation. The perturbation induces a reduction or even a collapse of the thermo-haline circulation. The latter is referred to as a "polar halocline catastrophe" (Bryan, 1986) and is characterized by the presence of a stable freshwater lid at the surface in high latitudes. This lid prevents water from sinking.

The reduction of deepwater formation results in a decrease of meridional heat transport, and cooling of the high latitudes follows. This cooling is located around the original sinking regions in the northern North Atlantic with amplitudes of $10\,^\circ$C or more. These amplitudes compare well with the paleoclimatic records, and the fact that the center of action is in the north Atlantic is consistent with paleoclimatic evidence. However, in all these models freshwater pulses cause an immediate response of the circulation: no time delay is observed. This is in disagreement with the most recent sea level reconstruction of the last deglaciation (Bard et al., 1996), where it is argued that the first meltwater pulse occurred about 1000 years before the onset of YD, that is, well before the major cooling. None of the present models is capable of simulating this postulated lag of the oceanic response. The sea level reconstruction, however, is only a globally integrated signal and does not indicate the location where this freshwater pulse was discharged into the ocean. It is thus interesting to note that the absence of a strong isotopic signal in the north Atlantic as well as other evidence suggests that the first meltwater event occurred around Antarctica (Clark et al., 1996).

Simulations with a simplified, coupled atmosphere-ocean climate model indicate that the discharge history strongly influences the transient behavior of the THC (Stocker and Wright, 1998). Two examples are shown in Figures 9.3 and 9.4. Both simulations contain a simple feedback mechanism between meltwater discharge and Atlantic meridional heat flux F. Whenever F exceeds 0.1×10^{15} W for the first time, the meltwater discharge into the north Atlantic starts and increases linearly in time. After the THC collapses and $F < 0.1 \times 10^{15}$ W, meltwater discharge stops (Figure 9.3) or starts decreasing linearly in time until it reaches zero (Figure 9.4). In other words, the freshwater anomaly is not prescribed but rather is closely tied to the circulation and thus to the delivery of heat to the high latitudes. This represents the simplest possible feedback mechanism between a climatic variable simulated by the model and the freshwater perturbation. Although we cannot give a quantitative justification for this choice, we note that a binge-purge mechanism was proposed for partly disintegrating ice sheets (MacAyeal, 1993a,b), with the effect that ice discharge rates increase progressively with time.

In principle, there are three possibilities for how the model can react to such a formulation. First, the meltwater discharge could induce a complete and permanent collapse of the THC with no further meltwater. Second, the THC could be weak and switch on and off on a short time scale, with an associated small meltwater discharge around a critical value. The third possibility, exhibited by the model, is the establishment of a quasi-periodic cycle of meltwater discharge, THC collapse, and recovery. The

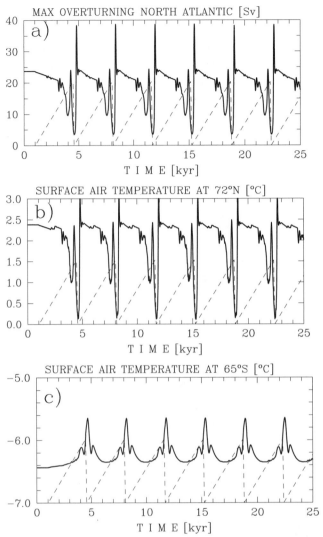

Figure 9.3. (a) Evolution of the maximum overturning in the North Atlantic as a result of a linearly increasing meltwater flux perturbation (dashed line). The thermo-haline circulation is weakening, initially linearly with the perturbation, but is interrupted by some significant, short-term reductions. The maximum meltwater discharge is 0.26 Sv. The thermo-haline circulation abruptly switches back on a few decades to centuries after the perturbation is stopped. (b) Evolution of the zonally averaged surface air temperature at 72°N as a result of changing meridional heat flux associated with the thermo-haline circulation. These changes are in phase with the reduction in thermohaline overturn. (c) The brief collapses of the North Atlantic thermo-haline circulation are associated with short warmings in the south. Amplitudes do not exceed 0.5 °C (from Stocker and Wright, 1998).

transient behavior is very different in the two cases. The sawtooth discharge results in gradual cooling punctuated by shorter cold spells in the north (Figure 9.3a). One such cycle looks surprisingly similar to the isotopic record of the Greenland ice core during the Bolling-Allerod-YD sequence. However, the coldest phase lasts for only a

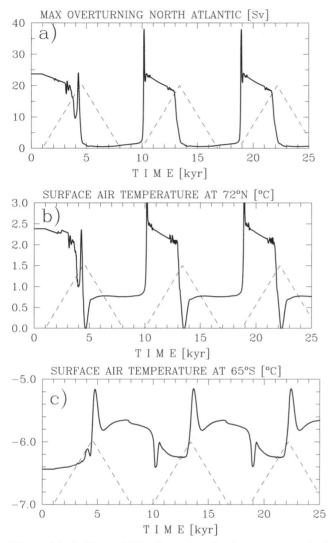

Figure 9.4. As Figure 9.3 but for a triangular freshwater perturbation. The prolonged melting through the time of the collapse of the thermo-haline circulation (a) results in an extended cold phase of a few thousand years in the north (b) with associated warming in the south (c). During the collapse (around the time of the maximum freshwater perturbation) surface air temperatures cool (warm) in the north (south) and relax back to intermediate values on a time scale of a few centuries. This is associated with equilibration of surface and intermediate ocean temperatures after the abrupt changes.

few decades, in contrast to the YD, which lasts for about 1200 years. Note that the south reacts with a smaller warming at the time of the full collapse (Figure 9.3c). This is the north–south seesaw.

The second scenario does show a longer time of collapsed THC, which is caused by sustained but decreasing meltwater fluxes after the collapsed THC (Figure 9.4). Even after the switch-off of the meltwater discharge, the circulation recovers only after about 2000 years. Note that in this case the cooling, preceding the abrupt cold period, is not punctuated by short cold spells (except at the beginning of the first cycle). Again, warming is simulated in the south when the Atlantic THC is switched off. These two examples illustrate that the reaction of the Atlantic THC depends very strongly on the temporal history of the freshwater flux, its amplitude as well as the location. Common to both experiments is the fact that the warming is always abrupt. This is the result of a convective feedback bringing warm and saline waters to the surface when convection begins in high latitudes (Wright and Stocker, 1991). These water masses then lose heat very quickly (note the overshoot in surface air temperature) and so enhance convection again. This positive feedback is not operating during the gradual cooling phase of the cycle because it is caused by an advective spin-down of the circulation (Bryan, 1986). This important feature – gradual cooling and abrupt warming – is common to all 24 Dansgaard/Oeschger events during the last glacial. Similar discharge profiles can induce very different responses depending, as well, on the initial state of the circulation. This implies that for realistic paleoclimatic simulations, we would require a much more detailed knowledge of the history of deglaciation and freshwater events during the last glacial. Not only the locations but also the temporal evolution of the volumes of freshwater should be reconstructed, if at all possible.

In addition to a qualitative check of the simulations of abrupt events with temperature reconstructions from ice cores, it is useful to investigate whether the simulations also reproduce other indicators of climate change. By coupling a prognostic biogeochemical component to the model (Marchal et al., 1998b), changes in the ^{13}C distribution and atmospheric CO_2 concentration can be simulated (Marchal et al., 1998a). It is found that changes of ^{13}C agree qualitatively with paleoceanographic reconstructions and that a long-standing controversy about the presence or absence of NADW during NADW in the Southern Ocean can be resolved. It turns out that ^{13}C is a faithful NADW indicator only in the North Atlantic, but that farther south, other effects (gas exchange and organic matter production) strongly influence the signal. Atmospheric CO_2, on the other hand, shows a small reduction at the beginning of a cold event but then increases steadily throughout the ongoing cold event (Marchal et al., 1999). This is consistent with CO_2 reconstructions from an Antarctic ice core during YD.

Although these results are very encouraging, it is disturbing that changes in the atmospheric concentration of radiocarbon, ^{14}C, during the termination of the last glacial have not yet been equally successfully modeled. The reconstruction clearly indicates increasing ^{14}C at the beginning of YD, with a peak about 200 years into YD (Hughen et al., 1998). During the following 1000 years – that is, during most of YD – ^{14}C decreases again. Model simulations, on the other hand, suggest that ^{14}C increases throughout YD and decreases only at the termination of YD (Mikolajewicz, 1996; Stocker and Wright, 1996). This is because of a sustained, full collapse of the THC. The data, however, strongly suggest that ventilation must already have recovered soon after the beginning of YD somewhere in the ocean, either in the Southern Ocean or by way of intermediate water formation in the Atlantic. Currently, this is not simulated by the available models.

9.5   Conclusions

Nonlinearities are a common feature of the Earth System. They occur in even the simplest formulation of planetary energy balance because of the dependence of albedo on surface air temperature. In the atmosphere-ocean system, nonlinearities can give rise to hysteresis behavior of the thermo-haline circulation. This is well documented by models. The existence of multiple equilibria also has implications for the evolution of future climate. Simulations by three-dimensional, coupled atmosphere-ocean general circulation models have indicated that the THC can be reduced, or even shut off, when atmospheric CO_2 concentrations continue to rise (Manabe and Stouffer, 1993, 1994). A thorough exploration of parameter space, however, is necessary because the global climate sensitivity (the global mean temperature increase for double CO_2, $\Delta T_{2\leftarrow}$) or the evolution of future CO_2 increase is poorly known.

Using a simplified coupled climate model, the results of Manabe and Stouffer (1994) could be reproduced, that is, the final CO_2 concentration in the atmosphere represents a threshold value beyond which the Atlantic THC shuts off (Figure 9.5). In addition, the rate of CO_2 increase, and hence the history of warming, involves threshold values (Stocker and Schmittner, 1997; Schmittner and Stocker, 1999; Stouffer and Manabe, 1999). A faster CO_2 increase leads to a faster warming of the surface of the ocean. Because downward mixing of heat is limited by ocean dynamics, the upper layers heat up more, and thus their density decreases correspondingly. If the density decrease is large enough, deepwater formation is reduced or inhibited altogether.

For a CO_2 increase of 1%/yr, typical of values in the decade 1980–90, and a climate sensitivity $\Delta T_{2\leftarrow} = 3.7\,^{\circ}C$ as in the model of Manabe and Stouffer (1993), the THC shuts off between 650 and 700 ppm (Figure 9.6). For current rates (about 0.5%/yr) and the most probable climate sensitivity of $\Delta T_{2\leftarrow} = 2.5\,^{\circ}C$, according to IPCC (1996), the critical threshold is much higher: between 1500 and 1600 ppm in this model. Based on results from many climate models and an understanding of the basic processes in the climate system, it is clearly not warranted to assume that the climate system would react in a linear fashion to perturbations irrespective of their amplitude. In particular, the dynamical coupling between atmosphere and ocean gives rise to multiple equilibria, which, in principle, imply the possibility of irreversible changes once thresholds are passed. One possibility of such a change is a shutdown of the Atlantic thermo-haline circulation as evidenced by an entire hierarchy of models. This would have profound implications for the climate in the North Atlantic region because it involves a large-scale modification of the circulation patterns and the heat balance in this ocean basin. Furthermore, it would affect the long-term uptake of CO_2 because NADW formation is a conduit of anthropogenic carbon into the deep sea (Joos et al., 1999).

One of the major tasks in climate modeling will be the investigation and assessment of such thresholds in the Earth System. Predicting such changes, that is, when and which thresholds will be crossed, is difficult, if not impossible, not primarily because of our limited understanding of the Earth System or limited computer resources but because the Earth is a truly nonlinear dynamical system (Lorenz, 1984).

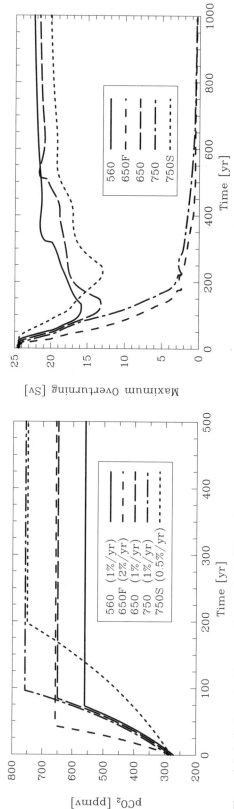

Figure 9.5. Left: Prescribed increase of atmospheric CO_2 until a maximum concentration is reached. The rates of increase are 1%/yr (exps. 560, 650, 750) and are slightly higher than the increase of CO_2 in the 1980s. Additional experiments are performed with 0.5%/yr (exp. 750S) and 2%/yr (exp. 650F). Right: Evolution of the overturning volume transport of the thermo–haline circulation in the North Atlantic. For an increase of 1%/yr the threshold value is between 650 ppm and 750 ppm. When that is passed, the thermo–haline circulation decreases and a new stable state is reached. The circulation recovers if the CO_2 increase is slower (exp. 750S), or it collapses if it is faster (exp. 650F) (from Stocker and Schmittner, 1997).

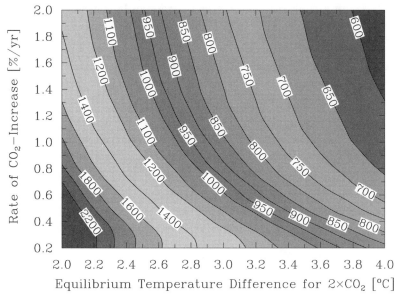

Figure 9.6. Threshold concentration of atmospheric CO_2 (in ppmv) beyond which the Atlantic thermo–haline circulation breaks down permanently as a function of the equilibration surface air temperature difference for a doubling of CO_2, $\Delta T_{2\leftarrow}$ (sensitivity of the climate model), and the rate of CO_2 increase in % per year. For the most likely climate sensitivity $\Delta T_{2\leftarrow} = 2.5\,°C$, IPCC, 1996) and the current rate of CO_2 increase of 0.5%/yr, the threshold lies between 1500 and 1600 ppmv in this model. As for any climate model, the actual values depend on model parameters (from Stocker and Schmittner, 1997). See also color plate section.

ACKNOWLEDGMENT

I thank the organizers of the workshop for a very stimulating week in a very special place. This chapter is dedicated to the late Hans Oeschger, whose continued interest in the Earth System was an inspiration to all. I was fortunate to exchange ideas and experience the vision of Hans during his last years in Bern. He was always a generous, enthusiastic, and discreet mentor. Support by the Swiss National Science Foundation is gratefully acknowledged.

REFERENCES

Bard, E., Hamelin, B., Arnold, M., Montaggioni, L., Cabioch, G., Faure, G., and Rougerie, F. (1996). Deglacial sea-level record from Tahiti corals and the timing of global meltwater discharge. *Nature*, **382**, 241–244.

Behl, R. J., and Kennet, J. P. (1996). Brief interstadial events in the Santa Barbara basin, NE Pacific, during the past 60 kyr. *Nature*, **379**, 243–246.

Blunier, T., Chappellaz, J., Schwander, J., Dällenbach, A., Stauffer, B., Stocker, T. F., Raynaud, D., Jouzel, J., Clausen, H. B., Hammer, C. U., and Johnsen, S. J. (1998). Asynchrony of Antarctic and Greenland climate change during the last glacial period. *Nature*, **394**, 739–743.

Blunier, T., Schwander, J., Stauffer, B., Stocker, T., Dällenbach, A., Indermühle, A., Tschumi, J., Chappellaz, J., Raynaud, D., and Barnola, J.-M. (1997). Timing of temperature variations during the last deglaciation in Antarctica and the atmospheric CO_2 increase with respect to the Younger Dryas event. *Geophys. Res. Lett.*, **24**, 2683–2686.

Bond, G., Broecker, W., Johnsen, S., McManus, J., Labeyrie, L., Jouzel, J., and Bonani, G. (1993). Correlations between climate records for North Atlantic sediments and Greenland ice. *Nature*, **365**, 143–147.

Bond, G., Heinrich, H., Broecker, W., Labeyrie, L., McManus, J., Andrews, J., Huon, S., Jantschik, R., Clasen, S., Simet, C., Tedesco, K., Klas, M., Bonani, G., and Ivy, S. (1992). Evidence for massive discharges of icebergs into the North Atlantic ocean during the last glacial period. *Nature*, **360**, 245–249.

Broecker, W. S. (1987). The biggest chill. *Natural History*, **96**, 74–82.

Broecker, W. S. (1997). Thermohaline circulation, the Achilles heel of our climate system: will man-made CO_2 upset the current balance? *Science*, **278**, 1582–1588.

Broecker, W. S. (1998). Paleocean circulation during the last deglaciation: a bipolar seesaw? *Paleoceanogr.*, **13**, 119–121.

Broecker, W. S., and Denton, G. H. (1989). The role of ocean-atmosphere reorganizations in glacial cycles. *Geochim. Cosmochim. Acta*, **53**, 2465–2501.

Broecker, W. S., Peteet, D. M., and Rind, D. (1985). Does the ocean-atmosphere system have more than one stable mode of operation? *Nature*, **315**, 21–25.

Bryan, F. (1986). High-latitude salinity effects and interhemispheric thermohaline circulations. *Nature*, **323**, 301–304.

Budyko, M. I. (1969). The effect of solar radiation variations on the climate of the earth. *Tellus*, **21**, 611–619.

Clark, P. U., Alley, R. B., Keigwin, L. D., Licciardi, J. M., Johnsen, S. J., and Wang, H. (1996). Origin of the first global meltwater pulse. *Paleoceanography*, **11**, 563–577.

Crowley, T. J. (1992). North Atlantic deep water cools the southern hemisphere. *Paleoceanography*, **7**, 489–497.

Dansgaard, W., Johnsen, S. J., Clausen, H. B., Dahl-Jensen, D., Gundestrup, N., Hammer, C. U., and Oeschger, H. (1984). North Atlantic climatic oscillations revealed by deep Greenland ice cores. In: *Climate Processes and Climate Sensitivity (Geophys. Monogr. Ser.)* [J. E. Hansen and T. Takahashi (eds.)]. Am. Geophys. Union, Washington, 288–298.

Dansgaard, W., Johnsen, S. J., Clausen, H. B., Dahl-Jensen, D., Gundestrup, N. S., Hammer, C. U., Hvidberg, C. S., Steffensen, J. P., Sveinbjörnsdottir, A. E., Jouzel, J., and Bond, G. (1993). Evidence for general instability of past climate from a 250-kyr ice-core record. *Nature*, **364**, 218–220.

Dansgaard, W., White, J. W. C., and Johnsen, S. J. (1989). The abrupt termination of the Younger Dryas climate event. *Nature*, **339**, 532–534.

Denton, G. H., and Hendy, C. H. (1994). Younger Dryas age advance of Franz Josef glacier in the southern alps of New Zealand. *Science*, **264**, 1434–1437.

Fanning, A. F., and Weaver, A. J. (1997). Temporal-geographical influences on the North Atlantic conveyor: Implications for the Younger Dryas. *Paleoceanography*, **12**, 307–320.

Gordon, A. L. (1986). Interocean exchange of thermocline water. *Journal of Geophysical Research*, **91**, 5037–5046.

Grimm, E. C., Jacobson G. L. Jr., Watts, W. A., Hanson, B. C. S., and Maasch, K. A. (1993). A 50,000-year record of climate oscillations from Florida and its temporal correlation with the Heinrich events. *Science*, **261**, 198–200.

Hughen, K. A., Overpeck, J. T., Lehman, S. C., Kashgarian, M., Southon, J., Peterson, L. C., Alley, R., and Sigman, D. M. (1998). Deglacial changes in ocean circulation from an extended radiocarbon calibration. *Nature*, **391**, 65–68.

Hughen, K. A., Overpeck, J. T., Peterson, L. C., and Trumbore, S. (1996). Rapid climate changes in the tropical Atlantic region during the last deglaciation. *Nature*, **380**, 51–54.

Hughes, T. M. C., and Weaver, A. J. (1994). Multiple equlibria of an asymmetric two-basin model. *Journal of Physical Oceanography*, **24**, 619–637.

Imbrie, J., et al. (1992). On the structure and origin of major glaciation cycles, 1. Linear responses to Milankovitch forcing. *Paleoceanography*, **7**, 701–738.

IPCC (1996): *Climate Change 1995: The Science of Climate Change*. Cambridge University Press, 572.

Joos, F., Plattner, G.-K., Stocker, T. F., Marchal, O.; and Schmittner, A. (1999). Global warming and marine carbon cycle feedbacks on future atmospheric CO_2. *Science*, **284**, 464–467.

Jouzel, J., Vaikmae, R., Petit, J. R., Martin, M., Duclos, Y., Stievenard, M., Lorius, C., Toots,

M., Mélières, M. A., Burckle, L. H., Barkov, N. I., and Kotlyakov, V. M. (1995). The two-step shape and timing of the last deglaciation in Antarctica. *Cliamte Dynamics*, **11**, 151–161.

Kennet, J. P., and Ingram, L. B. (1995). A 20,000-year record of ocean circulation and climate change from the Santa Barbara basin. *Nature*, **377**, 510–514.

Lehman, S. J., and Keigwin, L. D. (1992). Sudden changes in North Atlantic circulation during the last deglaciation. *Nature*, **356**, 757–762.

Lorenz, E. N. (1984). Irregularity: a fundamental property of the atmosphere. *Tellus, 36A*, 98–110.

MacAyeal, D. R. (1993a). Binge/purge oscillations of the Laurentide ice sheet as a cause of the North Atlantic's Heinrich events. *Paleoceanography*, **8**, 767–773.

MacAyeal, D. R. (1993b). A low-order model of the Heinrich event cycle. *Paleoceanography*, **8**, 767–773.

Macdonald, A. M., and Wunsch, C. (1996). An estimate of global ocean circulation and heat fluxes. *Nature*, **382**, 436–439.

Maier-Reimer, E., and Mikolajewicz, U. (1989). Experiments with an OGCM on the cause of the Younger Dryas. In: *Oceanography 1988* [A. Ayala-Castanares, W. Wooster and A. Yanez-Aranciba (eds.)]. UNAM Press, Mexico D.F., 207.

Manabe, S., and Stouffer, R. J. (1988). Two stable equilibria of a coupled ocean atmosphere model. *Journal of Climate*, **1**, 841–866.

Manabe, S., and Stouffer, R. J. (1993). Century-scale effects of increased atmospheric CO_2 on the ocean-atmosphere system. *Nature*, **364**, 215–218.

Manabe, S., and Stouffer, R. J. (1994). Multiple-century response of a coupled ocean-atmosphere model to an increase of atmospheric carbon dioxide. *Journal of Climate*, **7**, 5–23.

Manabe, S., and Stouffer, R. J. (1995). Simulation of abrupt climate change induced by freshwater input to the North Atlantic Ocean. *Nature*, **378**, 165–167.

Manabe, S., and Stouffer, R. J. (1997). Coupled ocean-atmosphere model response to freshwater input: comparison to Younger Dryas event. *Paleoceanography*, **12**, 321–336.

Marchal, O., Stocker, T. F., and Joos, F. (1998a). Impact of oceanic reorganizations on the ocean carbon cycle and atmospheric carbon dioxide content. *Paleoceanography*, **13**, 225–244.

Marchal, O., Stocker, T. F., and Joos, F. (1998b). A latitude-depth, circulation biogeochemical ocean model for paleoclimate studies. Model development and sensitivities. *Tellus*, **50B**, 290–316.

Marchal, O., Stocker, T. F., Joos, F., Indermühle, A., Blunier, T., and Tschumi, J. (1999). Modelling the concentration of atmospheric CO_2 during the Younger Dryas climate event. *Climate Dynamics*, **15**, 341–354.

Marotzke, J., Welander, P., and Willebrand, J. (1988). Instability and multiple equilibria in a meridional-plane model of the thermohaline circulation. *Tellus*, **40A**, 162–172.

Mikolajewicz, U. (1996). A meltwater-induced collapse of the "conveyor belt" thermohaline circulation and its influence on the distribution of ^{14}C and ^{18}O in the oceans. Hamburg, Max Planck Institut für Meteorologie: 1–25.

Mikolajewicz, U., Crowley, T. J., Schiller, A., and Voss, R. (1997). Modelling teleconnections between the North Atlantic and North Pacific during the Younger Dryas. *Nature*, **387**, 384–387.

Mikolajewicz, U., and Maier-Reimer, E. (1994). Mixed boundary conditions in ocean general circulation models and their influence on the stability of the model's conveyor belt. *Journal of Geophysical Research*, **99**, 22633–22644.

Model, F. (1950). Warmwasserheizung Europas. *Ber. Deut. Wetterdienst*, **12**, 51–60.

Oeschger, H., Beer, J., Siegenthaler, U., Stauffer, B., Dansgaard, W., and Langway, C. C. (1984). Late glacial climate history from ice cores. In: *Climate Processes and Climate Sensitivity (Geophys. Monogr. Ser.)* [J. E. Hansen and T. Takahashi (eds.)]. Am. Geophys. Union, Washington, 299–306.

Peteet, D. (1995). Global Younger Dryas? *Quaternary International*, **28**, 93–104.

Rahmstorf, S. (1995). Bifurcations of the Atlantic thermohaline circulation in response to changes in the hydrological cycle. *Nature*, **378**, 145–149.

Rintoul, S. R. (1991). South Atlantic interbasin exchange. *Journal of Geophysical Research*, **96**, 2675–2692.

Ruddiman, W. F., and McIntyre, A. (1981). The mode and mechanism of the last deglaciation: oceanic evidence. *Quaternary Research*, 16, 125–134.

Schiller, A., Mikolajewicz, U., and Voss, R. (1997). The stability of the North Atlantic thermohaline circulation in a coupled ocean-atmosphere general circulation model. *Climate Dynamics*, 13, 325–347.

Schmittner, A., and Stocker, T. F. (1999). The stability of the thermohaline circulation in global warming experiments. *Journal of Climate*, 12, 1117–1133.

Schmitz, W. (1995). On the interbasin-scale thermohaline circulation. *Reviews of Geophysics*, 33, 151–173.

Seidov, D., Sarnthein, M., Stattegger, K., Prien, R., and Weinelt, M. (1996). North Atlantic ocean circulation during the last glacial maximum and subsequent meltwater event: a numerical model. *Journal of Geophysical Research*, 101, 16305–16332.

Sellers, W. D. (1969). A global climate model based on the energy balance of the earth-atmosphere system. *Journal of Applied Meteorology*, 8, 392–400.

Singer, C., Shulmeister, J., and McLea, B. (1998). Evidence against a significant Younger Dryas cooling event in New Zealand. *Science*, 281, 812–814.

Steig, E. J., Brook, E. J., White, J. W. C., Sucher, C. M., Bender, M. L., Lehman, S. J., Morse, D. L., Waddington, E. D., and Clow, G. D. (1998). Synchronous climate changes in Antarctica and the North Atlantic. *Science*, 282, 92–95.

Stocker, T. F. (1998). The seesaw effect. *Science*, 282, 61–62.

Stocker, T. F., and Marchal, O. (2000). Abrupt climate change in the computer: Is it real? *Proc. U.S. Natl. Acad. Sci.*, 97, 1362–1365.

Stocker, T. F., and Schmittner, A. (1997). Influence of CO_2 emission rates on the stability of the thermohaline circulation. *Nature*, 388, 862–865.

Stocker, T. F., and Wright, D. G. (1991). Rapid transitions of the ocean's deep circulation induced by changes in surface water fluxes. *Nature*, 351, 729–732.

Stocker, T. F., and Wright, D. G. (1996). Rapid changes in ocean circulation and atmospheric radiocarbon. *Paleoceanography*, 11, 773–796.

Stocker, T. F., and Wright, D. G. (1998). The effect of a succession of ocean ventilation changes on radiocarbon. *Radiocarbon*, 40, 359–366.

Stocker, T. F., Wright, D. G., and Mysak, L. A. (1992). A zonally averaged, coupled ocean-atmosphere model for paleoclimate studies. *Journal of Climate*, 5, 773–797.

Stommel, H. (1958). The abyssal circulation. *Deep Sea Research*, 5, 80–82.

Stommel, H. (1961). Thermohaline convection with two stable regimes of flow. *Tellus*, 13, 224–230.

Stouffer, R. J., and Manabe, S. (1999). Response of a coupled ocean-atmosphere model to increasing atmospheric carbon dioxide: sensitivity to the rate of increase. *Journal of Climate*, 12, 2224–2237.

Taylor, K. C., Lamorey, G. W., Doyle, G. A., Alley, R. B., Grootes, P. M., Mayewski, P. A., White, J. W. C., and Barlow, L. K. (1993). The "flickering switch" of late Pleistocene climate change. *Nature*, 361, 432–436.

Vidal, L., Schneider, R., Marchal, O., Bickert, T., Stocker, T. F., and Wefer, G. (1999). Link between the North and South Atlantic during the Heinrich events of the last glacial period. *Climate Dynamics*, 15, 909–919.

Warren, B. A. (1981). Deep circulation of the world ocean. In: *Evolution of Physical Oceanography – Scientific Surveys in Honor of Henry Stommel* [B. A. Warren and C. Wunsch (eds.)]. MIT Press, 6–41.

Weaver, A. J., and Hughes, T. M. C. (1994). Rapid interglacial climate fluctuations driven by North Atlantic ocean circulation. *Nature*, 367, 447–450.

Welander, P. (1982). A simple heat-salt oscillator. *Dynamics of Atmospheres and Oceans*, 6, 233–242.

Wright, D. G., and Stocker, T. F. (1991). A zonally averaged ocean model for the thermohaline circulation, Part I: Model development and flow dynamics. *Journal of Physical Oceanography*, 21, 1713–1724.

Wright, D. G., and Stocker, T. F. (1993). Younger Dryas Experiments. In: *Ice in the Climate System* [W. R. Peltier (ed.)]. Springer-Verlag, Heidelberg, 395–416.

10 Simulations of the Climate of the Holocene

Perspectives Gained with Models of Different Complexity

J. E. KUTZBACH

ABSTRACT

The Earth's climate has changed significantly in the past 10,000 years. These climate changes were associated with changes of incoming solar radiation caused by orbital changes. The climatic response to the change in external forcing involved major adjustments of the atmosphere, ocean, cryosphere, and biosphere. Because these Earth System changes are well documented in time and space, there are opportunities to compare the observed behavior of the system to the results of simulations by numerical models.

This chapter reviews progress in climate simulations over a two-decade interval as models have been developed to increase the coupling among system components: first atmosphere models, then atmosphere-ocean-cryosphere models; and, recently, atmosphere-ocean-cryosphere-biosphere models. The scope of studies has also expanded from an initial emphasis on simulating the average climatic conditions of different epochs to a current emphasis on the simulation of climate variability at interannual to century time scales, and abrupt climatic changes, as a function of changing external forcing. In most instances the simulations have shown that ocean, cryosphere, and biosphere feedbacks can amplify the climate's response to changes of external forcing. Including more system components has generally led to improved agreement between observations and simulations. However, so far models have been unable to simulate the full magnitude, and the spatial and temporal structure, of Holocene climate change. These results also help underscore the importance of including all Earth System components in models being used to assess possible changes of climate in the future, changes that could be caused by human-related activities such as the burning of fossil fuels and deforestation.

10.1 Introduction

There are several reasons for studying atmosphere-ocean-biosphere interactions and the resultant climate during the Holocene: (1) observational studies and calculations with climate models indicate that changes in Earth's orbital geometry and the consequent precisely known changes in the seasonal and latitudinal distribution of solar radiation caused large climate changes; (2) the observed changes in the climate and the terrestrial biosphere during the Holocene were large in both spatial scale and magnitude, and therefore it should be possible to simulate these changes with current-generation climate models; and (3) data sets (terrestrial vegetation, lake status, indicators of sea-surface temperature and sea-ice conditions) are being assembled to permit increasingly detailed descriptions of the state of Holocene climates; these descriptions will aid in examination of mechanisms of climate change and geosphere-biosphere interactions, and in testing of the accuracy of model simulations.

During the Holocene, perihelion shifted from northern summer at 11 ka (11 thousand years ago) to northern fall at 6 ka to northern winter at 0 ka (Berger, 1978). Thus, the Northern Hemisphere seasonal insolation cycle was enhanced at 11 ka relative to 0 ka, and the month of largest positive insolation anomaly then moved from summer to late summer and into fall around 6 ka. Simultaneously, the Southern Hemisphere experienced a reduced seasonal insolation cycle at 11 ka, compared with 0 ka, and the month of largest negative insolation anomaly then moved into southern spring. The magnitude of these seasonal changes was as much as 25–30 Watts/m^2, representing changes of 7–8% in solar radiation at 11 ka. Averaged over the annual cycle, the global and latitudinal-average insolation was unchanged by these shifts in the season of perihelion. In contrast, the slightly increased tilt of the Earth's rotational axis at 11 ka, relative to modern, caused high-latitude insolation to increse in summer and produced a small annual-average increase of insolation of several Watts/m^2 near the poles and a slight decrease of insolation in the tropics; here also, the global average insolation was unchanged. The carbon dioxide concentration of the atmosphere was about 280 ppm at 11 ka and remained approximately constant at that level until about A.D. 1800, when the increase associated with burning of fossil fuels began (Raynaud et al., 1993). Some remnants of the large Northern Hemisphere ice sheets that had existed at the Last Glacial Maximum (LGM), about 21 ka, remained in North America and Europe at 11 ka (~30% of the LGM ice volume), but almost all this excess ice had melted by 6 ka.

This chapter will review the considerable array of Holocene climate simulations undertaken in the past 20 years. The review illustrates that in spite of progress, many questions remain unanswered about the magnitude, spatial structure, and temporal character of the response of the Earth System to the orbitally caused changes in incoming solar radiation. In some regions, terrestrial biosphere-climate interactions appear to amplify the climate change caused by the orbital forcing. Ocean-atmosphere interactions appear to amplify climate changes in some areas and reduce changes in other areas. In many regions, the changes in Holocene climate simulated by climate models are not as large as the observed changes. This discrepancy, now documented for Holocene climates, raises the possibility that current climate models may also underestimate future climate change.

Abrupt climate changes in the Holocene are not easily explained by orbital forcing alone. However, positive feedbacks in the atmosphere-biosphere-ocean system may have caused climate to change more rapidly, in response to slow changes in orbital forcing, than would otherwise be expected. With the advent of new climate model configurations permitting efficient century or multicentury simulations, it is now possible to examine whether different phases of the Holocene might have had different variability at interannual, decadal, or centennial time scales; these studies may provide useful perspectives as we consider the possible character of decade/century climate variability in the present and future.

10.2   Historical Notes

This chapter focuses primarily on simulations of Holocene climate with dynamical climate models, but it is important to recognize that energy budget modeling studies and observational studies pointed to the role of orbital forcing in Holocene climate change long before studies with dynamical climate models commenced. Using a greatly simplified energy budget climate model, Milankovitch (1920) found that northern summers could have been as much as 5 °C warmer and winters 5 °C colder when perihelion occurred in northern summer. Köppen and Wegener (1924), collaborating with Milankovitch, argued that the recession of the ice sheets and the occurrence of a thermal maximum in northern Europe in the early Holocene were linked to the orbitally forced maximum in northern summer insolation. However, this early support for the orbital theory was later challenged, and interest in this subject waned for several decades (see review in Kutzbach, 1983).

With the development of radiometric dating techniques, the close linkage between the timing of orbital changes and the timing of ice volume changes, as reflected in oxygen isotope records from marine sediment cores, firmly established that orbital changes were "pacemakers" of the glacial-interglacial cycles (Hays et al., 1976). At about the same time, there was renewed interest in the use of improved energy budget climate models to simulate the climate's response to orbital forcing. Suarez and Held (1976) showed that orbital changes could cause large changes in land temperature and sea-ice cover and, using a highly parameterized hydrologic model, large changes in snow cover in high latitudes (see also Schneider and Thompson, 1979). The strong sensitivity of high-latitude climate to orbital forcing was also demonstrated in the first simulations with general circulation models (Mitchell, 1977; see Mason, 1976). Somewhat later, general circulation models were used to demonstrate the sensitivity of tropical monsoon climates to orbital forcing (Kutzbach, 1981). Increasingly, observational studies also linked orbital forcing to high-latitude early Holocene warmth (Ritchie et al., 1983), mid-latitude warmth and dryness (Wright, 1987; Webb et al., 1987), and tropical monsoon enhancement (Prell, 1984a,b; Kutzbach and Street-Perrot, 1985; Wright et al., 1993). The subsequent evolution of these studies is described later.

10.3   Response of Climate to Orbital Forcing in Relation to Increasing Model Complexity

Climate models of different complexity have been used to study the response of the climate system to changes of insolation. This review describes both the similarities and the differences in results obtained from stepwise increases in model complexity. It also provides examples of the roles of individual system components: atmosphere, mixed-layer ocean, land surface and vegetation, and dynamical ocean.

10.3.1 Atmospheric Models with Prescribed SST and Sea Ice

The first simulations of the response of monsoon circulations to orbital forcing in the early Holocene (9 ka) used atmospheric dynamical models with all surface boundary conditions (except land surface temperature) fixed at modern values (Kutzbach, 1981; Kutzbach and Otto-Bliesner, 1982; Kutzbach and Guetter, 1986). The use of modern sea surface temperatures (SSTs) was based upon the assumption that SSTs would undergo only relatively small changes in response to orbital forcing because of the relatively larger thermal inertia of ocean compared with land. This assumption, in part also a necessity with early climate models, ignored the possible role of high-latitude sea-ice and snow-cover feedbacks and the role of changes in ocean upwelling in the Tropics (Kutzbach and Otto-Bliesner, 1982). This simple atmospheric model configuration showed that the enhanced Northern Hemisphere seasonal insolation cycle produced enhanced warming of the northern continents in summer ($2\,°C$ or more), enhanced cooling in winter ($2\,°C$ or more), and enhanced northern summer and winter monsoons. The summer monsoon enhancement exceeded that of winter because of the stronger water-vapor feedback in the warm season. Thus, summer rains increased more than winter rains decreased. Summer rainfall increases were of the order of 10% over that of the northern continents (Kutzbach and Otto-Bliesner, 1982). The magnitude of this summer monsoon response decreased as perihelion moved from summer (11 ka) into fall (6 ka) and past the maximum in the seasonal temperature cycle (Kutzbach and Guetter, 1986). Annual-average surface temperatures were unchanged, compared with modern, because summer warming was almost exactly balanced by winter cooling. Annual-average precipitation, however, was increased in northern tropical monsoon lands because of the nonlinearity of the water-vapor response. More-recent simulations have employed models of increased horizontal and vertical resolution, thereby allowing more-accurate descriptions of changes in jet stream structures, three-dimensional flow patterns, transient eddies, and model energetics (Hall and Valdes, 1997; de Noblet et al., 1996; Masson and Joussaume, 1997).

The Paleoclimatic Modeling Intercomparison Project (PMIP) has compared the results of simulations of 6 ka climate by 18 different general circulation models; the models all prescribed SSTs at modern values but employed improved parameterizations of soil moisture and land surface hydrology compared with earlier studies (Joussaume et al., 1999). All the simulations produced the general increase of temperature and summer monsoon precipitation over Northern Hemisphere land, proving that the overall response of climate models to orbital forcing is robust. However, the summertime warming of the Asian continent ranged from about $0.5\,°C$ to about $2\,°C$ among the 18 models, and the increase in northern summer monsoon rainfall ranged from near zero to about $2.5\ \mathrm{mm/day}$ (India) and $1.5\ \mathrm{mm/day}$ (Africa). The differences among the models were traced, in part, to differences in planetary albedo and cloud feedback processes. Net radiation increases at the top of the atmosphere over Asia in June, July, and August (JJA) ranged from $9\ \mathrm{W/m^2}$ to $16\ \mathrm{W/m^2}$ among the 18 models, compared with the increase in insolation of $\sim 22\ \mathrm{W/m^2}$ (Joussame et al., 1999). The task of assessing which of the simulations is most accurate, and why, remains.

10.3.2 Atmospheric Models Coupled to Swamp Oceans or Mixed-Layer Oceans

The next phase of model development added swamp oceans or mixed-layer oceans with simple thermodynamic sea-ice models, and improved soil moisture, runoff, and snow hydrology parameterizations (Mitchell, 1977; Kutzbach and Gallimore, 1988;

Mitchell et al., 1988; Liao et al., 1994; Phillips and Held, 1994). The simulations of Holocene climate made with these additional system components exhibited nonlinear effects of sea-ice and snow interactions. In high northern latitudes, the orbitally forced warmer summers and falls led to a later onset of sea-ice formation. With less sea ice and more open water, compared with modern, the surface albedo was lowered, and absorbed solar radiation was increased – a positive feedback. Especially over the Arctic, this effect produced additional warming in summer and fall, and in winter the ice was thinner and covered a smaller area. Spring melt-back was slightly delayed, but this effect was secondary. As a result of thinner ice and more leads, the flux of heat from ocean to atmosphere was increased, and winter temperatures were considerably warmer in higher northern latitudes than for modern conditions. The nonlinear and somewhat counterintuitive response of sea ice to the altered insolation forcing helped produce simulations that were in better agreement with observational evidence (Klimanov, 1984) than the earlier simulations with prescribed SSTs and sea ice. Atmospheric dynamical processes (horizontal advection and mixing by cyclones/anticyclones) were responsible for mixing some of the enhanced Arctic warmth onto the northern continents. Recent work illustrates an additional complexity of sea-ice response to orbital forcing. Compared with the strongly positive feedback exhibited by static thermodynamic sea-ice models, the addition of sea-ice dynamics to the models causes a small negative feedback that reduces the overall sensitivity of ice thickness to orbital forcing (Vavrus, 1999).

In middle and low latitudes, inserting an interactive soil moisture submodel in place of prescribed soil wetness triggered soil moisture feedbacks that acted to promote moisture recycling (a positive feedback). As a result, northern monsoon regions experienced larger precipitation increases due to orbital forcing than occurred in models with prescribed soil moisture (Gallimore and Kutzbach, 1989; Phillips and Held, 1994). Some northern continental interiors became even warmer and drier in the summer because of orbital forcing caused by increased evaporative losses and lowered soil moisture.

The maximum response of surface climate to orbital forcing clearly shifts forward in the annual cycle as perihelion shifts forward in season between 11 ka and 6 ka (Kutzbach et. al., 1998). Thus, the maximum *increase* in temperature and monsoon precipitation shifts from July at 11 ka to September at 6 ka; the total increase is less at 6 ka than at 11 ka.

10.3.3 Climate Models with Terrestrial Biosphere Interactions

Observational evidence indicates that significant changes of vegetation occurred in high, middle, and low latitudes during the Holocene (Wright et al., 1993). The boreal forest shifted northward, replacing tundra; grasslands expanded eastward in North America, replacing forest; and grasslands shifted northward in North Africa, replacing desert. Models have begun to be used to test how these changes in vegetation might influence the climate.

One-way Forcing: Vegetation Sensitivity Experiments. An indication of the potential response of the terrestrial biosphere to orbitally forced climate change has been gained by using the output of climate models – namely, the seasonal cycle of solar radiation, temperature, and precipitation – as input to vegetation models (Harrison et al., 1995; Kutzbach et al., 1998). The simulated biome distributions for 11 ka (early Holocene) and 6 ka (mid-Holocene) indicate (1) in the Arctic: northward expansion of boreal forest (taiga) replacing tundra; (2) in mid-latitude continental interiors: expansion of warm grassland/shrub replacing cool grassland/shrub; and (3) in the northern tropics: expansion of grassland and xerophytic woods/scrub replacing desert in N. Africa, and expansion of warmer and moister biome types in parts of SE and E. Asia. Although individual models differ from one another, these results were generally consistent among 10 climate models for 6 ka orbital forcing (Harrison et al., 1998).

Having confirmed that changes of climate cause vegetation change, what about changes in the opposite direction? Are the vegetation changes sufficiently large to affect climate? Sensitivity experiments with climate models suggest that this is so. In high latitudes, prescribed replacement of tundra by forest causes lowered springtime albedo because the trees extend above the snow whereas tundra remains snow-covered. This lower albedo of the forested landscape in late winter and early spring, relative to tundra, favors an increase in absorbed solar radiation and therefore additional warmth (Bonan et al., 1992). Thus, in the Holocene, if the orbitally forced increase in summer warmth caused forest to replace tundra, then the lowered albedo in the winter half year could cause additional warming – a positive feedback (Foley et al., 1994; TEMPO Members, 1996). In these forest-replacing-tundra sensitivity experiments, the additional warming due to vegetation change almost equaled the original warming due to orbital change alone. Similarly, if orbitally forced enhancements of summer monsoon precipitation in North Africa caused grasslands to invade desert, then the combination of lowered albedo and increased moisture recycling over the vegetated surface could have caused additional rainfall, a positive feedback (Street-Perrott et al., 1990; Kutzbach et al., 1996; Broström et al., 1998). Changes in soil texture and water-holding capacity could also have promoted increased moisture recycling.

Changes in the size of lakes and wetlands caused by climate change may also produce feedbacks on the atmosphere (Coe and Bonan, 1997; Broström et al., 1998; Carrington et al., in press).

Coupled Atmosphere-Biosphere Interaction. Because the results of one-way sensitivity experiments suggested that vegetation changes might exert strong positive feedbacks on climate, various groups have used coupled atmosphere-biosphere models to simulate the two-way interactions between climate and the terrestrial biosphere during the Holocene (Texier et al., 1997; Claussen and Gayler, 1997; Pollard et al., 1998; Ganopolski et al., 1998; Foley, et al., in press; Doherty et al., in press). Almost all these coupled models simulate expansion of northern forests and northern subtropical grasslands during the Holocene. The models differ both in the extent of vegetation change (for example, forest biome shifts of 100 km or 500 km) and in the magnitude of the climate change (for example, sub-Saharan rainfall increases of 10% or 50%). In most studies, the positive feedback of vegetation on the African monsoon still falls short of matching the observed major extension of monsoon rainfall and grasslands into the Sahara during the early and mid-Holocene; the one exception is Ganopolski et al. (1998), who find a very large positive feedback.

It is important for global change studies to determine the correct level of atmosphere-biosphere feedback. For example, the magnitude of high-latitude amplification of greenhouse warming may depend on the strength of this feedback. Studies of trace gas exchanges in northern forests and wetlands during the Holocene will also require improved models of climate/vegetation interactions.

10.3.4 Climate Models Coupled to Dynamical Ocean Models

The work mentioned in Sections 10.3.2 and 10.3.3 generally involved the coupling of atmospheric models to mixed-layer ocean models or the use of prescribed modern SSTs. Ocean currents, equatorial upwelling systems, subtropical gyre circulations, poleward heat transport processes, and the thermo-haline circulation are interacting features of the atmosphere-ocean-biosphere system, and the role of these atmosphere-ocean processes in the Holocene is beginning to receive greater attention. Within the Holocene, there is a significant amount of information about changes in sea surface temperatures (Ruddiman and Mix, 1993; Morley and Dworetzky, 1993), changes in upwelling, such as in the Arabian Sea (Prell, 1984a,b), and changes in surface conditions and sea ice in the northern North Atlantic (de Vernal et al., 1997).

Several studies have used coupled dynamical atmosphere/ocean models to simulate conditions at 11 ka (Liu et al., 1999) and 6 ka (Kutzbach and Liu, 1997; Hewitt and Mitchell, 1998; Bush, 1999; Otto-Bliesner, 1999). The common features of these simulations are the increased warmth of the Arctic, compared with modern, and the modified seasonal temperature cycle of the tropical and northern mid-latitude and subtropical oceans. SSTs respond rather directly to the insolation change, but with a lag of several months. In one model, changes of tropical ocean temperature further enhanced both the African and S. Asian summer monsoons (Hewitt and Mitchell, 1998); in a different model, the African monsoon was enhanced by SST increases in the North Atlantic, but the South Asian monsoon was somewhat weakened by the SST increases in the Indian Ocean (Liu et al., 1999). Bush (1999) reported an enhancement of the Asian monsoon, an enhancement of the Pacific Walker Circulation, and a lowering of ocean temperature in the eastern equatorial Pacific that resembled the La Niña phase of ENSO.

An experiment with a simplified atmosphere-ocean model coupled with a biosphere model showed that the ocean's response to high northern latitude warming at 6 ka included (1) a strong interaction between the effects of the expanding forest and the decreasing sea ice, with both ocean and vegetation feedbacks contributing to a large summertime warming of the Arctic of more than 3 °C, and (2) an increase in runoff from the northern continents that enhanced the freshwater flux to the North Atlantic, weakened the Atlantic thermo-haline circulation by about 10%, and warmed the ocean in high southern latitudes in agreement with some mid-Holocene climate estimates. In this model, the role of biosphere feedbacks dominated over ocean feedbacks in enhancing the monsoonal response to orbital forcing in the northern Tropics (Ganopolski et al., 1998).

In summary, orbitally forced polar processes can produce large changes in circum-Arctic land temperature, vegetation, snow cover, and runoff. These changes over the land can interact with the Arctic and northern North Atlantic ocean/sea-ice system and, in turn, possibly influence the thermo-haline circulation. Tropical monsoon systems respond to orbital forcing with varying feedbacks of biosphere and ocean. New modeling capabilities, when combined with increases in computer resources, should allow detailed study of these complex interactions.

10.3.5 Time-Dependent Simulations of Abrupt Climate Change

Studies of Holocene climates with three-dimensional climate models have employed relatively short "snapshot" simulations (10–50 years' duration) because of computational limitations. In contrast, time-dependent simulations of coupled climate systems with statistical-dynamical models have been run for hundreds of thousands of years and have provided major insights about the climate response to orbital forcing (and other forcing) at glacial/interglacial time scales (Berger et al., 1992, 1998; Berger, 1998). Recently, a statistical-dynamical atmospheric model coupled to a terrestrial biosphere model and simplified ocean basin model was used for a 9,000-year simulation of Holocene climate (Claussen et al., 1999). This simulation displayed an abrupt climate response to the gradually changing orbital forcing. The abrupt change at about 5.5 ka was traced to nonlinear vegetation feedback in the northern monsoon system as perihelion shifted from northern summer (9 ka) into northern fall, that is, beyond the monsoon maximum of the early Holocene. At about 5.5 ka, the vegetated Sahara, which had been formed by the orbitally forced monsoon enhancement of the early Holocene, reverted to desert, accompanied by an abrupt increase in surface albedo and an equally abrupt decrease in precipitation. Lake-level evidence from North Africa indicates an abrupt decrease in moisture at about this time. Meltwater pulses from the last stages of deglaciation could also have been a factor in abrupt changes of the thermo-haline circulation and climate in the early Holocene, such as at 8 ka (Street-Perrott and Perrott, 1990). With increasingly efficient models and more computer resources, time-dependent simulations of Holocene climate will become more common, and opportunities for studies of abrupt climate change will expand.

10.3.6 Decade/Century Variability as a Function of Mean Climate State

A question of great current interest is whether recent changes in the character of ENSO are related to changes in the mean climate caused by greenhouse warming. If the mean climate changes, will variability change? Changes of the character of El Niño are reported in simulations of climate with increased atmospheric concentrations of CO_2 (Knutson and Manabe, 1994; Meehl and Washington, 1996). Decade/century variability may also have changed in the Holocene, compared with present, because the external forcing and mean climate were different at 11 ka, 6 ka, and 0 ka (for example). High-resolution records (corals, laminated lake sediments, layered ice cores, tree rings) offer the possibility to observe such changes in variability.

Several studies have reported coupled-model simulations of Holocene climate variability (Hewitt and Mitchell, 1996; Otto-Bliesner, 1999; Clement et al., 1999). ENSO teleconnection patterns may have been different at 6 ka compared with present (Otto-Bliesner, 1999). Clement et al. (personal communication) show that the changed seasonal insolation cycle of the early Holocene might have suppressed El Niño events. Simulations with a coupled atmosphere-ocean model for 11 ka and 0 ka (Liu et al., 2000) showed that ENSO occurred in both time periods, and with similar spatial patterns, but the overall variability of eastern Pacific SST was reduced by 10–20% at 11 ka compared with modern. One possible explanation for the early Holocene reduction in ENSO variability is related to the enhanced Asian summer monsoon, which is forced by the enhanced summertime insolation at 11 ka. The strengthened monsoon helps to produce a stronger Walker cell over the equatorial Pacific and therefore increased equatorial easterlies, enhanced upwelling, and a colder equatorial ocean. The climatological bias toward colder conditions in the central and eastern equatorial Pacific could in turn favor less intense ENSO events (Liu et al., 2000). These first simulations of reduced interannual variability in the early Holocene appear to be somewhat consistent with preliminary paleoclimatic evidence that strong ENSO events developed only after 7 ka–6 ka (Sandweiss et al., 1996; Rodbell et al., 1999).

10.4   Conclusions

Observational records of Holocene climates provide a wealth of information about changes in the mean climate, abrupt changes, and changes in variability. A hierarchy of models is being used to study and help us understand the complex atmosphere-ocean-biosphere interactions that have occurred in response to orbital forcing. The results of these studies will help to inform us about the relative stability of Earth's climate and biosphere and its sensitivity to changes in external forcing, whether truly external (such as orbital) or human-related. Significant progress has been made in understanding how orbital forcing influences climate. However, this review serves to illustrate that climate models have so far been unable to simulate the full magnitude and spatial and temporal structure of Holocene climate change. Therefore, many challenges remain – both to improve the observational record and to improve the simulations and our understanding of the operative mechanisms.

ACKNOWLEDGMENT

The research on paleoclimates at the University of Wisconsin-Madison was funded in large part by grants from the National Science Foundation (NSF) Climate Dynamics Program and Earth System History Program, the Department of Energy, and modeling and computational resources from the NSF-sponsored National Center for Climatic Research.

REFERENCES

Berger, A. (1978). Long-term variations of daily insolation and Quaternary climatic change. Jour. of Atm. Sci., 35, 2362–2367.

Berger, A., Fichefet, Th., Gallée, H., Tricot, Ch., and van Ypersele, J. P. (1992). Entering the glaciation with a 2-D coupled climate model. Quat. Sci. Rev., 11, 481–493.

Berger, A. (1998). Milankovitch theory and climate. Rev. Geophys., 26, 624–657.

Berger, A., Loutre, M. F., and Galleé, H. (1998). Sensitivity of the LLN climate model to the astronomical and CO_2 forcings over the last 200 ky. Climate Dynamics, 14, 615–629.

Bonan, G. B., Pollard, D., and Thompson, S. L. (1992). Effects of boreal forest vegetation on global climate. Nature, 359, 716–718.

Broström, A., Coe, M., Harrison, S. P., Gallimore, R., Kutzbach, J. E., Foley, J., Prentice, I. C., and Behling, P. (1998). Land surface feedbacks and palaeomonsoons in northern Africa. Geophys. Res. Lett., 25(19), 3615–3618.

Bush, A. B. G. (1999). Assessing the impact of mid-Holocene insolation on the atmosphere-ocean system. Geophys. Res. Lett., 26(1), 99–102.

Carrington, D. P., Gallimore, R. G., and Kutzbach, J. E. Climate sensitivity to wetlands and wetland vegetation in Mid-Holocene North Africa. Climate Dynamics, in press.

Claussen, M., and Gayler, V. (1997). The greening of the Sahara during the mid-Holocene: results of an interactive atmosphere-biosphere model. Glob. Ecol. and Biogeog. Lett., 6, 369–377.

Claussen, M., Kubatzki, C., Brovkin, V., Ganopolski, A., Hoelzmann, P., and Hans-Joachim, P. (1999). Simulation of an abrupt change in Saharan vegetation in the mid-Holocene. Geophys. Res. Lett., 26(14), 2037–2040.

Clement, A. C., Seager, R., and Cane, M. A. (1999). Suppression of El Niño during the mid-Holocene by changes in the Earth's orbit. Unpublished manuscript.

Coe, M., and Bonan, G. (1997). Feedbacks between climate and surface water in Northern Africa during the middle-Holocene. Jour. of Geophys. Res., 102, 11,087–11,101.

de Noblet, N., Braconnot, P., Joussaume, S., and Masson, V. (1996). Sensitivity of simulated Asian and African summer monsoons to orbitally induced variations in insolation 126, 115, and 6 kBP. Climate Dynamics, 12(9), 589–603.

de Vernal, A., Rochon, A., Turon, J., and Matthiessen, J. (1997). Organic-walled dinoflagellate cysts: palynological tracers of sea-surface conditions in middle to high latitude marine environements. Geobios., 30, 905–920.

Doherty, R., Kutzbach, J. E., Foley, J., and Pollard, D. Fully-coupled climate/dynamical vegetation model simualtions over Northern Africa during the mid-Holocene. Climate Dynamics, in press.

Foley, J., Kutzbach, J. E., Coe, M. T., and Levis, S. (1994). Feedbacks between climate and boreal forests during the Holocene epoch. Nature, 371(6492), 52–54.

Foley, J., Levis, S., Costa, M., Doherty, R., and Pollard, D. Incorporating dynamic vegetation cover within global climate models. Ecol. App., in press.

Gallimore, R. G., and Kutzbach, J. E. (1989). Effects of soil moisture on the sensitivity of a climate model to earth orbital forcing at 9000 yr BP. Climatic Change, 14, 175–205.

Ganopolski, A., Kubatzki, C., Claussen, M., Brovkin, V., and Petoukhov, V. (1998). The influence of vegetation-atmosphere-ocean interaction on climate during the mid-Holocene. Science, 280, 1916–1919.

Hall, N. M., and Valdes, P. J. (1997). A GCM simulation of the climate 6000 years ago. Jour. of Clim., 10, 3–17.

Harrison, S. P., Kutzbach, J. E., Prentice, I. C., Behling, P. J., and Sykes, M. T. (1995). The response of Northern Hemisphere extratropical climate and vegetation to orbitally-induced changes in insolation during the last interglaciation. Quat. Res., 43, 174–184.

Harrison, S. P., Jolly, D., Laarif, F., Abe-Ouchi, A., Dong, B., Herterich, K., Hewitt, C., Joussaume, S., Kutzbach, J. E., Mitchell, J., de Noblet, N., and Valdes, P. (1998). Intercomparison of simulated global vegetation distributions in response to 6 kyr BP orbital forcing. Jour. of Clim., 11, 2721–2742.

Hays, J. D., Imbrie, J., and Shackleton, N. J. (1976). Variations in the Earth's orbit: Pacemaker of the ice ages. Science, 194, 1121–1132.

Hewitt, C. D., and Mitchell, J. F. B. (1996). GCM simulations of the climate of 6 kyr BP: mean changes and interdecadal variability. Jour. of Clim., 9, 3505–3529.

Hewitt, C. D., and Mitchell, J. F. B. (1998). A fully coupled GCM simulation of the climate of the mid-Holocene. Geophys. Res. Lett., 25(3), 361–364.

Joussaume, S., Taylor, K. E., Braconnot, P., Mitchell, J. F. B., Kutzbach, J. E., Harrison, S. P., Prentice, I. C., Broccoli, A. J., Abe-Ouchi, A., Bartlein, P. J., Bonfils, C., Dong, B., Guiot, J.,

Herterich, K., Hewitt, C. D., Jolly, D., Kim, J. W., Kislov, A., Kitoh, A., Loutre, M. F., Masson, V., McAvaney, B., McFarlane, N., de Noblet, N., Peltier, W. R., Peterschmitt, J. Y., Pollard, D., Rind, D., Royer, J. F., Schlesinger, M. E., Syktus, J., Thompson, S., Valdes, P., Vettoretti, G., Webb, R. S., and Wyputta, U. (1999). Monsoon changes for 6000 years ago: results of 18 simulations from the Paleoclimate Modeling Intercomparison Project (PMIP). Geophys. Res. Lett., 26(7), 859–862.

Klimanov, V. A. (1984). Paleoclimatic reconstructions based on the information statistical method. In *Late Quaternary Environments of the Soviet Union*, eds. H. E. Wright Jr. and C. W. Barnosky (Minneapolis, MN: University of Minnesota Press).

Knutson, T. R., and Manabe, S. (1994). Impact of increased CO_2 on simulated ENSO-like phenomena. Geophys. Res. Lett., 21, 2295–2298.

Köppen, W., and Wegener, A. (1924). Die Klimate des Quartärs. In *Die Klimate der Geo-logischen Vorzeit* (Berlin, Germany: Verlag Gebrüder Borntraeger).

Kutzbach, J. E. (1981). Monsoon climate of the early Holocene: Climate experiment with the Earth's orbital parameters for 9000 years ago. Science, 214, 59–61.

Kutzbach, J. E. (1983). Monsoon rains of the late pleistocene and early Holocene: patterns, intensity, and possible causes of changes. In *Variations in the Global Water Budget*, eds. A. Street-Perrott et al. Dordrecht, The Netherlands: D. Reidel Publishing Company.

Kutzbach, J. E., and Otto-Bliesner, B. L. (1982). The sensitivity of the African-Asian monsoonal climate to orbital parameter changes for 9000 yr B.P. in a low-resolution general circulation model. Jour. of the Atm. Sci., 39(6), 1177–1188.

Kutzbach, J. E., and Street-Perrott, F. A. (1985). Milankovitch forcing of fluctuations in the level of tropical lakes from 18 to 0 kyr BP. Nature, 317, 130–134.

Kutzbach, J. E., and Guetter, P. J. (1986). The influence of changing orbital parameters and surface boundary conditions on climate simulations for the past 18,000 years. Jour. of Atm. Sci., 43(16), 1726–1759.

Kutzbach, J. E., and Gallimore, R. G. (1988). Sensitivity of a coupled atmosphere/mixed-layer ocean model to changes in orbital forcing at 9000 years BP. Jour. of Geophys. Res., 93(D1), 803–821.

Kutzbach, J. E, Bonan, G., Foley, J., and Harrison, S. P. (1996). Vegetation and soil feedbacks on the response of the African monsoon to orbital forcing in the early to middle Holocene. Nature, 384, 623–626.

Kutzbach, J. E., and Liu, Z. (1997). Response of the African monsoon to orbital forcing and ocean feedbacks in the middle Holocene. Science, 278, 440–443.

Kutzbach, J. E., Gallimore, R., Harrison, S. P., Behling, P., Selin, R., and Laarif, F. (1998). Climate and biome simulations for the past 21,000 years. Quat. Sci. Rev., 17(6-7), 473–506.

Liao, X., Street-Perrott, F., and Mitchell, J. (1994). GCM experiments with different cloud parameterizations: comparisons with paleoclimatic reconstructions for 6000 years B.P. Paleoclimates – Data and Modeling, 1, 99–123.

Liu, Z., Gallimore, R., Kutzbach, J., Xu, X., Golubev, Y., Behling, P., and Selin, R. (1999). Modeling long term climate changes with equilibrium asynchronous coupling. Clim. Dynam., 15, 325–340.

Liu, Z., Jacob, R., Kutzbach, J., Harrison, S., and Anderson, J. (1999). Monsoon impact on El Niño in the early Holocene, PAGES, 7(2), 16–17.

Liu, Z., Kutzbach, J. E., and Wu, L. (2000). Modeling climate shift of El Niño variability in the Holocene. Geophys. Res. Letters, 27(15), 2265–2268.

Mason, B. J. (1976). Towards the understanding and prediction of climatic variations. Quat. Jour. of the Roy. Met. Soc., 102(433), 473–498.

Masson, V., and Joussaume, S. (1997). Energetics of the 6000 BP atmospheric circulation in boreal summer, from large scale to monsoon areas: a study with two versions of the LMD AGCM. Jour. of Clim., 10, 2888–2903.

Meehl, G. A., and Washington, W. M. (1996). El Niño-like climate change in a model with increased atmospheric CO_2 concentrations. Nature, 382, 56–60.

Milankovitch, M. (1920). *Théorie mathématique des phénomenes thermiques produits par la radiation solaite*. Gauthier-Villars, Paris, 338 pp.

Mitchell, J. F. B. (1977). Met 0 20 Tech. Note II/100 Meteorological Office, Bracknell.

Mitchell, J. F. B., Grahame, N. S., and Needham, K. H. (1988). Climate simulations for 9000 years before present: seasonal variations and the effect of the Laurentide Ice Sheet. Jour. of Geophys. Res., 93, 8283–8303.

Morley, J. J., and Dworetzky, B. A. (1993). Holocene temperature patterns in the South Atlantic. In *Global Climates since the Last Glacial Maximum*, eds. H. E. Wright Jr., J. E. Kutzbach, T. Webb, III, W. F. Ruddiman, F. A. Street-Perrott, and P. J. Bartlein (Minneapolis, MN: University of Minnesota Press), 125–135.

Otto-Bliesner, B. L. (1999). El Niño/La Niña and Sahel precipitation during the middle Holocene. Geophys. Res. Lett., 26, 87–90.

Phillips, P. J., and Held, I. M. (1994). The response of orbital perturbations in an atmospheric model coupled to a slab ocean. Jour. of Clim. 7, 767–782.

Pollard, D., Bergengren, J. C., Stillwell-Soller, L. M., Felzer, B., and Thompson, S. L. (1998). Climate simulations for 10000 and 6000 years BP using the GENESIS global climate model. Paleoclimates, 2, 183–218.

Prell, W. L. (1984a). Variation of monsoonal upwelling; a response to changing solar radiation. Climate Processes and Climate Sensitivity, Geophys. Monogr. Ser., vol. 29, AGU, Washington, D.C.

Prell, W. L. (1984b). Monsoonal climate of the Arabian sea during the late Quaternary; a response to solar radiation. In *Milankovitch and Climate*, eds. A. Berger, J. Imbrie, J. Hays, G. Kukla, and B. Saltzman (D. Reidel, Hingman, MA), 349–366.

Raynaud, D., Jouzel, J., Barnola, J. M., Chappelaz, J., Delmas, R., and Lorius, C. (1993). The ice record of greenhouse gases. Science, 259, 926–934.

Ritchie, J. C., Cwynar, L. C., and Spear, R. W. (1983). Evidence from north-west Canada for an early Holocene Milankovitch thermal maximum. Nature, 305, 126–128.

Rodbell, D. T., Seltzer, G. O., Anderson, D. M., Abbott, M. B., Enfield, D. B., and Newman, J. H. (1999). An ∼15,000-year record of El Niño – driven alluviation in southwestern Ecuador. Science, 283, 516–520.

Ruddiman, W. F., and Mix, A. C. (1993). The north and equatorial Atlantic at 9000 and 6000 yr BP. In *Global Climates since the Last Glacial Maximum*, eds. H. E. Wright, Jr., J. E. Kutzbach, T. Webb, III, W. F. Ruddiman, F. A. Street-Perrott, and P. J. Bartlein (Minneapolis, MN: University of Minnesota Press), 94–124.

Sandweiss, D., Richardson, J. B., Rieitz, E. J., Rollins, H. B., and Maasch, K. A. (1996). Geoarchaelogical evidence from Peru for a 5000 years B.P. onset of El Niño. Science, 273, 1531–1533.

Schneider, S. H., and Thompson, S. L. (1979). Ice ages and orbital variations: some simple theory and modeling. Quat. Res., 12, 188–203.

Street-Perrott, F. A., Mitchell, J. F. B., Marchand, D. S., and Brunner, J. S. (1990). Milankovitch and albedo forcing of the tropical monsoons: a comparison of geological evidence and numerical simulations for 9000 yr BP. Transactions of the Royal Society of Edinburgh, 81, 407–427.

Street-Perrott, F. A., and Perrott, R. A. (1990). Abrupt climate fluctuations in the tropics: the influence of Atlantic Ocean circulation. Nature, 343, 607–612.

Suarez, M. J., and Held, I. M. (1976). Modeling climatic response to orbital parameter variations. Nature, 263, 46–47.

TEMPO Members (1996). Potential role of vegetation feedback in the climate sensitivity of high-latitude regions: A case study at 6000 years B.P. Glob. Biogeochem. Cyc., 10(4), 727–736.

Texier, D., de Noblet, N., Harrison, S. P., Haxeltine, A., Joussaume, S., Jolly, D., Laarif, F., Prentice, I. C., and Tarasov, P. (1997). Quantifying the role of biosphere-atmosphere feedbacks in climate change: a coupled model simulation for 6000 years BP and comparison with palaeodata for northern Eurasia and northern Africa. Clim. Dyn., 13, 865–882.

Vavrus, S. J. (1999). The response of the coupled Arctic sea-ice atmosphere system to orbital forcing and ice motion at 6 kyr and 115 kyr BP. Jour. of Clim., 12, 873–896.

Webb, T. III, Bartlein, P. J., and Kutzbach, J. E. (1987). Climatic change in eastern North America during the past 18,000 years; comparisons of pollen data with model results. In *The Geology of North America: North America and adjacent oceans during the last deglaciation*, K-3 447–462. Boulder, CO: Geological Society of America.

Wright, H. E., Jr. (1987). Synthesis: the land south of the ice sheets. In *The Geology of North America: North America and adjacent oceans during the last deglaciation*, K-3 479–488. Boulder, CO: Geological Society of America.

Wright, H. E., Jr., Kutzbach, J. E., Webb, T. III, Ruddiman, W. F., Street-Perrott, F. A., and Bartlein, P. J., eds. (1993). *Global Climates since the Last Glacial Maximum* (Minneapolis, MN: University of Minnesota Press), 569 pp.

11 Interactions of Climate Change and the Terrestrial Biosphere

IAIN COLIN PRENTICE

ABSTRACT

The structure and function of the terrestrial biosphere are closely coupled to the atmosphere through multiple interactions involving physical changes to the land surface (biogeophysical feedback) and changes in the radiatively active gas composition of the atmosphere (biogeochemical feedback). Human activities are forcing large and pervasive changes in these interactions, making it more important than ever to understand the "natural" regulation of the atmosphere-terrestrial, biosphere-ocean system as it acted in the past as well as the consequences of human perturbations for the state and stability of the system in the future. Earth System Models are being developed for these purposes, although no model includes the full range of interactions that are known to occur. Study of paleoenvironmental records has yielded insights and information that can be used as a test of Earth System Models and as a stimulus to their improvement. For example, biogeophysical (vegetation-atmosphere) interactions as well as ocean–atmosphere interactions seem to be necessary for the correct simulation of major changes in vegetation and climate in response to orbital forcing during the Holocene. There is also considerable scope, as yet unexplored, to use paleorecords to test models with coupled climatic and biogeochemical components. The Last Glacial Maximum, in particular, poses fundamental challenges concerning the causation and radiative effects of low CO_2 and high eolian dust concentrations, and the biological effects and consequent feedbacks engendered by low CO_2. Fully prognostic models of the terrestrial biosphere (dynamical global vegetation models, or DGVMs) have been developed, partly in response to these challenges posed by paleorecords. DGVMs should form part of the design of Earth System Model experiments to test our ability to simulate changes in the system over annual to millenial time scales. Early applications of DGVMs to future climate scenarios predict a decline in the capacity of the terrestrial biosphere to take up the excess CO_2 released by human activities during the coming century. This predicted decline is a consequence of saturation of the biochemical response to CO_2 combined with enhanced release of soil carbon in a warmer world. Continued deforestation is also likely to reduce this capacity. These considerations suggest that it is important for CO_2 "stabilization" calculations to explicitly include the effects of changing climate and land use on terrestrial ecosystem carbon metabolism; however, there are still considerable differences among the results of DGVMs, and these differences must be resolved with independent data. Overall, our understanding of whole Earth System function is tantalizingly incomplete and is characterized by an array of interactions among different subsystems that include positive as well as negative feedbacks. Some particular aspects, however, are now relatively well understood, thanks to a combination of explicit numerical modeling and the systematic use of observational constraints in the contemporary and paleo domains.

11.1   Processes

Terrestrial ecosystems and climate interact through numerous biogeochemical and biogeophysical processes (Melillo et al., 1996). The term "interact" here means a two-way exchange, always including an *impact* of atmospheric conditions on the structure and function of ecosystems and a *feedback* of the structure and function of ecosystems on the atmosphere. Biogeochemical feedbacks influence climate by influencing the composition of the atmosphere, especially concentrations of greenhouse gases such as carbon dioxide (CO_2) and methane (CH_4). Biogeophysical feedbacks influence climate more directly, by altering the exchanges of energy, momentum, and water vapor between atmosphere and land – exchanges that are important in controlling the atmospheric circulation. This short overview begins by listing some of the atmosphere-biosphere interactions that are generally considered to be of potential importance, although the list is by no means exhaustive and many interactions are still poorly quantified.

- Exchanges of CO_2 and water vapor between the soil-vegetation system and the atmosphere are strongly coupled through stomatal behavior, and they are sensitive to variations in atmospheric conditions, including atmospheric CO_2 concentration itself. These exchanges are large enough that relatively small fluctuations in the annual balance of upward and downward fluxes can (a) have major consequences for the surface energy and water balance (Sellers et al., 1996, 1997), and (b) influence atmospheric CO_2 concentration measurably (Heimann, 1997). The particular effect of CO_2 concentration on the efficiency of photosynthesis results in a negative feedback, through which terrestrial ecosystems take up more carbon during periods of rapidly rising atmospheric CO_2 (as today) and would be expected to release carbon during periods of declining CO_2 (Taylor and Lloyd, 1992; Thompson et al., 1996). Interannual variability in climate also causes substantial year-to-year fluctuations in the annual rate of increase of CO_2 in the atmosphere. These fluctuations could in principle be explained by variations in ocean uptake or terrestrial uptake of CO_2. The relative magnitude of these contributions is not well established, but modeling and analytical studies indicate that a substantial fraction, at least, is attributable to variations in the carbon balance of the terrestrial biosphere (Bacastow, 1976; Kindermann et al., 1996; Heimann, 1997; Dettinger and Ghil, 1998). It therefore seems likely that the partitioning of carbon between the atmosphere and terrestrial ecosystems is also sensitive to climatic effects on ecosystem function on longer time scales.
- Methane (CH_4) production in wetland soils constitutes a component of heterotrophic respiration (HR) that occurs under waterlogged and therefore anaerobic conditions. Although a relatively small component of HR 2–5% in boreal wetlands (Christensen et al., 1996), this release of CH_4 is important from an Earth System point of view because CH_4 is a far more potent greenhouse gas than the

CO_2 that would otherwise be emitted under aerobic conditions (Schimel et al., 1996). Over periods of a year or longer, HR approximately balances net primary production (NPP), which is the integral over time of photosynthesis (CO_2 fixation by plants) minus autotrophic respiration (CO_2 release by plants). The annual imbalance of ± 2 Pg C yr^{-1} alluded to above is small compared with a global annual NPP that ≈ 60 Pg C yr^{-1}. Thus, atmospheric factors (CO_2 concentration and climate) that affect NPP, and all factors affecting the wetness of soils, can in principle affect the strength of the terrestrial CH_4 source and therefore the additional radiative forcing due to CH_4 (e.g., Hutchin et al., 1995; Walter et al., 1996; Potter, 1997).

- Some other carbon-containing reactive gases, originating mainly in terrestrial ecosystems, interact indirectly with CH_4 by affecting the oxidative capacity of the atmosphere, that is, the ability of HO_x (HO_2 and OH) radicals in the atmosphere to destroy a wide variety of gases, including CH_4. These key reactive gases include nonmethane hydrocarbons (NMHC) and carbon monoxide (CO), both gases whose natural sources are closely tied to ecosystem function (e.g., Jacob and Wofsy, 1988; Lerdau et al., 1997). NMHC is a side product of photosynthesis. CO is generated in fires, which are an integral component of the natural function of many ecosystem types, including savannas, grasslands, and coniferous forests (Bergamaschi et al., 1998).

- Cycling of carbon in ecosystems is naturally tied to the rate of nitrogen cycling, which in turn is dependent on atmospheric and soil conditions (e.g., Schimel et al., 1994). The rate of nitrogen cycling determines the emission rate of the biogenic trace gases nitrous oxide (N_2O) and NO_x (a shorthand for NO, NO_2, and other reactive compounds of N and O) from soils, and soil wetness determines which of these products is dominant (Firestone and Davidson, 1989; Khalil and Rasmussen, 1992; Bouwman et al., 1995; Davidson, 1995; Nevison et al., 1996). NO_x is also produced in biomass burning. N_2O is a potent and long-lived greenhouse gas (Schimel and Sulzman, 1995), whereas NO_x affects the oxidative capacity of the atmosphere by facilitating photochemical reactions involving CO and NMHC (e.g., Derwent, 1996).

- Because different types of plants, animals, and microorganisms are adapted to different climates, any climate change sustained over decades to centuries must bring about changes in the biological composition and structure of ecosystems (Woodward, 1987; Prentice, 1992; Webb and Bartlein, 1992). Changes in ecosystem structure (for example, between forest and nonforest ecosystems, or between evergreen and deciduous vegetation) influence physical properties of the land surface, most importantly albedo, surface roughness, canopy conductance, and rooting depth (Dickinson, 1992; Henderson-Sellers et al., 1993). By influencing net radiation and heat flux partitioning at the surface, these properties exert a strong control over the surface climate. In particular, warming at high northern latitudes shifts the forest limit poleward, reducing surface albedo (especially in late winter and spring before the snow melts) and thus generating a *positive* climate feedback of hemispheric extent (Bonan et al., 1992; Foley et al., 1994).

- Ecosystem structure also determines the potential source regions for mineral aerosol (dust) deflation, which takes place only at times and places where vegetation cover is minimal. Dust has a powerful radiative forcing effect on surface climate; the sign of this effect, however, depends on the albedo of the underlying marine or terrestrial surface (Tegen and Fung, 1995). Thus, changes in the atmospheric loading and spatial distribution of dust, associated with changes in the distribution of deserts, may affect the climate in a complex and as yet poorly understood manner (Kohfeld and Harrison, in press).

These processes do not act independently of one another. For example, changes in ecosystem structure depend on CO_2 exchange because the processes of establishment, growth, and mortality of individual organisms are all tied to carbon fixation. On the other hand, the current ecosystem structure influences CO_2 exchange because different types of plants respond differently to climatic conditions. Wetland ecosystem structure strongly influences the release of CH_4. Because the time for ecosystem structure to come into equilibrium with a changed climate is measured in hundreds of years, such interactions of structure and function can create long time lags and nonlinear behavior.

Further interactions have been speculated to exist between the marine and terrestrial biospheres, mediated through the atmosphere and/or river flows. For example, dust transferred from land to ocean provides a supply of the micronutrient iron, which may act to stimulate phytoplankton growth, either directly (Martin, 1990) or by promoting nitrogen fixation (Broecker and Henderson, 1998). Atmospheric dust loading also potentially influences chemical processes in the atmosphere (Dentener and Crutzen, 1993; Kohfeld and Harrison, in press).

The variety of conceivable interactions among the atmosphere, terrestrial biosphere, and ocean is evidently vast. To establish which processes may really be important, on various time scales, it is imperative to build numerical models that (a) incorporate current understanding of the processes and (b) are systematically evaluated and improved using the greatest possible variety of data sets describing relevant properties of the Earth System, both today and in the recent geological past. When models yield results consistent with the full range of observational constraints, particularly when simulating past Earth System states, we may have some confidence in analyzing model results to yield a more comprehensive picture of the operation of the system than could be obtained from observations alone.

11.2   Human Modifications

The role of the terrestrial biosphere as a mediator of physical and chemical processes in the Earth System was not fully appreciated until the 1990s. However, high-precision measurements of atmospheric CO_2 concentration have revealed not only the magnitude of the human perturbation but also how the terrestrial biosphere "breathes," generating a strong seasonal cycle and large variations in the annual net carbon balance (Heimann et al., 1998). Recognition of the human effect on the system led to the recognition of previously overlooked processes that are now realized to be an integral part of the system as it has functioned for millions of years.

All these processes are increasingly being interfered with by human actions, primarily fossil fuel burning and land use changes. It is important to understand the "natural" operation of these processes if we are to do a better job of understanding what may happen as a consequence of human actions today. Changes in the Earth System, documented in paleorecords of climate, vegetation, and atmospheric composition, provide a source of information about the natural operation of Earth System processes. We need to understand how the Earth System behaves in steady state and how it functions during periods of rapid climate change as documented in the paleorecord, in order to have confidence in our ability to project what will happen as human activities drive the system ever further from steady state (Broecker, 1997).

The following is a nonexhaustive summary of the changes being wrought by human activities on the atmosphere-biosphere processes discussed in the preceding section.

- The cycling of CO_2 between the atmosphere and the biosphere is being altered by deforestation, which returns additional CO_2 to the atmosphere (Houghton et al., 1987; Houghton, 1991; Schimel et al., 1995). On the other hand, forest regrowth on abandoned agricultural land and forest management changes, especially in the North Temperate Zone, may be contributing to the removal of CO_2 from the atmosphere (Kauppi et al., 1992; Melillo et al., 1995; Schimel et al., 1995). Fossil fuel emissions are still the major contributor to the rising atmospheric CO_2 content. The rising CO_2 itself is believed to be inducing a sink of CO_2 in terrestrial ecosystems because of disequilibrium between NPP and HR (Gifford, 1993; Lloyd and Farquhar, 1996; Tian et al., 1998; Kicklighter et al., in press). Other factors such as land use changes may also be contributing to the current sink. Analysis of recent CO_2 and O_2:N_2 ratio measurements suggests that extant ecosystems in the Tropics have been taking up about enough carbon to balance that released by deforestation, and an additional terrestrial carbon sink exists in the mid- to high latitudes of the Northern Hemisphere (Battle et al., 1996; Keeling et al., 1996; Heimann, 1997). Although the locations and magnitudes of terrestrial CO_2 sinks are difficult to estimate precisely and are controversial (e.g., compare Fan et al., 1998, with Rayner et al., in press), there is clearly a potential for both climate change and land use change to alter the terrestrial biosphere's capacity to act as a sink for CO_2 released by human activities.
- While additional CH_4 is released to the atmosphere from ruminant herds, landfills, and natural gas leakage, natural CH_4 sources are being diminished because of draining of wetlands for forestry or agriculture (Schimel et al., 1996). Atmospheric CH_4 now stands at an extremely high level relative to measurements for any preindustrial time and contributes almost one-fifth of the total greenhouse gas forcing; on the other hand, the growth rate of CH_4 in the atmosphere has slowed in recent years.
- Human activities strongly alter fire frequencies through deliberate burning or, in other regions, deliberate fire prevention. This has implications first of all for the CO_2 balance but also for the atmospheric budget of CO, NO_x, and other trace constituents that are released during fires. The net effect of current human actions

with regard to fire is difficult to estimate because fire affects virtually every aspect of biosphere-atmosphere interaction and because it is difficult to classify individual fire events unambiguously as natural or human-caused.

- The tie between carbon and nitrogen cycling is partially broken (a) because of the widespread use of nitrogen fertilizer (required to produce adequate protein to feed the human population), leading to release of both N_2O and NO_x and (b) because of NO_x emissions associated with fossil fuel combustion. Nitrogen deposition from the atmosphere (Holland et al., 1997) is believed to be a major cause of increased tree growth rates downstream from industrialized areas. This effect may be allowing additional carbon storage by these ecosystems and possibly contributing significantly to the terrestrial carbon sink (e.g., Townsend et al., 1994; Melillo et al., 1995). However, long-term effects of this nitrogen addition include damage symptoms, and the most affected ecosystems appear to be already N-saturated (Schulze et al., 1989). At present, far too little is known about the amount, type, and pattern of atmospheric nitrogen deposition, and about the ecosystem impacts.

- Deforestation is drastically modifying ecosystem structure in some regions, with potentially important consequences for the atmospheric circulation and regional climate (Henderson-Sellers et al., 1993). Because of the spatial complexity of human-caused deforestation, however, the feedbacks to climate may be very different from the simple ones predicted in the first sensitivity experiments on this topic.

- Agriculture and construction are contributing additional dust to the atmosphere, leading to a net negative radiative forcing that may be comparable in magnitude (though not in spatial pattern) to the effect of industrially generated sulphate aerosol (Tegen and Fung, 1995).

Current understanding of these impacts relies on a combination of geophysical observations (including remotely sensed measurements) and modeling. The need for numerical models, mentioned earlier in the context of understanding the "natural" operation of the geobiosphere, applies a fortiori to understanding the consequences of massive human involvement.

11.3 Models

Progress is being made toward the goal of building Earth System Models that include the main physical, chemical, and biological processes that interact in determining the state and change of the atmosphere, terrestrial biosphere, and oceans. Such modeling builds on the existence of three-dimensional physical models of the atmospheric and oceanic circulation, and on the existence of "fuzzy" laws governing the aggregate behavior of organisms (Prentice, 1998).

Earth System Models could be used as an aid to understanding the patterns seen in records of past environments and recent historical observations (including atmospheric gas measurements, remote-sensing measurements of photosynthetic activity, etc.). Today we are far from having models that include all the processes listed in Section 11.1. However, we do have a basic toolkit of models that describe the most basic ecosystem functions of CO_2, water, and energy exchange by terrestrial ecosystems and illustrate how these influence and are influenced by ecosystem structure and climate.

The recent evolution of global terrestrial biosphere models has taken place in two overlapping phases. In the first phase (about 1985–1995), two types of model evolved in parallel:

- Biogeography models (e.g., BIOME of Prentice et al., 1992) invoke plant-physiological controls on the distribution of plant forms as a way to predict the large-scale geographic distributions of biomes (tropical forests, savannas, boreal forests, deserts, grasslands, etc.).
- Biogeochemistry models (TEM of Melillo et al., 1993; Century of Parton et al., 1995; CARAIB of Warnant et al., 1994), and tens of others, invoke the mechanisms of photosynthesis, respiration, evapotranspiration, carbon and nitrogen allocation, and so on to predict the CO_2, water, and nitrogen metabolism of ecosystems.

The original biogeography models did not simulate ecosystem function, whereas biogeochemistry models did not predict where different types of ecosystem are found. Biogeochemistry models therefore have required a prescribed distribution of ecosystem types (VEMAP, 1995).

The second phase (about 1990 to present) has been characterized by a more comprehensive approach and has led to the development of the following types of models.

- Coupled biogeography-biogeochemistry models (e.g., BIOME3 of Haxeltine and Prentice, 1996; MAPSS of Neilson and Marks, 1994) predict the structure and function of ecosystems assumed to be in equilibrium with a given climate and soil regime, based on comparing the performance of different plant functional types in each environment.
- Dynamic global vegetation models (DGVMs, e.g., IBIS of Foley et al., 1996; LPJ of S. Sitch, B. Smith, and I. C. Prentice, unpublished) mimic the same processes but include the time-varying component by which ecosystem structure and function respond to changes in climate.
- At least one coupled DGVM-atmosphere model (IBIS-Genesis of Foley et al., 1998) fully incorporates ecosystem dynamics in the framework of an atmospheric general circulation model.
- Coupled DGVM-atmosphere-ocean dynamics models of intermediate complexity (Earth System Models of Intermediate Complexity, or EMICs, such as CLIMBER of Ganopolski et al., 1998a; Ganopolski et al., 1998b) incorporate simplified ecosystem dynamics at low spatial resolution with parameterized atmospheric and ocean models. This design allows long simulations, including the physical couplings among all three components.

This second phase of model development has also seen the first attempts to include the source strengths of various biogenic trace gases (nitrogen oxides, CH_4, NMHC, etc.) in the outputs of terrestrial biosphere models (e.g., Nevison et al., 1996; Potter, 1997). Progress is also being made toward the inclusion of biogeochemical processes into coupled DGVM-atmosphere models and EMICs.

Equilibrium models continue to have their uses, including simulations of past times where we have the data to test the performance of models under changed conditions, such as ice ages and periods of different Earth-Sun geometry (e.g., Harrison et al., 1998; Kutzbach et al., 1998), or as an indicator of the "potential" vegetation corresponding to any simulated future climate (Neilson, 1997). Much can also be learned from model experiments using equilibrium asynchronous coupling, or EAC (Claussen and Gayler, 1997; Kutzbach and Liu, 1997; Texier et al., 1997). In applying EAC to study atmosphere-biosphere interactions, one runs first a climate model and then a model to deduce the potential vegetation associated with that climate. Then one alters the land surface in the climate, model, and so on until the processes converge. This is a way to learn about the possible biogeophysical interactions mediated by vegetation structure without using a fully coupled model.

However, the pace of change in the atmospheric environment is already so fast that it far outstrips the response of ecosystem structure to the change (Prentice and Solomon, 1991; Melillo et al., 1996). Thus, realistic scenarios of the future must include models that can simulate transient effects. The system modeled is so complex that it appears to be capable of highly nonlinear effects that could in no way be predicted using equilibrium models.

11.4   The GAIM 6000 Yr BP Experiment

GAIM is the International Geosphere-Biosphere Programme (IGBP) activity on Global Analysis, Interpretation and Modeling (IGBP, 1994). One of the first tasks of GAIM, first proposed in 1992, was to use paleo data, climate models, and terrestrial biosphere models in concert to evaluate the possible importance of biogeophysical feedback (climate $\Rightarrow$ vegetation structure $\Rightarrow$ climate) in determining the global environment of the mid-Holocene, approximately 6000 years ago (6000 yr BP). This time period was chosen for the following reasons:

- It is unique as a geologically recent (therefore data-rich) period. On the one hand, the Earth-Sun geometry was distinctly different from that of today (a modified form of the high-northern-summer-insolation orbit that eventually halted the last Ice Age), but, on the other hand, unlike earlier times in the Holocene, the mid-latitude continental ice sheets were fully melted (Prentice and Webb, 1998).

- Its atmospheric CO_2 concentration was within the relatively preindustrial Holocene range, 260–280 ppm (Barnola et al., 1987; Indermuehle et al., 1999).

- Extensive climate modeling work was already under way, especially that using standard atmospheric model simulations performed by the world's leading climate modeling laboratories (Paleoclimate Modelling Intercomparison Project, PMIP: Joussaume et al., in press).

- Unequivocal evidence exists for a state of the terrestrial biosphere considerably different from that of today; for example, scrub and grassland vegetation and numerous lakes were distributed across most of what is now the Sahara desert, and most northern temperate and boreal forest zones and the circumpolar Arctic treeline were located farther north than present (Wright et al., 1993; Hoelzmann et al., 1998, Jolly et al., 1998; Tarasov et al., 1998; Prentice et al., in press).

This last point set a challenge: would it be possible to simulate an enormously expanded monsoon influence, big enough to account for the vegetation and lakes that were there? It was already suspected that the present Sahara desert is "self-sustaining" climatologically (Charney, 1975; Claussen, 1998), so it might be necessary to change the land surface conditions in a climate model in order to allow a green Sahara to be a stable condition (Street-Perrott and Perrott, 1990). On the other hand, it was already known in a qualitative sense, from atmospheric model experiments (e.g., Kutzbach and Street-Perrott, 1985; Kutzbach and Guetter, 1986; Mitchell et al., 1988), that the orbital conditions of the early to mid-Holocence do indeed lead climate models to predict an increase of the African monsoon.

To address this questions several sensitivity experiments (Kutzbach et al., 1996; Broström et al., 1997) and EAC simulations (Claussen and Gayler, 1997; Texier et al., 1997) have recently been performed. To summarize, it has been established that the biogeophysical feedback effect of a vegetated Sahara strongly enhanced the strength, the seasonal duration, and the northward extent of monsoon influence in northern Africa (Broström et al., 1997) compared with simulations based on orbital forcing of the atmosphere alone (Joussaume et al., in press). The picture is, however, further complicated by the likely contribution of sea surface temperature changes, brought about by altered surface wind fields (Kutzbach and Liu, 1997) and deep ocean circulation (Ganopolski et al., 1998a). Biogeophysical feedback amplified high-latitude warming at 6000 yr BP relative to present because of the greater energy absorption of forest compared with tundra; this amplification was greatly enhanced by the albedo feedback due to reduced sea ice in the Arctic Ocean (Foley et al., 1994; TEMPO, 1996; Ganopolski et al., 1998a). Thus, there may be synergistic effects of the ocean and terrestrial biosphere in modifying the atmospheric response in both regions (Berger, Chapter 8 in this volume).

The lesson for the future is that realistic simulation of climate must include the possibility of changes in the physical characteristics of the terrestrial biosphere, their interactions with the atmosphere, and the higher-order interactions among atmosphere, terrestrial biosphere, and ocean. Climate change may bring about changes in the terrestrial biosphere that could further modify the climate and in some respects amplify the climate change. On the other hand, there is a considerable potential for large-scale changes in land use to have unexpected climatic effects.

11.5   The Glacial World

The Last Glacial Maximum (LGM), about 21,000 years ago ($\approx$18,000 by the ^{14}C calendar) is another iconic period for paleoclimate modeling studies (Pinot et al., in press). Key features of the LGM include the following:

- Earth-Sun geometry close to present, while ice sheets (Fennoscandian, North American) were at a maximum and sea level at a minimum. Although ice-sheet variations are paced by the orbital variations, there is a phase lag; the LGM ice sheets were shaped by antecedent orbital conditions and were strongly out of equilibrium with the contemporary orbit.

- Concentrations of greenhouse gases CO_2, CH_4, and N_2O all at historic lows: CO_2 less than 200 ppm compared with 280 ppm in the late preindustrial Holocene (Barnola et al., 1987), CH_4 at about 400 ppb compared with 750 ppb (Blunier et al., 1995), N_2O at about 190 ppb compared with 270 ppb (Leuenberger and Siegenthaler, 1992).

- Like 6000 yr BP, the subject of an internationally standardized comparison of atmospheric models within PMIP (Pinot et al., in press).

- Evidence for a biosphere far more radically changed even than during the mid-Holocene; with greatly reduced or displaced forests in all latitudes, extensive tundra and steppes in mid-latitudes, and expanded deserts (Wright et al., 1993; Prentice et al., in press).

- A negative anomaly of $\delta^{13}C$ in the deep ocean (Duplessy et al., 1991), suggesting that total biosphere carbon content was on the order of 25% less than in the Holocene (Prentice and Sarnthein, 1993; Bird et al., 1994).

- Ice core, marine sediments, and terrestrial loess indicating many times greater atmospheric dust concentrations and fluxes than in the Holocene, especially at high latitudes (Petit et al., 1990; Steffensen, 1997; Mahowald et al., in press).

- Marine and terrestrial paleodata from the Tropics, remote from the ice sheet, showing widespread cooling. The tropical average was probably 2–3 °C less than present at modern sea level, but colder still in certain regions (Sonzogni et al., 1997; Farrera et al., in press). Terrestrial vegetation and snowline records from high elevations indicate a stronger cooling aloft – that is, a steeper than present lapse rate (Farrera et al., in press). There was much stronger cooling in the northern continents and high-latitude oceans, especially in winter (e.g., Webb et al., 1993; Weinelt et al., 1996; Peyron et al., 1998).

- An extraordinary level of "sub-Milankovitch" climate variability, on century to millenial time scales, expressed in the form of Dansgaard-Oeschger events, Heinrich events, and Bond cycles (Bond et al., 1992; Hammer et al., 1995; Broecker, 1997).

We do not yet know with any precision how the very large observed changes in the land surface might have contributed to determining the LGM climate through biogeophysical mechanisms of the kind discussed for 6000 yr BP, although a sensitivity experiment by Crowley and Baum (1997) suggested a significant and rather complex effect. Some attention, however, has been paid to the causes and effects of the high atmospheric dust loading at LGM. Possible causes include enhanced winds (Rea, 1994), reduced removal of dust in precipitation (Yung et al., 1996), and increased unvegetated source areas (Broecker and Henderson, 1998). The first two mechanisms are insufficient to fully simulate the enhanced dust loading at high latitudes (e.g., Joussaume, 1990; Andersen et al., 1998). However, the observed dust distribution and loading can be approached in an atmospheric transport model when all three mechanisms are allowed

(Mahowald et al., in press). Reduced precipitation and the physiological effect of low atmospheric CO_2 concentration contributed about equally to roughly double the area of deserts (including cold deserts in central and northeast Asia and south America) and to triple the total global flux of dust to the atmosphere, while the flux to the high latitudes was even more strongly enhanced (Mahowald et al., in press).

Simulations with atmosphere/mixed-layer ocean models have shown that the low atmospheric CO_2 concentration was important in maintaining cold conditions in the Tropics and Southern Hemisphere at the LGM (Broccoli and Manabe, 1987). However, the causes of the low CO_2 concentration are still unclear. In any event, the terrestrial biosphere was apparently not responsible. Several lines of evidence (Bird et al., 1994; Crowley, 1995; Peng et al., 1998) agree that terrestrial carbon storage must have been substantially less than present, thus requiring an oceanic mechanism to sequester an even larger amount of CO_2 than would be the case otherwise.

As for the radiative forcing effect of the simulated LGM dust field, even its sign is currently unclear: did atmospheric dust act to lock the Earth into a cold-climate regime (Harvey, 1988), or did it destabilize the ice sheets, thus contributing to the climate variability of the LGM (Overpeck et al., 1996)? The biogeochemical consequences have also yet to be quantified that is, could the additional dust help to maintain low atmospheric CO_2 concentrations by promoting phytoplankton productivity, as proposed by Martin (1990) and Broecker and Henderson (1998)?

Because our understanding of the LGM is still partial, it is hard to draw firm conclusions for the future (apart from future research needs!), but clearly a better understanding would be highly desirable. For example, data from the LGM combined with modeling studies could help us to gauge the possible magnitude of future tropical warming, the extent of additional warming to be expected in high mountain regions, the consequences of altered dust sources due to climatically or human-induced changes in the land surface, and ultimately – if the different radiative contributions of greenhouse gases, ice sheets, and dust can be disentangled – that much-disputed number, the sensitivity of the Earth's mean surface temperature to change in atmospheric CO_2.

11.6   Source or Sink of Carbon?

Since the advent of a worldwide network of measurement stations for atmospheric CO_2, much of our current understanding of the role of terrestrial ecosystems in the global carbon cycle comes from a combination of modeling with more-recent historical observations. Model experiments to date have been of two kinds: realistic attempts to reproduce the conditions of the last decades, and more-speculative attempts to project the future development of terrestrial ecosystems and their capacity to sequester CO_2 in a world increasingly warmed by anthropogenic CO_2. Experiments of the first type, invoking a wide range of observational constraints derived from the recent CO_2 record, are discussed by Heimann (Chapter 4 in this volume).

A standardized model exercise of the second type has been carried out by the research groups developing DGVMs (Cramer et al., in press). Both "historical" climate and "future" climate were obtained from a transient simulation of the coupled ocean-atmosphere general circulation model of the Hadley Centre, United Kingdom (Mitchell et al., 1995). This simulation assumes a continuing, steady exponential increase of atmospheric CO_2 concentration (or equivalent radiative forcing due to other greenhouse gases) and takes into account an associated scenario of changes in the global amount and distribution of sulphate aerosol. The DGVMs deliberately and artificially simulated a "natural" biosphere, dependent only on climate, soils, atmospheric CO_2, and its own recent history. The CO_2 scenario itself was a pure assumption, and there was no feedback from the CO_2 balance simulated by the DGVMs to the assumed time course of CO_2 concentration in the atmosphere. At the end of the experiment, the climate was instantaneously stabilized – a device to explore how far the simulated ecosystem state might be from equilibrium with the new climate.

Studies of the recent (observational) period help to constrain the response of terrestrial biosphere models to climate, on the time scales of interannual and to some extent interdecadal variability. On longer time scales, other processes may come into play. For example, changes in vegetation structure brought about by climate change may increase the potential for long-term carbon storage (over centuries) while transiently adding carbon to the atmosphere because of dieback of extant vegetation in some regions (the "carbon pulse" hypothesis of King and Neilson, 1992, and Smith et al., 1992). The processes involved are all explicitly simulated by DGVMs. However, the simulated behavior is essentially unconstrained by observations at the whole-system level. It may be possible to constrain such behavior by analysis of the paleorecord (especially by attempts to simulate the small-amplitude variability of CO_2 during the Holocene, as shown by recent high-resolution ice core records from Antarctica: Indermuehle et al., 1999). For the time being, the lack of constraint is an important caveat concerning DGVMs. Nevertheless, the DGVMs (which were generated from differing assumptions and perspectives) agreed in broad features. The carbon pulse hypothesis was not supported, but in all models the terrestrial sink for CO_2 (itself caused by the negative feedback mechanisms alluded to previously) began to level off or decline at around 50 years from present because of a combination of transient vegetation changes, the asymptotic nature of CO_2 fertilization, and the effect of warmer temperatures in stimulating HR (Cao and Woodward, 1998). The models differ in the magnitude and rate of the decline because of differences in the climate response of NPP (Cramer et al., in press).

One lesson from this admittedly stylized experiment was to draw attention to the heavy reliance currently placed on the continuation of a carbon sink in nonagricultural ecosystems. The simulations indicate that this sink is potentially vulnerable to climate change. A moment's reflection also leads to the conclusion that at least the tropical component of the terrestrial carbon sink probably cannot persist if conversion of forest to agriculture persists at its present rate. Thus, climate effects and potential land use changes must be considered in future attempts to estimate the CO_2 emissions reductions required, for example, to meet the objectives of the Kyoto Protocol. On the other hand, DGVMs must be subjected to more-rigorous tests that hopefully will reduce the uncertainty due to differences among models.

11.7   Concluding Remarks

We are presented with a tantalizing picture of the Earth System. On the one hand, interdisciplinary research involving a combination of geophysical observations and modeling appears to be providing firm explanations for some aspects of Earth System behavior as shown in recent observational records and paleo data for selected time periods. On the other hand, our understanding of the events shown in the paleo-record is seriously incomplete. Terrestrial biosphere processes have been shown to be an important component of Earth System dynamics, but only the most basic processes are as yet fully integrated into Earth System Models. "Predictions" of future climate and atmospheric CO_2 concentrations, made with various assumptions about future economic, technological, and environmental developments, continue to routinely ignore most of these processes. Thus, there is much to be done before some of the narratives pursued here can be translated into quantitative and testable statements about past global changes and into more-comprehensive tools for understanding the consequences of human activities.

Recognition of the significance of terrestrial biosphere in the climate system has further implications for future research priorities, and for our view of how the Earth System functions as a whole. If one generalization emerges from the bewildering variety of feedbacks that link the land to the atmosphere and ocean, it is the *absence* of any general rule favoring negative feedbacks that would stabilize the system. Such feedbacks exist, but there are positive, potentially destabilizing feedbacks in the ocean-atmosphere system (e.g., Stocker and Schmittner, 1997) as well as in the terrestrial biosphere-atmosphere system. Thus, no generalization can substitute for a thorough analysis of the processes, their incorporation into explicit numerical models, and the vigorous pursuit of observational constraints against which the models can be progressively honed.

REFERENCES

Andersen, K. K., Armengaud, A., and Genthon, C. (1998). Atmospheric dust under glacial and interglacial conditions. Geophys. Res. Lett., 25, 2281–2284.

Bacastow, R. B. (1976). Modulation of atmospheric carbon dioxide by the southern oscillation. Nature, 261, 116–118.

Barnola, J. M., Raynaud, D., Korotkevich, Y. S., and Lorius, C. (1987). Vostok ice core provides 160,000-year record of atmospheric CO_2. Nature, 329, 408–414.

Battle, M., Bender, M., Sowers, T., Tans, P. P., Butler, J. H., Elkins, J. W., Ellis, J. T., Conway, T., Zhang, N., Lang, P., and Clarke, A. D. (1996). Atmospheric gas concentrations over the past century measured in air from firn at the South Pole. Nature, 383, 231–235.

Bergamaschi, P., Brenninkmeijer, C. A. M., Han, M., Röckmann, T., Scharffe, D. H., and Crutzen, P. J. (1998). Isotope analysis based source identification for atmospheric CH_4 and CO sampled across Russia using the Trans-Siberian railroad. J. Geophys. Res., 103, 8227–8235.

Berger, A. The role of CO_2, sea level, and vegetation during the Milankovitch-forced glacial-interglacial cycles, Workshop on "Geosphere-Biosphere Interactions and Climate," Pontifical Academy of Sciences, Vatican City, 9–13 November 1998, Chapter 8 in this volume.

Bird, M. I., Lloyd, J., and Farquhar, G. D. (1994). Terrestrial carbon storage at the LGM. Nature, 371, 566.

Blunier, T., Chappellaz, J., Schwander, J., Stauffer, B., and Raynaud, D. (1995). Variations in atmospheric methane concentration during the Holocene epoch. Nature, 374, 46–49.

Bonan, G. B., Pollard, D. D., and Thompson, S. L. (1992). Effects of boreal forest vegetation on global climate. Nature, 359, 716–718.

Bond, G., Heinrich, H., Broecker, W., Labeyrie, L., McManus, J., Andrews, J., Huon, S., Jantschik, R., Clasen, S., Simet, C., Tedesco, K., Klas, M., Bonani, G., and Ivy, S. (1992). Evidence for massive discharges of icebergs into the North Atlantic ocean during the last glacial period. Nature, 360, 245–249.

Bouwman, A., van der Hoek, K., and Olivier, J. (1995). Uncertainties in the global source distribution of nitrous oxide. J. Geophys. Res., 100, 2785–2800.

Broccoli, A. J., and Manabe, S. (1987). The influence of contiental ice, atmospheric CO_2, and land albedo on the climate of the last glacial maximum. Clim. Dynam., 1, 87–99.

Broecker, W. S. (1997). Thermohaline circulation, the Achilles Heel of our climate system: will man-made CO_2 upset the current balance? Science, 278, 1582–1588.

Broecker, W. S., and Henderson, G. M. (1998). The sequence of events surrounding Termination II and their implications for the cause of glacial-interglacial CO_2 changes. Paleoceanography, 13, 352–364.

Broström, A., Coe, M., Harrison, S. P., Gallimore, R., Kutzbach, J. E., Foley, J., Prentice, I. C., and Behling, P. (1997). Land surface feedbacks and palaeomonsoons in northern Africa. Geophys. Res. Lett., 25, 3615–3618.

Cao, M., and Woodward, F. I. (1998). Dynamic responses of terrestrial ecosystem carbon cycling to global climate change. Nature, 393, 249–252.

Charney, J. G. (1975). Dynamics of deserts and drought in the Sahel. Q. J. Roy. Meteor. Soc., 101, 193–202.

Christensen, T., Prentice, I. C., Kaplan, J., Haxeltine, A., and Sitch, S. (1996). Methane flux from northern wetlands and tundra: an ecosystem source modelling approach. Tellus, 48B, 652–661.

Claussen, M., and Gayler, V. (1997). The greening of the Sahara during the mid-Holocene: results of an interactive atmosphere-biome model. Global Ecol. Biogeogr., 6, 369–377.

Claussen, M. (1998). On multiple solutions of the atmosphere-vegetation system in present-day climate. Glob. Change Biol., 4, 549–559.

Cramer, W., Boudeau, A., Woodward, I. I., Prentice, I. C., Betts, R. A., Brorlin, V., Cox, P. M., Fisher, V., Foley, J., Friend, A. D., Unchank, C., Lomas, M. R., Ramanlenthy, N., Sitch, S., Smith, B., White, A., and Young-Molling, C. Dynamic Responses of global terrestrial ecosystems to changes in CO_2 and climate. Global Change Biology, in press.

Crowley, T. J. (1995). Ice-age terrestrial carbon changes revisited. Global Biogeochem. Cy., 9, 377–389.

Crowley, T. J., and Baum, S. K. (1997). Effect of vegetation on an ice-age climate model simulation. J. Geophys. Res., 102(D14), 16463–16480.

Davidson, E. A. (1995). Linkages between carbon and nitrogen cycling and their implications for storage of carbon in terrestrial ecosystems, ed. G. M. Woodwell and F. T. Mackenzie, *Biotic feedbacks in the global climatic system*, pp. 219–230, Oxford University Press, New York – Oxford.

Dentener, F. J., and Crutzen, P. J. (1993). Reaction of N_2O_5 on tropospheric aerosols – impact on the global distributions of NO_x, O_3, and OH. J. Geophys. Res.-Atmos. 98(D4), 7149–7163.

Derwent, R. G. (1996). The influence of human activities on the distribution of hydroxyl radicals in the troposhpere. Phil. Trans. Roy. Soc. Lond., 354, 1–30.

Dettinger, M. D., and Ghil, M. (1998). Seasonal and interannual variations of atmospheric CO_2 and climate. Tellus, 50B, 1–24.

Dickinson, R. E. (1992). Land surface, ed. K. E. Trenberth, Climate system modeling, pp. 149–171, Cambridge University Press, Cambridge.

Duplessy, J. C., Bard, E., Arnold, M., Shackleton, N. J., Duprat, J., and Labeyrie, L. (1991). How fast did the ocean-atmosphere system run during the last deglaciation? Earth Planet. Sc. Lett., 103, 27–40.

Fan, S., Gloor, M., and Mahlman, J. (1998). A large terrestrial carbon sink in North America implied by atmospheric and oceanic carbon dioxide data and models. Science, 282, 442–446.

Farrera, I., Harrison, S. P., Prentice, I. C., Ramstein, G., Guiot, J., Bartlein, P. J., Bonnefille, R., Bush, M., Cramer, W., Grafenstein, U., von Holmgren, K., Hooghiemstra, H., Hope, G., Jolly, D., Lauritzen, S. E., Ono, Y., Pinot, S., Stute, M., and Yu, G. (in press). Tropical palaeoclimates at the Last Glacial Maximum: A new synthesis of terrestrial data. Clim. Dynam.

Firestone, M. K., and Davidson, E. A. (1989). Microbiological basis of NO and N_2O production and consumption in soil, ed. M. O. Andreae and D. S. Schimel, *Exchange of trace gases between terrestrial ecosystems and the atmosphere*, pp. 7–21, Wiley, New York.

Foley, J. A., Kutzbach, J. E., Coe, M. T., and Levis, S. (1994). Feedbacks between climate and boreal forests during the Holocene epoch. Nature, 371, 52–54.

Foley, J. A., Prentice, I. C., Ramankutty, N., Levis, S., Pollard, D., Sitch, S., and Haxeltine, A. (1996). An integrated biosphere model of land surface processes, terrestrial carbon balance, and vegetation dynamics. Global Biogeochem. Cy., 10, 603–628.

Foley, J. A., Levis, S., and Prentice, I. C. (1998). Coupling dynamic models of climate and vegetation, Glob. Change Biol., 4(5), 561–579.

Ganopolski, A., Kubatzki, C., Claussen, M., Brovkin, V., and Petoukhov, V. (1998a). The influence of vegetation-atmosphere-ocean interaction on climate during the Mid-Holocene. Science, 280, 1916–1919.

Ganopolski, A., Rahmstorf, S., Petoukhov, V., and Claussen, M. (1998b). Simulation of modern and glacial climates with a coupled global model of intermediate complexity. Nature, 391, 351–356.

Gifford, R. M. (1993). Implications of CO_2 effects on vegetation for the global carbon budget, ed. M. Heimann, The global carbon cycle, pp. 165–205, Proceedings of the NATO Advanced Study Institute, Il Ciocco, Italy.

Hammer, C. U., Clausen, H. B., Dansgaard, W., Neftel, A., Kristinsdottir, P., and Johnson, E. (1995). Continuous impurity analysis along the Dye 3 deep core. Geophysical Monographs, 33, 90–94.

Harrison, S. P., Jolly, D., Laarif, F., Abe-Ouchi, A., Dong, B., Herterich, K., Hewitt, C., Joussaume, S., Kutzbach, J. E., Mitchell, J., de Noblet, N., and Valdes, P. (1998). Intercomparison of simulated global vegetation distribution in response to 6kyr B.P. orbital forcing. J. Climate, 11, 2721–2742.

Harvey L. D. D. (1988). Climatic impact of ice-age aerosols. Nature, 334, 333–335.

Haxeltine, A., and Prentice, I.C. (1996). BIOME3: an equilibrium terrestrial biosphere model based on ecophysiological constraints, resource availability, and competition among plant functional types. Global Biogeochem. Cy., 10, 693–709.

Heimann, M. (1997). A review of the contemporary global carbon cycle and as seen a century ago by Arrhenius and Hägbom. Ambio, 26, 17–24.

Heimann, M., Esser, G., Haxeltine, A., Kaduk, J., Kicklighter, D. W., Knorr, W., Kohlmaier, G. H., McGuire, A. D., Melillo, J., Moore III, B., Otto, R. D., Prentice, I. C., Sauf, W., Schloss, A., Sitch, S., Wittenberg, U., and Würth, G. (1998). Evaluation of terrestrial carbon cycle models through simulations of the cycle of atmospheric CO_2: First results of a model intercomparison study. Global Biogeochem. Cy., 12(1), 1–24.

Henderson-Sellers, A., Dickinson, R. E., Durbidge, T. B., Kennedy, P. J., McGuffie, K., and Pitman, A. J. (1993). Tropical deforestation: modeling local- to regional-scale climate change. J. Geophys. Res., 98, 7289–7315.

Hoelzmann, P., Jolly, D., Harrison, S. P., Laarif, F., Bonnefille, R., and Pachur, H.-J. (1998). Mid-Holocene land-surface conditions in northern Africa and the Arabian peninsula: A data set for the analysis of biogeophysical feedbacks in the climate system. Global Biogeochem. Cy., 12, 35–51.

Holland, E. A., Braswell, B. H., Lamarque, J. F., Townsend, A., Sulzman, J., Muller, J. F., Dentener, F., Brasseur, G., Levy, H., Penner, J. E., and Roelofs, G. J. (1997). Variations in the predicted spatial distribution of atmospheric nitrogen deposition and their impact on carbon uptake by terrestrial ecosystems. J. Geophys. Res. Atmos., 102(D13), 15849–15866.

Houghton, R. A., Boone, R. D., Fruci, J. R., Hobbie, J. E., Melillo, J. M., Plam, C. A., Peterson,

B. J., Shaver, G. R., Woodwell, G. M., Moore, B., Skole, D. L., and Myers, N. (1987). The flux of carbon from terrestrial ecosystems to the atmosphere in 1980 due to changes in land use: Geographic distribution of the global flux. Tellus, 39B, 122–139.

Houghton, R. A. (1991). Tropical deforestation and atmospheric carbon dioxide. Climatic Change, 19, 99–118.

Hutchin, P. A., Press, M. C., Lee, J. A., and Ashenden, T. W. (1995). Elevated concentrations of CO_2 may double methane emissions from mires. Glob. Change Biol., 1, 125–128.

IGBP (1994). IGBP global modeling and data activities 1994–1998. Global Change Report, No. 30.

Indermuehle, A., Stocker, T. F., Joss, F., Fischer, H., Smith, H. J., Wahlen, M., Deck, B., Mastroianni, D., Tschumi, J., Blunier, T., Meyer, R., and Stauffer, B. (1999). Holocene carbon-cycle dynamics based on CO_2 trapped in ice at Taylor Dome, Antarctica. Nature, 398, 121–126.

Jacob, D., and Wofsy, S. (1988). Photochemistry of biogenic emissions over the Amazon forest. J. Geophys., 93, 1477–1486.

Jolly, D., Prentice, I. C., Bonnefille, R., Ballouche, A., Bengo, M., Brenac, P., Buchet, G., Burney, D., Cazet, J.-P., Cheddadi, R., Edorh, T., Elenga, H., Elmoutaki, S., Guiot, J., Laarif, F., Lamb, H., Lezine, A.-M., Maley, J., Mbenza, M., Peyron, O., Reille, M., Reynoud-Farrera, I., Riollet, G., Ritchie, J. C., Roche, E., Scott, L., Ssemmanda, I., Straka, H., Umer, M., van Campo, E., Vilimumbalo, S., Vincens, A., and Waller, M. (1998). Biome reconstruction from pollen and plant macrofossil data for Africa and the Arabian peninsula at 0 and 6000 years. J. Biogeogr., 25, 1007–1027.

Joussaume, S. (1990). Three-Dimensional simulations of the atmospheric cycle of desert dust particles using a general circulation model. J. Geophys. Res., 95(D2), 1909–1941.

Joussaume, S., Taylor, K. E., Braconnot, P., Mitchell, J., Kutzbach, J., Harrison, S. P., Prentice, I. C., Abe-Ouchi, A., Bartlein, P. J., Bonfils, C., Broccoli, A. J., Dong, B., Guiot, J., Herterich, K., Hewitt, C., Jolly, D., Kim, J. W., Kislov, A., Kitoh, A., Masson, V., McAvaney, B., McFarlane, N., de Noblet, N., Peltier, W. R., Peterschmitt, J. Y., Pollard, D., Rind, D., Royer, J. F., Schlesinger, M. E., Syktus, J., Thompson, S., Valdes, P., Vettoretti, G., Webb, R. S., and Wyputta, U. (in press). Monsoon changes for 6000 years ago: results of 18 simulations from the Paleoclimate Modeling Intercomparison Project (PMIP), Geophys. Res. Lett.

Kauppi, P. E., Mielkäinen, K., and Kuusela, K. (1992). Biomass and carbon budget of European forests. Science, 256, 70–74.

Keeling, R. F., Piper, S. C., and Heimann, M. (1996). Global and hemispheric CO_2 sinks deduced from changes in atmospheric O_2 concentration. Nature, 381(6579), 218–221.

Khalil, M. A. K., and Rasmussen, R. A. (1992). The global sources of nitrous oxide. J. Geophys. Res., 97, 14651–14660.

Kicklighter, D. W., Bruno, M., Dönges, S., Esser, G., Heimann, M., Helfrich, J., Ift, F., Joos, F., Kadku, J., Kohlmaier, G. H., McGuire, A. D., Melillo, J. M., Meyer, R., Moore III, B., Nadler, A., Prentice, I. C., Sauf, W., Schloss, A. L., Sitch, S., Wittenberg, U., and Wirth, G. (in press). A first order analysis of the potential of CO_2 fertilization to affect the global carbon budget: A comparison of four terrestrial biosphere models, Tellus.

Kindermann, J., Wurth, G., Kohlmaier, G. H., and Badeck, F. W. (1996). Interannual variation of carbon exchange fluxes in terrestrial ecosystems. Global Biogeochem. Cy., 10(4), 737–755.

King, G. A., and Neilson, R. P. (1992). The transient response of vegetation to climate change: a potential source of CO_2 to the atmosphere. Water Air Soil Pollution, 64, 365–383.

Kohfeld, K. E., and Harrison, S. P. (in press). How well can we simulate past climates? Evaluating models using global palaeoenvironmental datasets. Quaternary Sci. Rev.

Kutzbach, J. E., and Street-Perrott, F. A. (1985). Milankovitch forcing of fluctuations in the level of tropical lakes from 18 to 0 kyr BP. Nature, 317, 130–134.

Kutzbach, J. E., and Guetter, P. J. (1986). The influence of changing orbital parameters and surface boundary conditions on climate simulations for the past 18,000 years. J. Atmospheric Science, 43, 1726–1759.

Kutzbach, J. E., Bonan, G. B., Foley, J. A., and Harrison, S. P. (1996). Vegetation and soils feedbacks on the African monsoon response to orbital forcing in the Holocene. Nature, 384, 623–626.

Kutzbach, J. E., and Liu, Z. (1997). Response of the African monsoon to orbital forcing and ocean feedbacks in the Middle Holocene. Science, 278, 440–443.

Kutzbach, J., Gallimore, R., Harrison, S., Behling, P., Selin, R., and Laarif, F. (1998). Climate and biome simulations for the past 21,000 years. Quaternary Sci. Rev., 17, 473–506.

Lerdau, M., Litvak, M., Palmer, P., and Monson, R. (1997). Controls over monoterpene emissions from boreal forest conifers. Tree Physiol., 17, 563–569.

Leuenberger, M., and Siegenthaler, U. (1992). Ice-age atmospheric concentration of nitrous oxide from an Antarctic ice core. Nature, 360, 449–451.

Lloyd, J., and Farquhar, G. D. (1996). The CO_2 dependence of photosynthesis, plant growth responses to elevated atmospheric CO_2 concentrations, and their interaction with soil nutrient status. I. General principles and forest ecosystems. Funct. Ecol., 10(1), 4–32.

Mahowald, N., Kohfeld, K. E., Hansson, M., Balkanski, Y., Harrison, S. P., Prentice, I. C., Schulz, M. and Rodhe, H. (in press). Dust sources and deposition during the last glacial maximum and current climate: a comparison of model results with paleodata from ice cores and marine sediments. J. Geophys. Res.

Martin, J. (1990). Glacial-interglacial CO_2 change: the iron hypothesis. Paleoceanography, 5, 1–13.

Melillo, J. M., McGuire, A. D., Kicklighter, D. W., Moore III, B., Vörösmarty, C. J., and Schloss, A. L. (1993). Global climate change and terrestrial net primary production. Nature, 363, 234–240.

Melillo, J. M., Kicklighter, D. W., McGuire, A. D., Peterjohn, W. T., and Newkirk, K. (1995). Global change and its effects on soil organic carbon stocks, Dahlem Conference Proceedings, pp. 175–189, John Wiley & Sons, New York.

Melillo, J. M., Prentice, I. C., Farquhar, G. D., Schulze, E.-D., and Sala, O. E. (1996). Terrestrial biotic response to environmental change and feedbacks to climate, ed. J. T. Houghton, L. G. M. Filho, B. A. Callander, N. Harris, A. Kattenberg, and K. Maskell, Climate Change 1995. *The science of climate change*, pp. 449–481, Cambridge University Press, Cambridge.

Mitchell, J. F. R., Grahame, N. S., and Needham, K. J. (1988). Climate simulations for 9000 years before present: seasonal variations and effect of the Laurentide ice sheet. J. Geophys. Res., 93, 8283–8303.

Mitchell, J. F. B., Johns, T. C., Gregory, J. M., and Tett, S. F. B. (1995). Climate response to increasing levels of greenhouse gases and sulphate aerosols. Nature, 376, 501–504.

Neilson, R. P., and Marks, D. (1994). A global perspective of regional vegetation and hydrologic sensitivities from climatic change. J. Veg. Sci., 5(5), 715–730.

Neilson, R. P. (1997). Simulated changes in vegetation distribution under global warming, ed. R. T. Watson, M. C. Zinyowera, and R. H. Moss, *The regional impacts of climate change*, pp. 441–456, Cambridge University Press, Cambridge.

Nevison, C. D., Esser, G., and Holland, E. A. (1996). A global model of changing N_2O emissions from natural and perturbed soils. Climatic Change, 32(3), 327–378.

Overpeck, J., Rind, D., Lacis, A., and Healy, R. (1996). Possible role of dust-induced regional warming in abrupt climate change during the last glacial period. Nature, 384, 447–449.

Parton, W. J., Scurlock, J. M. O., Ojima, D. S., Schimel, D. S., and Hall, D. O. (1995). Impact of climate-change on grassland production and soil carbon worldwide. Glob. Change Biol., 1, 13–22.

Peng, C. H., Guiot, J., and Campo, E. V. (1998). Estimating changes in terrestrial vegetation and carbon storage: Using palaeoecological data and models. Quaternary Sci. Rev., 17(8), 719–735.

Petit, J. R., Mounier, L., Jouzel, J., Korotkevich, Y. S., Kotyakov, V. I., and Lorius, C. (1990). Paleo-climatological and chronological implications of the Vostok core dust record. Nature, 343, 56–58.

Peyron, O., Guiot, J., Cheddadi, R., Tarasov, P., Reille, M., de Beaulieu, J. L., Bottema, S., and Andreu, V. (1998). Climate reconstruction in Europe for 18,000 yr B. B. from pollen data. Quaternary Res., 49, 183–196.

Pinot, S., Ramstein, G., Harrison, S. P., Prentice, I. C., Guiot, J., Joussaume, S., and Stute, M. (in press). PMIP models and data comparison over tropical areas for the Last Glacial Maximum. Clim. Dynam.

Potter, C. S. (1997). An ecosystem simulation model for methane production and emission from wetlands. Global Biogeochem. Cy., 11, 495–506.

Prentice, I. C., and Solomon, A. M. (1991). Vegetation models and global change, ed. R. S. Bradley, *Global changes of the past*, pp. 365–384, UCAR/Office for Interdisciplinary Earth Studies, Boulder, Colorado.

Prentice, C., Cramer, W., Harrison, S. P., Leemans, R., Monserud, R. A., and Solomon, A. M. (1992). A global biome model based on plant physiology and dominance, soil properties and climate. J. Biogeogr., 19, 117–134.

Prentice, I. C. (1992). Climate change and long-term vegetation dynamics, ed. D. C. Glenn-Lewin, R. A. Peet, and T. Veblen, *Plant succession: Theory and prediction*, pp. 293–339, Chapman and Hall, London.

Prentice, I. C., and Sarnthein, M. (1993). Self-regulatory processes in the biosphere in the face of climate change, ed. J. Eddy and H. Oeschger, *Global changes in the perspective of the past*, pp. 29–38, Wiley, Chichester.

Prentice, I. C. (1998). Ecology and the Earth System., ed. H.-J. Schellnhuber and V. Wenzel, *Earth system analysis*, pp. 219–240. Berlin: Springer, Potsdam Institute for Climate Impact Research.

Prentice, I. C., and Webb III, T. (1998). BIOME 6000: reconstructing global mid-Holocene vegetation patterns from palaeoecological records. J. Biogeogr., 25, 997–1005.

Prentice, I. C., Jolly, D., and BIOME 6000 participants. Mid-Holocene and glacial-maximum vegetation geography of the northern continents and Africa. J. Biogeogr., in press.

Rayner, P. J., Enting, I. G., Francey, R. J., and Langenfelds, R. (in press). Reconstructing the recent carbon cycle from atmospheric CO_2, $\delta^{13}C$ and O_2/N_2 observations, Tellus.

Rea, D. K. (1994). The paleoclimatic record provided by eolian deposition in the deep sea: the geologic history of wind. Rev. Geophys., 32, 159–195.

Schimel, D. S., Braswell, B. H., Holland, E. A., McKeown, R., Ojima, D. S., Painter, T. H., Parton, W. J., and Townsend, A. R. (1994). Climatic, edaphic and biotic controls over carbon and turnover of carbon in soils. Global Biogeochem. Cy., 8, 279–293.

Schimel, D. S., Enting, I. G., Heimann, M., Wigley, T. M. L., Raynaud, D., Alves, D., and Siegenthaler, U. (1995). CO_2 and the carbon cycle, ed. J. T. Houghton, L. G. M. Filho, J. Bruce, H. Lee, B. A. Callander, E. Heites, N. Harris, and K. Maskell, *Climate change 1994: Radiative forcing of climate change and an evaluation of the IPCC IS92 emission scenarios*, Cambridge University Press, Cambridge.

Schimel, D. S., and Sulzman, E. (1995). Variability in the earth climate system: Decadal and longer timescales. Rev. Geophys., 33, 873–882.

Schimel, D., Alves, D., Enting, I., Heimann, M., Joos, F., Raynaud, D., Wigley, T., Prather, M., Derwent, R., Ehhalt, D., Fraser, P., Sanhueza, E., Zhou, X., Jonas, P., Charlson, R., Rodhe, H., Sadasivan, S., Shine, K. P., Fouquart, Y., Ramaswamy, V., Solomon, S., Srinivasan, J., Albritton, D., Derwent, R., Isaksen, I., Lal, M., and Wuebbles, D. (1996). Radiative forcing of climate change, ed. J. T. Houghton, L. G. M. Filho, B. A. Callander, N. Harris, A. Kattenberg, and K. Maskell, *Climate change 1995: The Science of climate change*, pp. 69–131, Cambridge University Press, Cambridge.

Schulze, E. D., Devries, W., and Hauhs, M. (1989). Critical loads for nitrogen deposition on forest ecosystems. Water Air Soil Pollution, 48, 451–456.

Sellers, P. J., Bounoua, L., Collatz, G. J., Randall, D. A., Dazlich, D. A., Los, S. O., Berry, J. A., Fung, I., Tucker, C. J., Field, C. B., and Jensen, T. G. (1996). Comparison of radiative and physiological effects of doubled atmospheric CO_2 on climate. Science, 271, 1402–1406.

Sellers, P. J., Dickinson, R. E., Randall, D. A., Betts, A. K., Hall, F. G., Berry, J. A., Collatz, G. J., Denning, A. S., Mooney, H. A., Nobre, C. A., Sato, N., Field, C. B., and Henderson-Sellers,

A. (1997). Modeling the exchanges of energy, water, and carbon between continents and the atmosphere. Science, 275, 502–509.

Smith, T. M., Shugart, H. H., Bonan, G. B., and Smith, J. B. (1992). Modeling the potential response of vegetation to global climate change. Adv. Ecol. Res., 22, 93–116.

Sonzogni, C., Bard, E., Rostek, F., Dollfus, D., Rosell, M. A., and Eglinton, G. (1997). Temperature and salinity effects on alkenone ratios measured in surface sediments from the Indian Ocean. Quaternary Res., 47, 344–355.

Steffensen, J. P. (1997). The size distribution of microparticles from selected segments of the Greenland Ice Core Project ice core representing different climatic periods. J. Geophys. Res., 102, 26755–26763.

Stocker, T. F., and Schmittner, A. (1997). Influence of CO_2 emission rates on the stability of the thermohaline circulation. Nature, 388, 862–865.

Street-Perrott, F. A., and Perrott, R. A. (1990). Abrupt climate fluctuations in the tropics: the influence of Atlantic Ocean circulation. Nature, 343, 607–611.

Tarasov, P. E., Webb III, T., Andreev, A. A., Afanaseva, N. B., Berezina, N. A., Bezusko, L. G., Blyakharchuk, T. A., Bolikhovskaya, N. S., Cheddadi, R., Chernavskaya, M. M., Chernova, G. M., Dorofeyuk, N. I., Dirksen, V. G., Elina, G. A., Filimonova, L. V., Glebov, F. Z., Guiot, J., Gunova, V. S., Harrison, S. P., Jolly, D., Khomutova, V. I., Kvavadze, E. V., Osipova, I. M., Panova, N. K., Prentice, I. C., Saarse, L., Sevastyonov, D. V., Vokova, V. S., and Zernitskaya, V. P. (1998). Present-day and mid-Holocene biomes reconstructed from pollen and plant macrofossil data from the former Soviet Union and Mongolia. J. Biogeogr., 25, 1029–1053.

Taylor, J. A., and Lloyd, J. (1992). Sources and sinks of atmospheric CO_2. Aust. J. Bot., 40, 407–418.

Tegen, I., and Fung, I. (1995). Contribution to the atmospheric mineral aerosol load from land surface modification. J. Geophys. Res., 100, 18707–19726.

TEMPO Members (Bartlein, P. J., Foley, J. A., Harrison, S. P., Hostetler, S., Kutzbach, J. E., Liu, Z., Prentice, I. C., and Webb, T.) (1996). The potential role of vegetation feedback in the climate sensitivity of high-latitude regions: a case study at 6000 years before present. Global Biogeochem. Cy., 10, 727–736.

Texier, D., de Noblet, N., Harrison, S. P., Haxeltine, A., Jolly, D., Joussaume, S., Laarif, F., Prentice, I. C., and Tarasov, P. (1997). Quantifying the role of biosphere-atmosphere feedbacks in climate change: coupled model simulation for 6000 years BP and comparision with paleodata for northern Eurasia and northern Africa. Clim. Dynam., 13, 865–882.

Thompson, M. V., Randerson, J. T., and Malmstrom, C. M. (1996). Change in net primary production and heterotrophic respiration: How much is necessary to sustain the terrestrial carbon sink? Global Biogeochem. Cy., 10, 711–726.

Tian, H. Q., Melillo, J. M., and Kicklighter, D. W. (1998). Effect of interannual climate variability on carbon storage in Amazonian ecosystems. Nature, 396, 664–667.

Townsend, A. R., Braswell, B. H., Holland, E. A., and Penner, J. E. (1994). Spatial and temporal patterns in potential terrestrial carbon storage resulting from deposition of fossil fuel derived nitrogen. Ecol. Appl., 6, 806–814.

VEMAP Members. (1995). Vegetation/ecosystem modeling and analysis project: comparing biogeography and biogeochemistry models in a continental-scale study of terrestrial ecosystem responses to climate change and CO_2 doubling. Global Biogeochem. Cy., 9, 407–437.

Walter, B. P., Heimann, M., Shannon, R. D., and White, J. R. (1996). A process based model to derive methane emission from natural wetland. Geophys. Res. Lett., 23, 3731–3734.

Warnant, P., Francois, L., and Strivay, D. (1994). CARAIB – A global model of terrestrial biological productivity. Global Biogeochem. Cy., 8(3), 255–270.

Webb III, T., and Bartlein, P. J. (1992). Global changes during the last 3 million years: climatic controls and biotic responses. Annu. Rev. Ecol. Syst., 23, 141–173.

Webb III, T., Bartlein, P. J., Harrison, S. P., and Anderson, K. H. (1993). Vegetation, lake levels and climate in eastern North America for the past 18,000 years, ed. H. E. Wright, Jr,

J. E. Kutzbach, T. Webb III, W. F. Ruddiman, F. A. Street-Perrott, and P. J. Bartlein, *Global changes since the Last Glacial Maximum*, pp. 415–467, University of Minnesota Press, Minneapolis.

Weinelt, M., Sarnthein, M., Pflaumann, U., Schulz, H., Jung, S., and Erlenkeuser, H. (1996). Ice-free Nordic seas during the last glacial maximum? Potential sites of deepwater formation. Paleoclimates, 1, 1–4.

Woodward, F. I. (1987). *Climate and plant distribution*, Cambridge University Press, Cambridge.

Wright, H. E., Kutzbach, J. E., Webb III, T., Ruddiman, W. F., Street-Perrott, F. A., and Bartlein, P. J., eds. (1993). *Global changes since the Last Glacial Maximum*, University of Minnesota Press, Minneapolis.

Yung, Y. L., Lee, T., Wang, C.-H., and Shieh, Y.-T. (1996). Dust: A diagnostic of the hydrologic cycle during the last glacial maximum. Science, 271, 962–963.

PART FOUR

INFORMATION FROM THE PAST

12 The Record of Paleoclimatic Change and Its Greenhouse Implications

W. R. PELTIER

Perhaps the most important contribution of paleoclimate analysis to the ongoing debate concerning our future in the "greenhouse" is that associated with the issue of rapid climate change. The ultra-low-frequency variability derivative of the 100 kyr cycle of the late Pleistocene is now a rather well understood consequence of the nonlinear response of the system to orbital insolation forcing (e.g., see Tarasov and Peltier, 1997, 1999 for detailed discussion and Berger, Chapter 8 in this volume), but no similar degree of understanding may be claimed for the millennium and shorter-period variability that is evident in the $\delta^{18}0$ atmospheric temperature proxy from the Summit, Greenland, ice cores during the interval of time known as Oxygen Isotope Stage 3. This is a critical gap, especially inasmuch as it has been suggested that these Dansgaard-Oeschger (D-O) oscillations might also be induced under the enhanced greenhouse conditions that we expect will be characteristic of the future.

The only explicit theory of the D-O oscillation in the current literature is that based on a highly reduced model of the global thermo-haline circulation (Sakai and Peltier, 1995, 1996, 1997, 1999), in which it is shown that the THC should "fibrillate" in response to application of a supercritical freshwater forcing applied to the high latitudes of the North Atlantic basin. In this model the D-O oscillation occurs through a simple Hopf bifurcation when the buoyancy flux is somewhat weaker than would be required to arrest the deep circulation completely. That such an enhancement of the high-latitude buoyancy flux might be expected under ice-age conditions follows from the fact that the basin was then bounded on the west and east by the massive Laurentide and Fennoscandian/Barents Sea ice sheets, respectively. As evidenced by the episodic occurrence of Heinrich events (Heinrich, 1988), during which intense discharges of icebergs from these ice sheets apparently covered the high-latitude North Atlantic with massive amounts of freshwater, the idea that enhanced buoyancy flux might indeed have been characteristic of the stage 3 period is not at all unreasonable. The enhanced buoyancy of the surface water is not associated with increased P−E (Precipitation minus evaporation), as could well be the case in the warmer "enhanced greenhouse" world of the future, but rather with the enhanced surface "run-off" carried by the berg-flux.

The idea that such high-latitude freshwater forcing could be a major determinant of climate system variability has of course become the prevalent view in connection with the Younger Dryas climatic reversal that occurred during the Last Glacial Maximum (LGM)-to-Holocene transition (e.g., Broecker and Denton, 1989, and see Broecker, Chapter 5 in this volume). High-latitude foraminifera-based sea surface salinity (SSS) reconstructions by Duplessy et al. (1991), and others subsequently, have indeed established that high-latitude Atlantic SSS was strongly depressed during glacial conditions in the regions where North Atlantic deep water (NADW) forms today (see Chapter 14 by Duplessy in this volume). An important recommendation for future research is that more-detailed analysis with more fully anticulated models be performed to test the robustness of the above referenced theory of the D-O oscillation. Because the ice-mechanical instability through which Heinrich events are generated appears to be entrained to the D-O oscillation, it would appear necessary to develop a sound theory of the latter in order to better understand the details of the former.

The issue of the mechanism(s) underlying rapid climate change is not one that we might reasonably expect to be entirely settled on the basis of models; the level of complexity involved is simply too high for us to be overly sanguine of success through this means alone. What is also required is a concerted program of data analysis, perhaps focused on the most recent 2000 years of Earth history but including further work on the YD event. In this, and on the basis of the most recent literature, it would seem important to pay particular attention to the relative phasing of events in the Northern and Southern Hemispheres (see Chapter 5 [Broecker] and Chapter 15 [Jouzel] in this volume). That an out-of-phase relationship existed during the YD now seems to be well established on the basis of the intercomparison of accurately time-constrained data from the Greenland and Vostok Antarctica ice cores. One possible explanation of such a relationship between the hemispheres posits a complementarity of the deepwater production rates between them. When NADW production rate is high, the Northern Hemisphere is warm. When this rate falls and the Northern Hemisphere is cooled in consequence, the rate of Antarctic bottom water (AABW) production increases to compensate, and the Southern Hemisphere is thereby warmed. Broecker has suggested that support for this view is found in the ^{14}C age of abyssal waters. In an enhanced program focused on the most recent 2000 years of Earth history, an appropriate primary target might be the so-called Little Ice Age during which there occurred a distinct maximum in the extent of mountain glaciers. Although the cost would be high for the sufficiently high frequency sampling of the Vostok core to support such a focused effort, this should nevertheless be carefully considered.

Paleoclimate inferences also continue to provide important constraints on the mechanics of the global carbon cycle. In this regard, however, a significant further gap in understanding continues to be that related to the mechanism(s) responsible for the glacial-to-interglacial variation of atmospheric carbon dioxide concentration. The recently constructed models of the 100 kyr ice-age cycle, which are successful in explaining this phenomenon as a nonlinear response to orbital insolation forcing (Tarasov and Peltier, 1997, 1999), employ the Vostok measurements of the variation of atmospheric CO_2 concentration from Eemean to Holocene to enhance the insolation signal and strongly suggest that the observed $[CO_2]$ depression during the glacial period is required in order to understand the continental ice volumes that are inferred to have been characteristic of the LGM epoch based on the observed sea level depression. There is still no entirely satisfactory mechanistic model that can correctly predict the observed variation of $[CO_{2-}]$ as a combined effect of the solubility pump and the biological pump. The magnitude of the problems that continue to bedevil our understanding of biospheric feedbacks on climate is also clearly evident in the fundamental disagreements that have arisen in connection with the identification of particular continental surfaces as being either sources or sinks of this trace gas in the modern climate system (see in particular Chapter 4 by Heimann in this volume).

An important issue in paleoclimatology remains outstanding, providing a cogent reminder of the extent to which the climate system continues to provide challenges to the scientific imagination. This issue relates to the theory of ice-age occurrence in the more remote past. Recent analysis has focused on the late pre-Cambrian glaciation that occurred approximately 700 million years ago. Paleomagnetic evidence has been widely construed to suggest that this glaciation occurred not at high latitude but rather at extremely low latitude. The debate concerning mechanism has recently come to be polarized between two extreme views. In one, the so-called snowball Earth hypothesis, the entire surface of the Earth is imagined to have become ice-covered (e.g., as most recently discussed by Hoffman et al., 1998). In the other view, the obliquity of the orbit is imagined to have been so large at that time that only the lowest latitudes were susceptible to glacial advance (Williams et al., 1998; see also Kasting, Chapter 13, in this volume). Neither of these hypotheses has yet been subjected to particularly rigorous test, although the snowball hypothesis is gaining intellectual momentum (e.g., see Hyde et al., 2000). This most recent work, however, which was based on the neoproteriozoic continental reconstruction of Dalziel (1997), strongly suggests the fully glaciated "hard snowball" scenario of Haffman et al. (1998) to be as unlikely as the high obliquity scenario of Williams et al. is unnecessary. The detailed GCM reconstructions of the climate of that era discussed in Hyde et al. do lead to glaciation of the continental fragments that then existed at low latitude, but this occurs in the presence of a substantial equatorial refugium of open water.

REFERENCES

Broecker, W. S., and Denton, G. (1989). The role of ocean-atmosphere reorganizations in glacial cycles. Geochim. Cosmochim. Acta., 53, 2465–2501.

Dalziel, I. W. D. (1997). Neoproteriozoic-Paleozoic geography and tectonics: Review, hypothesis, environmental speculation. GSA Bulletin, 109(1), 16–42.

Duplessy, J.-C., Labeyrie, L., Juillet-Leclerc, A., Maitre, F., Dupart, J., and Sarnthein, M. (1991). Surface salinity reconstruction of the North Atlantic Ocean during the last glacial maximum. Oceanol. Acta. 14, 311–323.

Heinrich, H. (1988). Origin and consequences of cyclic ice rafting in the northeast Atlantic Ocean during the past 130,000 years. Quat. Res., 29, 142–152.

Hoffman, P. F., Kaufman, A. J., Halverson, G. P., and Schrag, D. E. (1998). Carbon isotopes in Namibean rocks – new argument for the "Snowball Earth." Science, 281, 1342–1346.

Hyde, W. T., Crowley, T. J., Baum, S. K., and Peltier, W. R. (2000). Neoproteriozoic "Snowball Earth" simulations with a coupled climate ice-sheet model. *Nature*, 405, 425–429.

Sakai, K., and Peltier, W. R. (1995). A simple model of the Atlantic thermolahine circulation: internal and forced variability with paleoclimatological implications. J. Geophys. Res., 100, 13455–13479.

Sakai, K., and Peltier, W. R. (1996). A multi-basin reduced model of the global thermohaline circulation: paleoceanographic analyses of the origins of ice-age climate variability. J. Geophys. Res., 101, 22535–22561.

Sakai, K., and Peltier, W. R. (1997). Dansgaard-Oeschger oscillations in a coupled atmosphere-ocean climate model. J. Climate, 10, 949–970.

Sakai, K., and Peltier, W. R. (1999). A dynamical systems model of the Dansgaard-Oeschger oscillation and the origin of the Bond cycle. J. Climate, 12, 2238–2255.

Tarasov, L., and Peltier, W. R. (1997). Terminating the 100 kyr ice-age cycle. J. Geophys. Res., 100, 21665–21693.

Tarasov, L., and Peltier, W. R. (1999). The impact of thermo-mechanical ice-sheet coupling on a model of the 100 kyr ice-age cycle. J. Geophys. Res.

Williams, D. M., Kasting, J. F., and Frakes, L. A. (1998). Low-latitude glaciation and rapid changes in the Earth's obliquity explained by obliquity-oblateness-feedback. Nature, 396, 453–455.

13   Long-Term Stability of Earth's Climate

The Faint Young Sun Problem Revisited

JAMES F. KASTING

Although the main focus of this chapter is to assess our understanding of climate on decadal-to-millenial time scales, it is also sometimes useful to step back and see what we know on very long ones (millions to billions of years). One reason for doing so was illustrated by a question that was raised after this paper was presented. The questioner asked, "Wouldn't the radiative forcing from atmospheric CO_2 increases saturate once CO_2 reached a certain value, say, 2–4 times the preindustrial level?" This question is relevant to long-term anthropogenic global warming because it is conceivable that fossil fuel burning could eventually raise atmospheric CO_2 levels to 8–10 times the preindustrial value if we consume the bulk of the available fossil fuel within the next few hundred years. My answer to this question was an emphatic, "No!" Climate model calculations performed for the early Earth have shown that the radiative forcing from CO_2 remains roughly log-linear (1.5–4.5 °C for each doubling of CO_2 and $\sim$10 °C for each factor of 10 increase in CO_2) up to a factor of at least 1000 increase in atmospheric CO_2 levels. The reason is that, as the strongest CO_2 absorption bands – the 15 μm band in particular – become saturated, weaker absorption bands at other thermal-infrared (IR) wavelengths begin to become important. This fact is well known to planetary scientists who have tried to simulate the greenhouse effect of Venus's dense CO_2 atmosphere, and it is incorporated in climate models of the early Earth that rely on high CO_2 abundances to offset reduced solar luminosity early in the Earth's history. It is occasionally overlooked by climate modelers who are mainly interested in the response of Earth's climate to more modest (factor of 2) changes in atmospheric CO_2 levels. I mention this because it is one example of how developing an understanding of climate in its broadest sense can shed light on important points that might otherwise be missed.

13.1   The Faint Young Sun Problem

As hinted at above, the fundamental problem in understanding Earth's climate history on billion-year time scales has to do with the faintness of the young Sun. Standard solar evolution models (e.g., Gough, 1981; Gilliland, 1989) predict that the Sun was $\sim$30% less luminous at the time when it entered the main sequence, some 4.5 billion years ago, and that it has brightened more or less linearly with time up to the present day. The reason has to do with the conversion of hydrogen to helium in the Sun's core by nuclear fusion. As the core becomes richer in He, it becomes denser, it contracts and heats up, and the rate of nuclear fusion increases. A convenient analytic formula for expressing the change in solar luminosity with time is (Gough, 1981)

$$\frac{S}{S_0} = \frac{1}{1 + 0.4(t/t_0)} \tag{1}$$

where $t =$ time in billions of years before the present, or Ga, $t_0 = 4.6$ Ga, and S_0 is the present solar constant, 1370 W/m^2. This change in luminosity with time is illustrated by the solid curve in Figure 13.1.

This change in solar luminosity with time – although small by comparison with the much larger changes in luminosity that will occur after the Sun leaves the main sequence, about 5 billion years hence – is large enough to have an appreciable effect on Earth's climate. The basic equation for planetary energy balance can be written as

$$\sigma T_e^4 = \frac{S}{4}(1 - A) \tag{2}$$

where T_e is Earth's effective radiating temperature, A is the planetary albedo ($\cong 0.3$), and σ is the Stefan–Boltzmann constant, 5.67×10^{-8} W/m^2/K^4. Plugging in the numbers for the modern Earth yields $T_e = 255$ K. This is lower than the observed mean surface temperature, T_s, by about 33 K. The difference between the two numbers is attributed to the greenhouse effect of the atmosphere. Most of that warming is caused by just two IR-active gases: H_2O and CO_2.

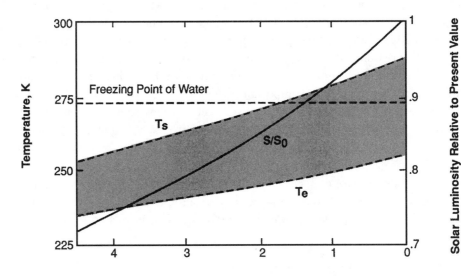

Figure 13.1. Diagram illustrating the faint young Sun problem. The solid curve represents solar luminosity relative to today's value. T_e and T_s are the Earth's effective radiating temperature and surface temperature, respectively. A constant atmospheric CO_2 level and fixed tropospheric humidity were assumed in the calculation. (After Kasting et al., 1988.)

If one extrapolates back to 4.6 Ga, using Equations (1) and (2) and holding the planetary albedo constant, one finds that T_e should have been lower than today by about 22 K. If atmospheric CO_2 had remained constant, T_s should have dropped by an even larger amount, $\sim$33 K, because the amount of water vapor in the atmosphere would have declined as the surface cooled off. The predicted changes in T_e and T_s with time are shown by the dashed curves in Figure 13.1. The surface temperature calculations were performed using a one-dimensional radiative-convective climate model (Kasting and Ackerman, 1986) using a fixed tropospheric relative humidity distribution and neglecting any kind of cloud feedback. The results are similar to those found by Sagan and Mullen (1972), who used a somewhat simpler climate model. In both calculations, T_s drops below the freezing point of water prior to $\sim$2.0 Ga, implying a totally ice-covered Earth. Geologists know, however, that this is not what actually happened on the early Earth, inasmuch as there is evidence for life back to at least 3.5 Ga (Schopf, 1993) and evidence for water-lain sedimentary rocks back to almost 4.0 Ga (Sagan and Mullen, 1972; Schopf, 1983). The conflict between the geologic record and the predictions of such simple climate models has come to be called the "faint young Sun paradox."

The faint young Sun paradox is not a paradox at all, of course, if either Earth's albedo was lower or its greenhouse effect was higher in the past. A greatly lower albedo is possible, in principle, if cloud cover was substantially reduced at that time. This has been suggested several times (Henderson-Sellers, 1979; Rossow et al., 1982), but it has yet to be shown to be very likely. General circulation model (GCM) studies with a swamp ocean predicted a 20% decrease in cloudiness as a consequence of the Earth's faster rotation (Jenkins, 1993; Jenkins et al., 1993), but a more recent study by the same author (Jenkins, 1996) indicates that this effect goes away if sea surface temperatures are specified. A much colder Earth might conceivably have had fewer clouds, but any reduction in planetary albedo would likely have been offset by an increase in highly reflective sea ice. So cloud cover changes remain an improbable solution to the faint young Sun problem.

Fortunately, there is another, more intuitive way of solving this problem, namely, by postulating that the atmospheric greenhouse effect was higher in the past. Several different infrared-active gases might, in principle, have been involved. Sagan and Mullen themselves (1972) suggested ammonia, NH_3. Subsequent investigations (Kuhn and Atreya, 1979; Kasting, 1982) have shown, though, that ammonia would have been rapidly photolyzed to N_2 and H_2, unless it was protected by a UV shield (Sagan and Chyba, 1997). I shall return to the question of possible UV screens in a moment. For now, let us simply note that ammonia is less likely to have contributed substantially to the greenhouse effect on the early Earth than are several other gases.

13.2 CO_2 and the Carbonate-Silicate Cycle

One greenhouse gas that was almost certainly abundant on the early Earth is carbon dioxide, CO_2. CO_2 is the second most important greenhouse gas in the present atmosphere (after H_2O) and, as mentioned at the outset, can provide more than enough warming to solve the faint young Sun problem. (Note that water vapor, by itself, cannot do so because it is always close to its saturation point. Hence, H_2O acts as a feedback mechanism, rather than as a forcing mechanism, for climate change.) The idea that high CO_2 concentrations could have compensated for low solar luminosity was first suggested by Hart (1978), although it remained for Owen et al. (1979) to demonstrate this with a detailed radiative-convective climate model. Subsequent calculations by Kuhn and Kasting (1983), Kasting et al. (1984), and Kiehl and Dickinson (1987) have confirmed this conclusion and have demonstrated reasonable agreement between the different climate models. CO_2 could, in fact, have overcompensated for the faint young Sun and produced a very warm early Earth. Walker (1985) suggested that the CO_2 partial pressure on an ocean-covered early Earth could have been as high as 10 bars for the first several hundred million years of the planet's history. According to Kasting and Ackerman (1986), this would have produced a mean surface temperature of 80–90 °C. Although this sounds extremely warm, there is no evidence to prove that this could not have been the case. Note that the oceans would not have been anywhere near boiling at this temperature because the total atmospheric pressure was considerably higher than 1 bar. Indeed, the oceans would not have boiled until surface temperature reached the critical point (374 °C) because the atmosphere-ocean system works much like a pressure cooker.

It is not only the large potential for greenhouse warming by CO_2 that makes this gas a good candidate for solving the faint young Sun problem. There are good, independent reasons for expecting that CO_2 levels should have been high on the early Earth. On long ($>10^6$ year) time scales, atmospheric CO_2 concentrations are controlled primarily by the carbonate-silicate cycle (Figure 13.2). CO_2 dissolves in rainwater, forming carbonic acid (H_2CO_3). Carbonic acid dissolves silicate rocks on the continents in a process called *silicate weathering*. The by-products of silicate weathering,

THE CARBONATE-SILICATE CYCLE

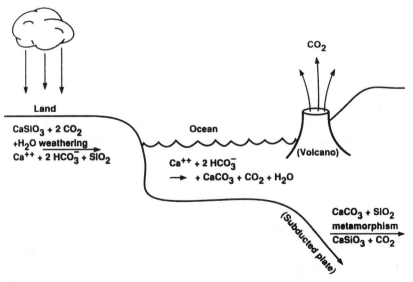

Figure 13.2. Diagram illustrating the carbonate-silicate cycle.

including calcium (Ca^{++}) and bicarbonate (HCO_3^-) ions, are carried by rivers down to the ocean, where certain organisms, such as planktonic foraminifera, use them to make shells of calcium carbonate ($CaCO_3$). When these organisms die, some of the calcium carbonate collects in sediments on the ocean floor. These carbonate sediments are carried away from the mid-ocean ridge axes by the process of seafloor spreading. At some plate boundaries, the seafloor is subducted and the carbonate sediments are carried down into the upper mantle. When this happens, the carbonate sediments are heated, and carbonate minerals are converted back into silicate minerals (carbonate metamorphism), releasing gaseous CO_2 in the process. This CO_2 is returned to the atmosphere by volcanism. The rate of this cycle is sufficient to recycle all of the CO_2 in the combined atmosphere-ocean system in about half a million years. (On long time scales, it is legitimate to consider the atmosphere and the ocean as one combined reservoir for CO_2. The ocean contains most of the carbon – about 60 times the amount that resides in the atmosphere.)

As pointed out originally by Walker et al. (1981), the carbonate-silicate cycle provides a natural solution to the faint young Sun problem. If Earth's oceans were to have frozen over early in the planet's history because of low solar luminosity, the hydrologic cycle would have come to a virtual standstill, and silicate weathering on land would have ceased. Carbonate metamorphism would have continued, however, inasmuch as there is a huge reservoir of carbonate rocks – the equivalent of 60–80 bars of CO_2 – stored in Earth's crust. Thus, on a frozen Earth, volcanic CO_2 should have accumulated in the atmosphere until the greenhouse effect became large enough to melt the ice. As we shall see below, this sequence of global glaciation followed by CO_2 build-up and melting may actually have occurred two or three times during Earth's history. During most of Earth's history, however, the negative feedback inherent in this cycle should have prevented this from happening. The reduction in the silicate weathering rate as Earth's surface cooled would have driven atmospheric CO_2 concentrations higher, warming the climate and preventing further ice build-up. The evidence for this is that Earth has remained habitable, and usually ice-free, throughout most of its history.

If one accepts the logic outlined above, then it is possible to turn the argument around and to ask how much CO_2 would have been required to compensate for reduced solar luminosity at different times in Earth's history, assuming that CO_2 and H_2O were the two dominant greenhouse gases. The answer is shown in Figure 13.3 (from Kasting, 1992). At 4.5 Ga, a minimum of 0.1 bars of CO_2, or ~300 times the present atmospheric level (PAL), would have been needed to compensate for a 30% reduction in solar luminosity. On the other hand, CO_2 concentrations may actually have been much higher than this (10 bars?) at that time. Somewhat tighter constraints on pCO_2 can be derived during glacial periods, when both upper and lower limits on surface temperature can be estimated. The limits shown in Figure 13.3 correspond to global mean temperatures between 5 °C and 20 °C during the Paleoproterozoic (2.2–2.3 Ga) and Neoproterozoic (0.6–0.8 Ga) glaciations (Kasting, 1992). Keep in mind, though, that these estimates assume that CO_2 and H_2O were the only important greenhouse gases (which may not have been true prior to ~2.2 Ga), and they do not account for transient excursions in CO_2 that may have occurred during global glaciations.

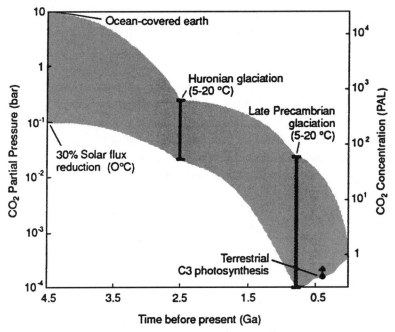

Figure 13.3. Atmospheric CO_2 levels required to compensate for reduced solar luminosity in the past. The right-hand scale is the CO_2 concentration in units of PAL, or "times the present atmospheric level." (After Kasting, 1992.)

13.3   Other Greenhouse Gases: CH_4

Although CO_2 could, in principle, have solved the faint young Sun problem by itself, there are reasons for suspecting that other greenhouse gases were involved. One of these reasons stems from a study of paleosols (ancient soils) by Rye et al. (1995). Rye et al. examined several paleosols ranging in age from 2.2 Ga to 2.8 Ga and looked for the presence of the mineral siderite, $FeCO_3$. According to their analysis, if atmospheric pCO_2 was above a certain value, siderite should have formed as these soils were weathered. None of the paleosols they looked at, though, contained even a trace of this mineral. Rye et al. concluded that atmospheric CO_2 concentrations during this time interval were a factor of 20 or more lower than the minimum concentration predicted by the climate model (Figure 13.3). (The actual upper limit on pCO_2 from the paleosol data depends on surface temperature because the equilibrium reactions involved in siderite formation are temperature-dependent.)

Now, the absence of siderite from a soil profile is not conclusive evidence that atmospheric CO_2 levels were low, inasmuch as one can think of other reasons (e.g., subsequent dissolution) that this mineral might have disappeared. However, there are other reasons for suspecting that CO_2 and H_2O were not the only important greenhouse gases at this time. Most analyses of the rise of atmospheric oxygen suggest that a major increase in O_2 levels occurred at about 2.0–2.2 Ga (Cloud, 1972; Walker et al., 1983; Kasting, 1993; Holland, 1994). (For an alternative viewpoint, see Ohmoto, 1996, 1997.) If Walker et al. (1983) and Kasting (1993) are correct, then the atmosphere was essentially anoxic prior to 2.2 Ga. Under such circumstances, the abundance of certain reduced gases, CH_4 and H_2 in particular, could have been much higher than today. H_2 does not absorb radiation at thermal infrared wavelengths and, hence, would not have directly affected climate, but CH_4 does absorb and might therefore have been important. This fact was recognized by Kiehl and Dickinson (1987), who pointed out that the presence of significant CH_4 concentrations in the early atmosphere could have allowed CO_2 levels to be lower without violating the geologic constraints on Earth's climate. They performed radiative-convective temperature calculations for various combinations of CH_4, CO_2, and solar flux. Since that time, my students and I have performed additional calculations with a climate model that contains a more up-to-date database of methane absorption lines (Pavlov et al., 2000). We have also performed photochemical modeling (Brown, 1999) to estimate how high atmospheric CH_4 levels might have been during the period (2.2–2.8 Ga) when the paleosol constraints apply.

Most of the methane that is in the atmosphere today was produced by methanogenic bacteria that live in anaerobic environments such as rice paddies, wetlands, and the intestines of ruminants (especially cows). These bacteria make their metabolic living by using hydrogen to reduce CO_2

$$CO_2 + 4H_2 \rightarrow CH_4 + 2H_2O$$

and by similar reactions involving acetate or formate as the electron donor. On an anoxic early Earth, atmospheric H_2 concentrations should have been high enough – 10^{-4} atm or more (Kasting, 1993) – that methanogenic bacteria could have lived directly at Earth's surface. (The minimum threshold for metabolism in methanogenic bacteria is $\sim$65 μatm [Lovley, 1985]. However, typical Michaelis-Menton K_m coefficients for growth are about 4–8 μM dissolved H_2, which corresponds to H_2 partial pressures of 5.5×10^{-3} to 1.1×10^{-2} atm [Zinder, 1993, p. 143].) Furthermore, methanogens are believed to have originated very early during biological evolution, based on sequencing of their 16S ribosomal RNA (Woese and Fox, 1977). Hence, there is every reason to expect that they were abundant during the Archean Era (Walker, 1977; Kasting et al., 1983; Walker et al., 1983). Additional evidence for the past presence of methanogens comes from isotopically light (-60‰!) kerogens in rocks dated at 2.8 Ga. Such extreme fractionations are thought to have been produced by a two–step process involving methanogenesis followed by assimilation of isotopically light CH_4 by methanotrophic bacteria (Hayes, 1983, 1994).

It is difficult to predict what the methane flux on the Archean Earth might have been, but one can use photochemical models to examine the consequences of different assumptions. Figure 13.4 shows the results of calculations performed by L. Brown as part of her thesis work here at Penn State (Kasting and Brown, 1998; Brown, 1999). The model used was a one-dimensional photochemical model similar to that of Kasting et al. (1983), Zahnle (1986), and Kasting (1990), but with updated hydrocarbon rate constants. The solid curve in Figure 13.4 shows the calculated photochemical destruction rate of CH_4 as a function of atmospheric CH_4 mixing ratio. Most of the destruction of CH_4 occurs either by photolysis at Ly α (121.6 nm) or by reaction with OH. Because there is little photochemical production of CH_4 within the atmosphere, this destruction rate

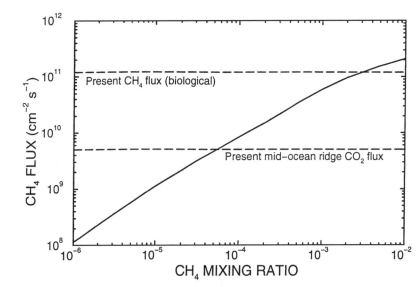

Figure 13.4. CH_4 destruction rate as a function of atmospheric CH_4 mixing ratio, as calculated with a photochemical model. The CH_4 source must equal the destruction rate in order to maintain a steady state. The dashed lines show the present biological CH_4 source and a possible abiotic CH_4 source from a highly reduced mantle. (After Kasting and Brown, 1998.)

must be balanced by an influx of methane from surface sources. The two dashed lines show possible biotic and abiotic surface fluxes of CH_4. The upper line is the present biological CH_4 flux, $\sim 3 \times 10^{13}$ mol yr^{-1}, or 1×10^{11} CH_4 molecules cm^{-2} s^{-1} (Watson et al., 1990). The lower line is the present abiotic source of carbon from mid-ocean ridge outgassing, $\sim 1.5 \times 10^{12}$ mol yr^{-1} (DesMarais and Moore, 1984; Marty and Jambon, 1987). Most of this carbon is emitted as CO_2 at present, but it could have been released as CH_4 in the past if the upper mantle was more reduced at that time (Kasting and Brown, 1998). The abiotic source was probably not important during the Archean because most of the methane should have been produced biologically.

Figure 13.4 shows that a CH_4 flux comparable to the present biological methane flux could have produced an atmospheric CH_4 mixing ratio of $\sim 3 \times 10^{-3}$. By comparison, the present atmospheric CH_4 mixing ratio is ~ 1.6 ppm, or 1.6×10^{-6}. The difference arises because the photochemical lifetime of CH_4 is about 2000 times longer in a weakly reduced primitive atmosphere than it is today ($\sim 20,000$ years, as compared with 12 years). A caveat should be added here: This high CH_4 concentration implies a correspondingly high escape rate of hydrogen to space if hydrogen was escaping at the diffusion-limited rate (Walker, 1977). The methanogenic biota might have run out of H_2 unless hydrogen escape was actually slower than this. It seems likely, however, that this was indeed the case because the exospheric temperature was probably low and the escape was energy-limited (Kasting and Brown, 1998). So an Archean atmosphere containing 0.1% CH_4 is entirely plausible.

Suppose that the Archean atmosphere was indeed methane-rich. What would have been the implications for climate? Recent radiative-convective model calculations by our group here at Penn State (Pavlov et al., 2000) are shown in Figure 13.5. The solid curves

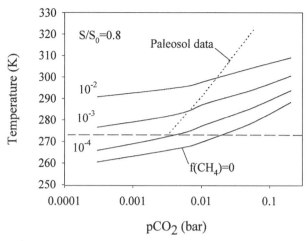

Figure 13.5. Global mean surface temperature as a function of atmospheric CO_2 and CH_4 concentrations, as calculated with a radiative-convective climate model. The solar flux is assumed to be 80% of its present value. The dotted curve represents the upper limit on pCO_2 obtained from paleosols (Rye et al., 1995). (After Pavlov et al., 1999.)

show calculated mean surface temperatures at 2.8 Ga for different amounts of CH_4 and CO_2. According to Equation (1), the solar flux was about 80% of its present value at that time. The dashed line shows the freezing temperature of water, and the dotted curve represents the upper limit on CO_2 derived from paleosols (Rye et al., 1995). The figure shows that, for a methane-free atmosphere, the amount of CO_2 required to keep the surface temperature above freezing is considerably higher than the limit imposed by the paleosol data. For $f(CH_4)$ greater than a few times 10^{-4}, however, the disagreement disappears. For example, an atmosphere with $f(CH_4) = 3 \times 10^{-3}$ – the concentration predicted for the present-day methane production rate – could have maintained a mean surface temperature within a few degrees of the present value, 288 K, even if atmospheric CO_2 concentrations were no higher than today.

I conclude that, if methane was abundant in the Archean atmosphere, CO_2 concentrations may have been well below those shown in Figure 13.3. Indeed, the silicate-weathering feedback should have worked to ensure this: high CH_4 concentrations should have led to warmer surface temperatures, which in turn should have led to faster rates of silicate weathering and increased loss of CO_2.

I should add one caveat and one corollary. The caveat is this: Photochemical modeling suggests that methane can be polymerized to higher hydrocarbons, forming Titan-like smog, when the C/O ratio in the atmosphere is greater than ~1 (Zahnle, 1986; Brown, 1999). For the atmospheres discussed here, this condition is equivalent to having $f(CH_4)/f(CO_2) > 1$. Thus, the calculations shown in the upper left of Figure 13.5, which do not account for the presence of such haze, may not be realistic. The presence of such a haze would have affected Archean climate both by absorbing incoming solar radiation and by increasing the magnitude of the greenhouse effect (Sagan and Chyba, 1997). To determine what the net effect on surface temperature would have been requires detailed modeling.

The corollary, which was actually anticipated some time ago (Walker et al., 1983; Kasting et al., 1983), is that the rise of O_2 around 2.2 Ga could have caused a significant drop in global surface temperature by wiping out the methane component of the atmospheric greenhouse. This, in turn, may have triggered the first episode of widespread glaciation in Earth's history during the Paleoproterozoic. Indeed, in at least one locality (the Huronian sequence of Canada), glacial deposits are underlain by sedimentary rocks that appear to have been deposited under reducing conditions and are overlain by rocks that were deposited under oxidizing conditions (Roscoe, 1969, 1973; Walker et al., 1983). This is entirely consistent with the hypothesis outlined above, in which methane played a substantial role in warming the Archean climate.

13.4   The Mystery of Low-Latitude Precambrian Glaciation

This brings me to my last topic: the issue of low-latitude Precambrian glaciation. On the basis of paleomagnetic data, it has been argued for many years that at least some of the glacial deposits from the Neoproterozoic (0.85–0.65 Ga) were formed at low latitudes (Frakes, 1979; Zhang and Zhang, 1985; Embleton and Williams, 1986; Williams, 1993, and other references therein). Recently, Evans et al. (1997) and Williams and Schmidt (1997) have suggested that the Paleoproterozoic glaciation ($\sim$2.2 Ga) may also have occurred at low latitudes. If these analyses are correct, then glacial climates during the Precambrian must have been markedly different from those of the Phanerozoic. Three different possibilities come to mind to explain the observations: (1) Earth's magnetic field was aligned at a large angle to its spin axis; (2) Earth's obliquity was much higher than today, so the equator was colder than the poles; or (3) the entire Earth was glaciated from the poles down to the Tropics. This last idea has been termed the "snowball Earth hypothesis" (Kirshvink, 1992; Hoffman et al., 1998).

A change in the alignment of Earth's magnetic field would, of course, make the problem of low-latitude glaciation disappear because rocks that are believed to have crystallized near the equator might actually have formed near one of the poles. Proponents of this idea might point out that both Neptune and Uranus have magnetic fields that are tilted with respect to their spin axes (Beatty and Chaikin, 1990, Ch. 3). Their fields also have a large quadrupole moment in addition to the dipole contribution. However, these giant planets are very different in structure from Earth; in particular, their magnetic fields are probably generated by convection of metallic hydrogen throughout a large portion of their mantles. By contrast, Earth's predominantly dipole magnetic field is generated by convection in the liquid outer core. It has higher moments as well, but these higher-order terms have largely decayed by the time one reaches the Earth's surface. And there is no indication from the recent past that it is possible for the dipole component of the field to be offset from the spin axis by much more than it is today, about 11 degrees. So it is completely ad hoc to suggest that this happened during the Precambrian. One must look elsewhere to explain the evidence for low-latitude glaciation.

The second idea – the high-obliquity hypothesis – was suggested many years ago by the Australian geologist George Williams (1975) and has been advocated by him ever since (Williams, 1993, and references therein). Its chief strength is that it would allow the poles to remain ice-free at the same time that the Tropics were glaciated. (At obliquities above ~54°, the equator receives less annually averaged solar insolation than do the poles [Ward, 1974].) This avoids the main problem with the snowball Earth hypothesis, namely, explaining how life could have made it through such a catastrophe. With that thought in mind, my student Darren Williams performed a series of calculations to see whether it is conceivable that Earth's obliquity could have changed (Williams et al., 1998). The one physical mechanism that might have caused this to happen is so-called "climate friction" – the feedback between Earth's obliquity and its oblateness (Rubincam, 1993; Bills, 1994). Earth's oblateness changes as ice accumulates at the poles, or at the equator during a postulated high-obliquity phase. If this happens in a particular phase relationship with Earth's normal 2.5° obliquity oscillation, then the obliquity can gradually drift toward either higher or lower values. We found that if the retreat of the ice lagged the peak solar insolation by less than ~20° or more than ~200°, the obliquity change could be negative, that is, in the proper direction needed to make George Williams's hypothesis work. Such an obliquity change could also explain the 5° inclination of the lunar orbit with respect to the ecliptic plane, which has been a thorn in the side of celestial mechanicians for many years (Goldreich, 1966; Rubincam, 1975). Since the time our calculations were performed, however, a new theory has been put forward that may explain the Moon's orbit (Touma and Wisdom, 1998). Inasmuch as it is also difficult to imagine why the phase lag should be so long in the climate friction mechanism, I now consider the high-obliquity hypothesis to be rather improbable. It remains, however, as one conceivable alternative to the theory outlined next.

13.5   The Snowball Earth Hypothesis

A third way of explaining the evidence for low-latitude glaciation is the snowball Earth hypothesis. This idea has been around for some time (Kirshvink, 1992), but it has gained new credibility from the recent study by Hoffman et al. (1998). Despite assertions to the contrary (Jenkins and Scotese, 1998), there is no reason why global glaciation could not have occurred. If the polar ice sheets were ever to have advanced far enough equatorward, Earth's albedo would have become very high, and freezing of the remaining parts of the ocean is to be expected. This point is illustrated by energy-balance climate model calculations by Caldeira and Kasting (1992) (Figure 13.6). In this type of climate model, the equator-to-pole temperature gradient is simulated by parameterizing latitudinal heat transport by the atmosphere and oceans in the form of a diffusion equation, with a diffusion coefficient chosen to match present-day conditions. The solid curves in Figure 13.6 represent stable solutions for the sine of the ice-line latitude in this model, whereas dashed curves represent unstable solutions. Inspection of the rightmost curve, for $pCO_2 = 3 \times 10^{-4}$ bar (the present value), shows that the ice line is unstable for latitudes less than ~30° (sine latitude $\cong 0.5$). If the ice line moves equatorward of this point, it will advance to the stable, completely ice-covered solution at the bottom of the diagram. The critical latitude at which the instability sets in depends on the details of the climate model, in particular the assumed albedo of snow and sea ice (0.66 in the Caldeira and Kasting model). Crowley and Baum's (1993) climate model

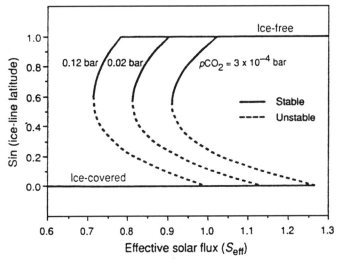

Figure 13.6. Ice-line latitude as a function of solar flux and atmospheric CO_2 level, as calculated by an energy-balance climate model. The solid curves represent stable solutions; dashed curves represent unstable solutions. (After Caldeira and Kasting, 1992.)

is apparently stable with ice sheets down to 25°, but it, too, should become unstable at some point, given similar assumptions about the high albedo of snow and ice.

Global glaciation can be reversed by the build-up of CO_2 in Earth's atmosphere. According to Figure 13.6, an accumulation of 0.12 bars of CO_2 would cause the climate system to revert to an ice-free state, given present solar luminosity. Somewhat higher CO_2 levels might have been required during the Late Precambrian, when the solar flux was only ∼94% of its present value (Equation 1). At a CO_2 outgassing rate of 8×10^{12} mol/yr (Holland, 1978), 0.12 bars of CO_2 could accumulate in ∼3 million years, assuming no transfer of CO_2 between the atmosphere and oceans. Considerably higher atmospheric CO_2 concentrations might have been required to reverse the early Proterozoic glaciation, when solar luminosity was only 84% of its present value. Indeed, Caldeira and Kasting (1992) predicted that global glaciation might have been irreversible during the first part of Earth history because of the formation of CO_2 ice clouds. This prediction now appears to have been incorrect, though, because such clouds should warm the surface, rather than cool it (Forget and Pierrehumbert, 1997).

The preceding discussion begs the question of how global glaciation could have been triggered in the first place, given the negative feedback on climate provided by the carbonate-silicate cycle. The reason it could have happened is that the ocean surface can freeze much faster than the atmospheric CO_2 level can respond to it. A simple calculation using Equation 2 shows that if the planetary albedo were increased from 0.3 to 0.6 (its approximate value in the Caldeira and Kasting snowball Earth model), the planetary energy deficit would be sufficient to freeze the topmost kilometer of the oceans within less than a thousand years. Thus, what could have happened in the Late Precambrian is that the polar ice sheets crept slowly down toward the critical latitude as a consequence of a gradual drawdown of atmospheric CO_2. The drawdown might have been caused by supercontinent breakup and a concomitant increase in organic carbon burial on newly created continental shelves, as suggested by Hoffman et al. (1998). Perhaps a better explanation, though, is the suggestion by Marshall et al. (1988): clustering of the continents near the equator allowed silicate weathering to proceed in spite of a generally cold global climate. CO_2 was drawn down very low, perhaps near present-day concentrations, and the polar ice caps advanced to the critical latitude. When they reached that point, even a short-lived climatic perturbation, such as a volcanic eruption, could have caused the remaining tropical oceans to freeze over.

What would conditions have been like during such a global glaciation? This is a difficult question and one that deserves consideration because it has important implications for the survival of the biota. If the ice had remained clean, then the planetary albedo should have been ~0.62 (Caldeira and Kasting, 1992), and Equation 2 predicts that $T_e \cong 220$ K (compared with 255 K today). At the equator, solar insolation is ~25% higher than the global average, but even there T_e would have been only ~230 K. The atmospheric greenhouse effect was, if anything, smaller than today because of the lower amount of water vapor. I shall be generous and assume it was 25 K. Then, even at the equator, the mean annual surface temperature was only ~255 K. The thickness of the ice covering the oceans would have been limited by the geothermal heat flux (~0.06 W/m^2), which would have had to escape through the ice by conduction (Bada et al., 1994). The conductive heat flux, in turn, is

$$F = \frac{\lambda \Delta T}{\Delta z} \tag{3}$$

where λ ($= 2$ W/m/K) is the thermal conductivity of ice, ΔT is the temperature difference across the ice layer, and Δz is its thickness. Assuming that the bottom of the ice was at 271 K (the freezing point of seawater), the ice thickness at the equator should have been ~530 m. Everywhere else, the ice would have been thicker.

Here lies the paradox of the snowball Earth model. How did life make it through such a catastrophe? Half a kilometer of ice would have been totally opaque to sunlight, so photosynthetic organisms could not have survived beneath the ice surface. The continents would also have had subfreezing climates, so it is not obvious how life could have survived there. Bacteria (or Archea) that live in subsurface environments such as the Columbia River basalt (Stevens and McKinley, 1995) or within the mid-ocean ridge vent systems would not even know that the glaciation had occurred, but such organisms are phylogenetically distinct from most modern forms of life, so it is highly unlikely that life re-evolved from them at this late stage in Earth history.

The solution to this paradox may lie in local, geothermal-rich environments such as Iceland, where heat from Earth's interior is concentrated in a particular area. Iceland itself is basically a piece of the Mid-Atlantic spreading ridge that sticks up above the ocean surface. Photosynthetic organisms may have survived the climate catastrophe by living in warm pools in volcanic areas such as this.

Alternatively, the answer to the paradox may have to do with the cleanness of the ice. After the oceans had frozen over, evaporation would have been shut off and snowfall over the continents would have dropped to minuscule levels. Ice and snow covering the continents would have been worn away by ablation, exposing dry, dusty soil underneath.

The dust would have been picked up by the winds and deposited globally, darkening the ice. The darker the ice, the higher the surface temperature, and the thinner the ice cover would have been, according to the argument just given. This process is self-limiting, however: if the sea-ice cover had begun to melt back substantially, evaporation would have been reestablished, fresh snow would have fallen, and the planetary albedo would have crept upward once again. Thus, it seems unlikely that dust by itself could have ended the climate catastrophe: the Earth would have remained glaciated until volcanic CO_2 built up in the atmosphere. If anything, the time scale for recovery might have been lengthened by the dust, inasmuch as transient melting of ice would have allowed some atmospheric CO_2 to dissolve in the oceans.

These ideas are all speculative, of course. The snowball Earth hypothesis requires more testing before it can be said to be well established. In particular, it would be nice to acquire accurate age dates and paleolatitudes on as many Neoproterozoic and Paleoproterozoic glacial deposits as possible and thereby confirm whether the glaciations were synchronous on all the continents, as the theory would predict. (The high obliquity hypothesis, by contrast, would make a distinctly different set of predictions.) This is a fascinating topic for future research because it poses interesting questions for biological evolution here on Earth and because it may also have implications for the habitability of other planets within or outside our own solar system.

13.6   Conclusions

Earth has remained habitable throughout most of its history despite a large increase in solar luminosity over that time. The most likely solution to the faint young Sun problem involves higher concentrations of atmospheric CO_2 in the past, possibly bolstered by high concentrations of CH_4 prior to ~ 2.2 Ga. Elimination of the methane component of the greenhouse by the rise of atmospheric O_2 at this time may have triggered the Paleoproterozoic glaciation. The stabilizing feedback between atmospheric CO_2 concentrations and climate is not foolproof because the time required for the oceans to freeze is much shorter than the time required for volcanic CO_2 to build up in the atmosphere. Thus, transient global glaciations are possible and may well have occurred during both the early and the late Proterozoic. Life, however, is evidently resilient to even a catastrophe of this magnitude. This suggests that the chances of finding life elsewhere in the galaxy may be even greater than previously thought.

REFERENCES

Bada, J. L., Bigham, C., and Miller, S. L. (1994). Impact melting of frozen oceans on the early Earth: Implications for the origin of life. *Proc. Nat. Acad. Sci.*, 91, 1248–1250.

Beatty, J. K., and Chaikin, A. (1990). *The New Solar System* (Cambridge, MA: Sky Publishing Corp.).

Bills, B. G. (1994). Obliquity-oblateness feedback: Are climatically sensitive values of the obliquity dynamically unstable? *Geophys. Res. Lett.*, 21, 177–180.

Brown, L. L. (1999). *Photochemistry and Climate on Early Earth and Mars*. Ph.D. dissertation, Penn State University.

Caldeira, K., and Kasting, J. F. (1992). Susceptibility of the early Earth to irreversible glaciation caused by carbon dioxide clouds. *Nature*, 359, 226–228.

Cloud, P. E. (1972). A working model of the primitive Earth. *Amer. J. Sci.*, 272, 537–548.

Crowley, T. J., and Baum, S. K. (1993). Effect of decreased solar luminosity on Late Precambrian ice extent. *J. Geophys. Res.*, 98, 16,723–716,732.

DesMarais, D. J., and Moore, J. G. (1984). Carbon and its isotopes in mid-oceanic basaltic glasses. *Earth Planet. Sci. Lett.*, 69, 43–57.

Embleton, B. J., and Williams, G. E. (1986). Low paleolatitude of deposition for the Late Precambrian periglacial varvites in South Australia: Implications for paleoclimatology. *Earth Planet. Sci. Lett.*, 79, 419–430.

Evans, D. A., Beukes, N. J., and Kirshvink, J. L. (1997). Low-latitude glaciation in the Proterozoic era, *Nature*, 386, 262–266.

Forget, F., and Pierrehumbert, R. T. (1997). Warming early Mars with carbon dioxide clouds that scatter infrared radiation. *Science*, 278, 1273–1276.

Frakes, L. A. (1979). *Climates Throughout Geologic Time* (New York: Elsevier), 310 pp.

Gilliland, R. L. (1989). Solar evolution. *Global Planet. Change*, 1, 35–55.

Goldreich, P. (1966). History of the lunar orbit, *Rev. Geophys.*, 4, 411–439.

Gough, D. O. (1981). Solar interior structure and luminosity variations. *Solar Phys.*, 74, 21–34.

Hart, M. H. (1978). The evolution of the atmosphere of the Earth. *Icarus*, 33, 23–39.

Hayes, J. M. (1983). Geochemical evidence bearing on the origin of aerobiosis, a speculative hypothesis. In *Earth's Earliest Biosphere: Its Origin and Evolution*, ed. Schopf, J. W. (Princeton, New Jersey: Princeton University Press), 291–301.

Hayes, J. M. (1994). Global methanotrophy at the Archean-Proterozoic transition. In *Early Life on Earth*, ed. Bengtson, S. (New York: Columbia University Press), 220–236.

Henderson-Sellers, A. (1979). Clouds and the long term stability of the Earth's atmosphere and climate. *Nature*, 279, 786–788.

Hoffman, P. F., Kaufman, A. J., Halverson, A. J., and Schrag, D. P. (1998). A Neoproterozoic snowball Earth. *Science*, 281, 1342–1346.

Holland, H. D. (1978). *The Chemistry of the Atmosphere and Oceans* (New York: Wiley), 351 pp.

Holland, H. D. (1994). Early Proterozoic atmospheric change. In *Early Life on Earth*, Bengtson, S. ed. (New York: Columbia University Press), 237–244.

Jenkins, G. S. (1993). A general circulation model study of the effects of faster rotation, enhanced CO_2 concentrations, and solar forcing: Implications for the faint young Sun paradox. *J. Geophys. Res.*, 98, 20,803–20,811.

Jenkins, G. S. (1996). A sensitivity study of changes in Earth's rotation rate with an atmospheric general circulation model. *Global and Planet. Change*, 11, 141–154.

Jenkins, G. S., Marshall, H. G., and Kuhn, W. R. (1993). Precambrian climate: the effects of land area and Earth's rotation rate. *J. Geophys. Res.*, 98, 8785–8791.

Jenkins, G. S., and Scotese, C. (1998). Comment on "The snowball Earth hypothesis." *Science*, 282, 1644–1645.

Kasting, J. F. (1990). Bolide impacts and the oxidation state of carbon in the Earth's early atmosphere. *Origins of Life*, 20, 199–231.

Kasting, J. F. (1993). Earth's early atmosphere. *Science*, 259, 920–926.

Kasting, J. F. (1992). Proterozoic climates: the effect of changing atmospheric carbon dioxide doncentrations. In *The Proterozoic Biosphere: A Multidisciplinary Study*, eds. Schopf, J. W., and Klein, C. (Cambridge: Cambridge University Press), 165–168.

Kasting, J. F. (1982). Stability of ammonia in the primitive terrestrial atmosphere. *J. Geophys. Res.*, 87, 3091–3098.

Kasting, J. F., and Ackerman, T. P. (1986). Climatic consequences of very high CO_2 levels in the Earth's early atmosphere. *Science*, 234, 1383–1385.

Kasting, J. F., and Brown, L. L. (1998). Setting the stage: the early atmosphere as a source of biogenic compounds. In *The Molecular Origins of Life: Assembling the Pieces of the Puzzle*, ed. Brack, A. (New York: Cambridge University Press), pp. 35–56.

Kasting, J. F., Pollack, J. B., and Crisp, D. (1984). Effects of high CO_2 levels on surface temperature and atmospheric oxidation state of the early Earth. *J. Atmos. Chem.*, 1, 403–428.

Kasting, J. F., Zahnle, K. J., and Walker, J. C. G. (1983). Photochemistry of methane in the Earth's early atmosphere. *Precambrian Res.*, 20, 121–148.

Kasting, J. F., Pollack, J. B., and Toon, O. B. (1988). How climate evolved on the terrestrial planets. *Scientific American*, 258, no. 2, 90–97.

Kiehl, J. T., and Dickinson, R. E. (1987). A study of the radiative effects of enhanced atmospheric CO_2 and CH_4 on early Earth surface temperatures. *J. Geophys. Res.*, 92, 2991–2998.

Kirschvink, J. L. (1992). Late Proterozoic low-latitude global glaciation: the snowball Earth. In *The Proterozoic Biosphere: A Multidisciplinary Study*, eds. Schopf, J. W., and Klein, C. (Cambridge: Cambridge University Press), 51–52.

Kuhn, W. R., and Atreya, S. K. (1979). Ammonia photolysis and the greenhouse effect in the primordial atmosphere of the Earth. *Icarus*, 37, 207–213.

Kuhn, W. R., and Kasting, J. F. (1983). The effects of increased CO_2 concentrations on surface temperature of the early Earth. *Nature*, 301, 53–55.

Lovley, D. R. (1985). Minimum threshold for hydrogen metabolism in methanogenic bacteria. *Appl. Environ. Microbiol.*, 49, 1530–1531.

Marshall, H. G., Walker, J. C. G., and Kuhn, W. R. (1988). Long-term climate change and the geochemical cycle of carbon. *J. Geophys. Res.*, 93, 791–802.

Marty, B., and Jambon, A. (1987). $C/^3He$ in volatile fluxes from the solid Earth: Implications for carbon geodynamics. *Earth Planet. Sci. Lett.*, 83, 16–26.

Ohmoto, H. (1996). Evidence in Pre-2.2 Ga paleosols for the early evolution of atmospheric oxygen and terrestrial biota. *Geology*, 24, 1135–1138.

Ohmoto, H. (1997). When did the Earth's atmosphere become oxic? *Geochemical News*, 93 (Fall), 13.

Owen, T., Cess, R. D., and Ramanathan, V. (1979). Early Earth: An enhanced carbon dioxide greenhouse to compensate for reduced solar luminosity. *Nature*, 277, 640–642.

Pavlov, A. A., Kasting, J. F., Brown, L. L., Rages, K. A., and Freedman, R. (2000). Greenhouse warming by CH_4 in the atmosphere of early Earth. *J. Geophys. Res.*, 105, 11981–11990.

Roscoe, S. M. (1969). Huronian rocks and uraniferous conglomerates in the Canadian shield. *Geol. Surv. Can. Pap.*, 68–40.

Roscoe, S. M. (1973). The Huronian supergroup: a Paleophebian succession showing evidence of atmospheric evolution. *Geol. Soc. Can. Spec. Pap.*, 12, 31–48.

Rossow, W. B., Henderson-Sellers, A., and Weinrich, S. K. (1982). Cloud feedback: A stabilizing effect for the early Earth? *Science*, 217, 1245–1247.

Rubincam, D. P. (1993). The obliquity of Mars and "climate friction." *J. Geophys. Res.*, 98, 10,827–10,832.

Rubincam, D. P. (1975). Tidal friction and the early history of the Moon's orbit. *J. Geophys. Res.*, 80, 1537–1548.

Rye, R., Kuo, P. H., and Holland, H. D. (1995). Atmospheric carbon dioxide concentrations before 2.2 billion years ago. *Nature*, 378, 603–605.

Sagan, C., and Chyba, C. (1997). The early faint Sun paradox: Organic shielding of ultraviolet-labile greenhouse gases. *Science*, 276, 1217–1221.

Sagan, C., and Mullen G. (1972). Earth and Mars: Evolution of atmospheres and surface temperatures. *Science*, 177, 52–56.

Schopf, J. W. (1983). *Earth's Earliest Biosphere: Its Origin and Evolution* (Princeton, New Jersey: Princeton University Press).

Schopf, J. W. (1993). Microfossils of the early Archean Apex Chert: New evidence for the antiquity of life. *Science*, 260, 640–646.

Stevens, T. O., and McKinley, J. P. (1995). Lithoautotrophic microbial ecosystems in deep basalt aquifers. *Science*, 270, 450–454.

Touma, J., and Wisdom, J. (1998). Resonances in the early evolution of the Earth-Moon system. *Astron. J.*, 115, 1653–1663.

Walker, J. C. G. (1985). Carbon dioxide on the early Earth. *Origins of Life*, 16, 117–127.

Walker, J. C. G. (1977). *Evolution of the Atmosphere* (New York: Macmillan).

Walker, J. C. G., Hays, P. B., and Kasting, J. F. (1981). A negative feedback mechanism for the long-term stabilization of Earth's surface temperature. *J. Geophys. Res.*, 86, 9776–9782.

Walker, J. C. G., Klein, C., Schidlowski, M., Schopf, J. W., Stevenson, D. J., and Walter, M. R. (1983). Environmental evolution of the Archean-Early Proterozoic Earth. In *Earth's Earliest Biosphere: Its Origin and Evolution*, ed. Schopf, J. W. (Princeton, New Jersey: Princeton University Press), 260–290.

Ward, W. R. (1974). Climatic variations on Mars. 1. Astronomical theory of insolation. *J. Geophys. Res.*, 79, 3375–3386.

Watson, R. T., Rodhe, H., Oeschger, H., and Siegenthaler, U. (1990). Greenhouse gases and aerosols. In *Climate Change: The IPCC Scientific Assessment*, eds. Houghton, J. T., Jenkins, G. J., and Ephraums, J. J. (Cambridge: Cambridge University Press), 1–40.

Williams, D. M., Kasting, J. F., and Frakes, L. A. (1998). Low-latitude glaciation and rapid changes in the Earth's obliquity explained by obliquity-oblateness feedback. *Nature*, 396, 453–455.

Williams, G. E. (1993). History of the Earth's obliquity. *Earth Sci. Rev.*, 34, 1–45.

Williams, G. E. (1975). Late Precambrian glacial climate and the Earth's obliquity. *Geological Magazine*, 112, 441–465.

Williams, G. E., and Schmidt, P. W. (1997). Paleomagnetism of the Palaeoproterozoic Gowganda and Lorraine formations, Ontario: Low paleolatitude for Huronian glaciation. *Earth Planet. Sci. Lett.*, 153, 157–169.

Woese, C. R., and Fox, G. E. (1977). Phylogenetic structure of the prokaryotic domain: the primary kingdoms. *Proc. Natl. Acad. Sci. USA*, 74, 5088.

Zahnle, K. J. (1986). Photochemistry of methane and the formation of hydrocyanic acid (HCN) in the Earth's early atmosphere. *J. Geophys. Res.*, 91, 2819–2834.

Zhang, H, and Zhang, W. (1985). Paleomagnetic data, Late Precambrian magnetostratigraphy, and tectonic evolution of eastern China. *Precambrian Res.*, 29, 65–75.

Zinder, S. H. (1993). Physiological ecology of methanogens. In *Methanogenesis: Ecology, Physiology, Biochemistry, and Genetics*, ed. Ferry, J. G. (New York, Chapman and Hall), 128–206.

14 Physical and Chemical Properties of the Glacial Ocean

J. C. DUPLESSY

ABSTRACT

During the last climatic cycle, sea surface temperature and salinity changed in the whole ocean, noticeably in the North Atlantic, which is an area of deep water formation. This change resulted in major variations of the thermo-haline circulation and the CO_2 cycle. This chapter describes the main mechanisms responsible for these changes: (a) insolation forcing responsible for changes in the climate system exhibiting periodicities of 19, 23, and 41 kyr and (b) interaction between the cryosphere and the ocean, which resulted in rapid (centennial) climatic changes.

The paleoclimatic record shows that the thermo-haline circulation and the Earth's climate are extremely sensitive to minor perturbations, such as freshwater discharge at high latitude into the North Atlantic Ocean. Under pure natural conditions, the geological data indicate that warm times are associated with active thermo-haline circulation, whereas reduced thermo-haline circulation prevailed during glacial times. If the global warming induced by human activities would result in enhanced precipitation over the North Atlantic, the resulting salinity changes might act as a massive iceberg discharge occurring during the glaciation. This would reduce the mean flux of deep water formed annually, the rate of thermo-haline circulation, and the oceanic heat flux brought to the Norwegian Sea, and it would deeply affect the European climate.

14.1 Introduction

Over the past 20 years, micropaleontological and stable isotope analysis of deep sea sediments has shown alternations between glacial and interglacial conditions, which are manifested by waning and waxing of large continental ice sheets. They coincide with changes in temperature, salinity, and circulation in the ocean, as well as changes in temperature and humidity over the continents. Geological data also provides strong support for the Milankovitch theory, which relates climatic variations to insolation changes due to changes in the Earth's orbital parameters (Hays et al., 1976). This mechanism leads primarily to changes in the intensity of the seasonal cycle, whereas the mean annual insolation budget is rather constant. This shows that the climate system is extremely sensitive to minor perturbations of the atmosphere energy budget.

Paleoclimate data were also the first to point to the ability of the Earth's climate system to switch abruptly between significantly different climatic modes. The coupling between the ocean, the atmosphere, and the ice is probably responsible for these variations. Because the atmosphere has a rather short time constant, the oceanic thermo-haline circulation variability is one major process that can be linked to long-term climatic changes, from decades to millennia. Geological data do not suggest that the thermo-haline circulation is a self-oscillating, perhaps self-regulating, mechanism. On the contrary, they show that instabilities of the ice sheets, ice shelves, and sea ice are playing a fundamental role in driving the thermo-haline circulation. They raise the problem of the impact of precipitation changes over the northern North Atlantic on the flux of deep water formed over decades to millennia. The significance and abruptness of past climatic changes are heightened by the fact that they cannot be studied using instrumental data because the instrumental period is too short, and because the origin of the changes is poorly understood. It is quite plausible that the mechanisms responsible for past abrupt changes might act again in response to the trace-gas-induced warming of the next century. In this chapter, I concentrate on the reconstruction of ocean circulation variations during the last glaciation, on evidence of the main mechanisms that are responsible for water mass and heat transport changes, and on their impact on ocean chemistry, atmospheric composition, and climate. Special attention is paid to the coupling between the oceans, the atmosphere, and the cryosphere. Finally, I review the major uncertainties associated with these reconstructions and their consequences on our ability to understand the future behavior of the climate system.

14.2   Physical Properties and Circulation of the Glacial Ocean

The oceanic general circulation is forced by wind stress and exchange of heat and freshwater with the atmosphere, because it is highly sensitive to density differences maintained by salt and heat in the areas of deep and intermediate water formation. Therefore, paleoclimatologists have developed various methods to generate quantitative estimates of past sea surface temperature and salinity and to reconstruct the major features of the global deep water circulation.

14.2.1 Sea Surface Temperature (SST)

An accurate reconstruction of past SST variations is a prerequisite to paleo-oceanographic studies. During the past decades, two methods have been commonly used to estimate past SSTs based on planktonic foraminiferal composition: the Imbrie and Kipp transfer function (Imbrie and Kipp, 1971) and the modern analog technique (Hutson, 1977). They usually provide similar estimates within statistical errors. This result gives strong support to the validity of the CLIMAP reconstruction of SST during the Last Glacial Maximum (LGM) (CLIMAP, 1981; Prell, 1985). Basically, it shows that high latitudes experienced major cooling (up to $10\,^{\circ}C$) and large sea-ice cover, while tropical areas experienced only minor SST drops (Figure 14.1). The cooling was stronger close to equatorial upwellings, which were more active than today because of enhanced trade winds.

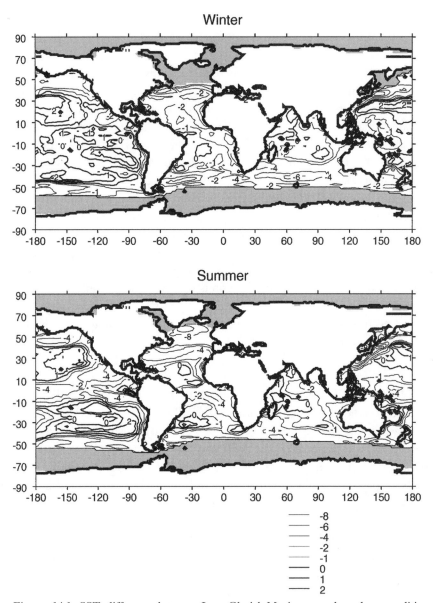

Figure 14.1. SST difference between Last Glacial Maximum and modern conditions for Northern Hemisphere winter and summer (CLIMAP, 1981).

Recently, several studies suggested that the CLIMAP reconstruction may be erroneous both in very cold (Norwegian Sea) and very warm (tropical ocean) areas. Using a new transfer function and a large set of Norwegian-Greenland sea cores, Weinelt (1996) suggested that SSTs were higher than those reconstructed by CLIMAP and that these seas were ice-free during glacial summer. These revised SST estimates are probably too high, because the transfer function used in this study does not allow us to estimate accurately very low SST (see Weinelt's Figure 4, page 289), but the high abundance of foraminifera in deep-sea sediments deposited during the last glaciation supports the hypothesis of seasonally ice-free Nordic seas, allowing a high summer foraminiferal productivity.

In the Tropics, fossil populations of foraminifera exhibit only small variations during the last climatic cycles. They were interpreted as an indication that the tropical ocean experienced only a minor cooling during the last glaciation. This result is challenged by selective measurements performed on corals: Measurements on a fossil coral indicate that 10,300 years ago mean annual SSTs near Vanuatu in the southwestern Pacific Ocean were about 5 °C cooler than today and that seasonal variations in SST were larger (Beck et al., 1997). Similarly, $^{18}O/^{16}O$ and Sr/Ca measurements performed in Atlantic corals collected along the slopes of Barbados indicate that SSTs were 5 °C cooler than today 18,000–19,000 years ago, during the LGM (Guilderson et al., 1994). Such significant coolings during the glaciation are in agreement with several continental evidences, noticeably the glacial depression of snowlines in tropical mountains (Hawaii, Colombia, East Africa, and New Guinea) and vegetation changes (Bonnefille et al., 1992; Bonnefille et al., 1990; Rind and Peteet, 1985). However, it should be stressed that most continental vegetation changes in the Tropics may also be explained by the low CO_2 content of the atmosphere and a minor air cooling (Jolly and Haxeltine, 1997). In addition, the validity of the Sr/Ca paleothermometer, which is generally used for detecting a large SST drop in the Tropics, is not yet fully established. Other comparisons relying on micropaleontological transfer functions or alkenone paleothermometry show a good agreement between low-altitude-continental and marine-shelf temperature estimates, and both indicate minor tropical cooling (Bard et al., 1997; Rostek et al., 1993; Van Campo et al., 1990). However, marine SST estimates, which were generated at proximity of the African continent (Van Campo et al., 1990), are slightly colder than those of CLIMAP, which were obtained in the open ocean (CLIMAP, 1981). This suggests that foraminifera may be faithful indicators of past SSTs but that the sediment core coverage originally used by CLIMAP should be enlarged in tropical oceans in order to obtain a precise reconstruction.

In conclusion, the CLIMAP (1981) reconstruction still provides the best maps of SST available for the LGM, although some inaccuracies probably exist. Model simulations have been performed with tropical SSTs colder than CLIMAP estimates as boundary conditions. These simulations exhibit land climates colder than those computed with CLIMAP SSTs (Webb et al., 1997). This result has been put forward in support of a revision of the CLIMAP reconstruction of the Last Glacial Maximum. However, this argument rests on a very simple interpretation of continental vegetation changes, an interpretation that does not take into account the impact of the low atmospheric CO_2 content. Meaningful results could be obtained only by coupling atmospheric general circulation models with models that include enough physiology to compute the response of the continental vegetation to atmospheric CO_2 changes. Sensitivity experiments have also been made to simulate the impact of the uncertainties of the CLIMAP reconstruction on our understanding of the glacial ocean circulation. They are discussed later.

14.2.2 Sea Surface Salinity (SSS)

To compute the density of a water mass requires both temperature and salinity. Whereas SST can be estimated directly from counts of foraminiferal shells present in marine sediment, no similar micropaleontological method is available to generate salinity estimates. However, a geochemical method can be used and is fully discussed by Duplessy et al. (1991). In fact, the $^{18}O/^{16}O$ ratio of ocean surface water is closely tied to the salinity: the higher the salinity, the higher the $^{18}O/^{16}O$ ratio, because the $H_2^{16}O$ molecule evaporates faster than the $H_2^{18}O$ molecule (Epstein, 1953; Craig and Gordon, 1965). This results in an empirical relationship between sea surface salinity and $^{18}O/^{16}O$ ratio, a relationship that is approximately linear. Its slope depends on the local precipitation and evaporation rates as well as mixing with deeper water (Craig and Gordon, 1965), but it is rather constant ($\pm 20\%$) in the open ocean. The geochemical method for estimating past SSS rests on the fact that the $^{18}O/^{16}O$ ratio of surface water is recorded in the $^{18}O/^{16}O$ ratio of planktonic foraminifera (Duplessy et al., 1991.)

The $^{18}O/^{16}O$ ratio of foraminifera, like that of other calcareous organisms, depends on both the $^{18}O/^{16}O$ ratio and the temperature of the seawater in which these organisms have grown: on the one hand, the $^{18}O/^{16}O$ ratio of $CaCO_3$ directly reflects variations of the $^{18}O/^{16}O$ ratio of the ambient water; on the other hand, the oxygen isotope fractionation between $CaCO_3$ and water increases by about 0.25% for each degree the water is cooled (Epstein et al., 1953; Shackleton, 1974). This is expressed by the paleotemperature equation

$$T = 16.9 - 4.38 \times (\delta^{18}O_{carbonate} - \delta^{18}O_{water}) \\ + 0.10 \times (\delta^{18}O_{carbonate} - \delta^{18}O_{water})^2$$

in which $\delta^{18}O$ is the $^{18}O/^{16}O$ ratio difference from that in a reference sample. This equation is valid in the whole range of seawater temperature (from $-1\,^{\circ}C$ to $30\,^{\circ}C$) and reflects the isotopic separation between ^{18}O and ^{16}O during carbonate deposition. Its validity therefore rests on the isotope thermodynamics and is independent of climate and time. It was first tentatively used to determine quantitatively the range of oceanic temperature fluctuations in the Quaternary (Emiliani, 1955). A precise interpretation of the variations of the foraminiferal $\delta^{18}O$ in deep-sea sediments is difficult, however, because both the seawater temperature and $\delta^{18}O_{water}$ vary with the climate. Indeed, continental ice sheets are depleted in ^{18}O, and during a glaciation the water remaining in the ocean becomes richer in ^{18}O (i.e., the $\delta^{18}O_{water}$ increases). This isotopic enrichment is entirely reflected in the calcitic test of foraminifera. Moreover, because the temperature at which the shell formed decreases, the $\delta^{18}O_{carbonate}$ increases. Thus, heavy isotopic ratios indicate cold climate. Conversely, temperature increase and ice cap melting will be reflected by light $\delta^{18}O_{carbonate}$ values.

The paleotemperature equation should be valid for both benthic foraminifera that live at the sediment-water interface and planktonic foraminifera that live in the upper part of the water column. $\delta^{18}O$ measurements performed on benthic foraminifera have been used to reconstruct ice volume changes and to estimate past deep water temperature variations (Labeyrie et al., 1987; Shackleton, 1974, 1987). The interpretation is more complex for planktonic foraminifera, which may live at different seasons and depths in the water column (Emiliani, 1954). Duplessy et al. (1991) used core-top analyses to demonstrate that, under modern conditions, the isotopic temperatures determined from the $\delta^{18}O$ of planktonic foraminiferal shells are linearly linked to the observed summer SSTs. However, this is true only within a limited and well-defined temperature range. This is characteristic of each species and is termed the "optimum temperature range."

By solving the paleotemperature equation, the $\delta^{18}O$ of planktonic foraminifera can therefore be used to derive an estimate of the water $\delta^{18}O$ in the past, provided that summer SST is independently estimated by a transfer function applied to microfaunal counts (Duplessy et al., 1992). This method has been used to generate past sea surface salinity estimates during the last deglaciation in the North Atlantic Ocean (Duplessy et al., 1993; Duplessy et al., 1992). It should be stressed that salinity and seawater $\delta^{18}O$ are both tracers of surface water hydrology only. Estimates of past seawater $\delta^{18}O$ can be obtained with an accuracy of $\pm 0.3\%o$ if past SSTs are determined with an accuracy of $\pm 1\,°C$. The accuracy of past salinity estimates also depends sharply on the accuracy of our knowledge of the relationship between sea surface water salinity and $\delta^{18}O$, a relationship that itself depends on local rates of evaporation, precipitation, runoff, and mixing with underlying waters. Usually the slope linking salinity to $\delta^{18}O$ varies between 2 and 3 (Duplessy et al., 1991; Kallel et al., 1997), so the accuracy of single salinity estimates is no better than $0.6\%o$. Replicate analyses are therefore necessary to improve the accuracy of such salinity estimates.

A map of sea surface salinity reconstruction for the North Atlantic during the LGM exhibits high values in the northern subtropical waters and a decreasing trend toward high latitudes (Figure 14.2). A marked gradient coincided with the thermal polar front, except in the middle Atlantic between 20°W and 40°W. Surface waters in the Norwegian-Greenland seas exhibited values significantly lower than those of today. The lowest values were found close to Greenland and the European continent, near the edge of the sea ice, and also in the Bay of Biscay, which received freshwater and ice-sheet meltwater carried by European rivers. The salinity gradient between high and low latitudes was larger than that of today, something that requires either a stronger freshwater atmospheric transport from the Tropics to the subpolar ocean, a weaker tropical water transport by surficial oceanic currents toward high latitudes, or a combination of both.

14.2.3 Density

Estimates of SST and SSS can be used to establish a T/S diagram for surface waters. Isodensity lines can be calculated following Cox et al. (1970). This allows us to compare the density of surface water during winter cooling with that of deep waters calculated from a global budget of the salinity and deep water temperature estimates derived from benthic foraminifera $\delta^{18}O$ values (Labeyrie et al., 1992).

During LGM, because of the density barrier resulting from the very low deep water temperature, salinity values higher than $35.5\%o$ would be required to produce the densities of both intermediate and deep waters by winter cooling within the open ocean (Figure 14.3). The only area with sufficiently high surface water salinity/density

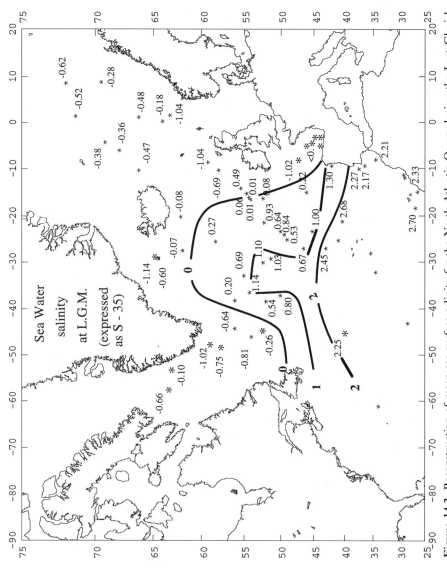

Figure 14.2. Reconstruction of mean sea surface salinity for the North Atlantic Ocean during the Last Glacial Maximum (Duplessy et al., 1991).

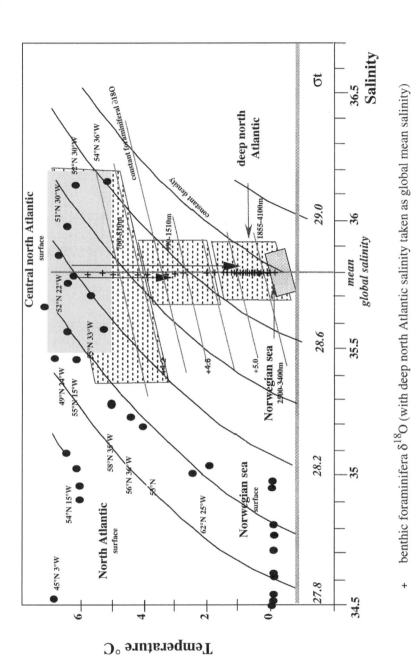

+ benthic foraminifera δ¹⁸O (with deep north Atlantic salinity taken as global mean salinity)

● planktic foraminifera : Sea water temperature derived from Transfer Functions
  and salinity calculated using the planktic foraminifera δ¹⁸O and temperature estimates

Figure 14.3. Distribution of the different water masses in the North Atlantic and Norwegian Sea during the Last Glacial Maximum on a T, S, σt diagram. σt = (d−1)*1000, where d is the density of the surface water. Benthic foraminifera δ¹⁸O are plotted at the intersection between the corresponding equilibrium isotopic fractionation lines and the estimated mean North Atlantic salinity. Surface water values are independently calculated (see text). The potential areas of deep water formation are associated with the vertical arrows, which simulates the temperature changes during winter cooling (from Labeyrie et al., 1992).

is relatively small, between 49°N and 54°N, and 20°W to 40°W, about midway between North America and Europe, south of Iceland (Duplessy et al., 1991). Salinities were too low in the Norwegian Sea to produce a significant amount of deep water, even by sea-ice freezing and brine formation during peak winter conditions. However, during the course of the glaciation, the location of deep and intermediate sources varied within the North Atlantic Ocean. In particular, during massive iceberg discharges, which resulted in a strong salinity decrease of North Atlantic surface water in the 40°N–55°N latitudinal band, brine formation was active and resulted in surface water sinking and reaching equilibrium density at 1 to 2 km depth (Veum et al., 1992; Vidal et al., 1998).

14.2.4 Deep Water Flow Lines

The distribution of $\delta^{13}C$ in the ocean is controlled principally by photosynthesis and mineralization of organic carbon, and by mixing between water masses of different isotopic composition (Broecker and Peng, 1982). Photosynthesis in surface water preferentially extracts ^{12}C from seawater, causing the enrichment in ^{13}C of surface water total dissolved CO_2. When all the nutrients are consumed by phytoplankton, surface waters have lost about 10% of their original ΣCO_2 (and are about 10% poorer in ΣCO_2 than deep waters). Because the carbon that has been removed by phytoplankton is strongly impoverished in ^{13}C ($\delta^{13}C \sim -20\%$o), the surface reservoir is enriched by about 2‰ over the mean $\delta^{13}C$ of the deep water.

Deep water masses today form at high latitudes. At their sites of formation, the main deep water masses, the North Atlantic deep water (NADW) and the Antarctic bottom water (AABW) are depleted in nutrients and enriched in ^{13}C relative to mean ocean water. After leaving the surface layers of the ocean, $\delta^{13}C$ of deep water masses changes only through two processes: remineralization of organic carbon and mixing between water masses with different $\delta^{13}C$ values. Because remineralization of organic carbon consumes oxygen at depth and releases CO_2 depleted in ^{13}C, there is a close relationship between the oxygen dissolved in deep waters and their $\delta^{13}C$. As the oxidation of organic matter regenerates nutrients (phosphate, nitrate), there is also a close relationship between deep water nutrient content and their $\delta^{13}C$. Kroopnick (1984) demonstrated that the distribution of $\delta^{13}C$ in deep water delineates the general distribution of water masses and that modern $\delta^{13}C$ gradients in the deep ocean follow the net flow of the deep waters from the Atlantic to the Pacific (Figure 14.4a).

Some species of benthic foraminifera (genus *Cibicides*) closely record the modern $\delta^{13}C$ distribution in the world ocean, providing a good proxy for the reconstruction of the past $\delta^{13}C$ gradients in the ocean (Duplessy et al., 1984; Duplessy and Shackleton, 1985). It is assumed that this was true also in the past. However, in addition to effects of deep water circulation, temporal changes in benthic foraminiferal $\delta^{13}C$ occur because of global changes in the distribution of carbon between the ocean and various reservoirs such as continental biosphere or organic carbon in shelf sediments (Boyle and Keigwin, 1982; Broecker and Peng, 1982; Shackleton, 1977). Any individual record of benthic foraminiferal $\delta^{13}C$ is therefore a complicated combination of global effects, circulation effects, local productivity effects, and changes in mixing ratios between the principal deep water masses (Curry et al., 1988).

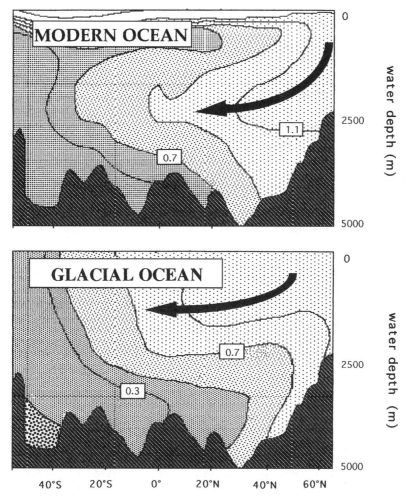

Figure 14.4. Reconstruction of the changes in $\delta^{13}C$ (expressed as deviation from the international PDB standard) across the Atlantic Ocean for the modern period (adapted from Kroopnick, 1984).

Other indicators may be used to estimate the nutrient content of ocean waters in the past. For instance, profiles of cadmium (Cd) dissolved in surface and deep waters are closely similar to those for phosphate (Broecker and Peng, 1982). Cd and P are removed efficiently from the surface ocean by organisms; these elements are incorporated into organic debris that sink. During their fall in the water column, these debris decompose partly in a nearly stoichiometric fashion. The general circulation of the ocean is then superimposed on this one-dimensional cycle to create signature variations in the chemical composition of oceanic water masses. Boyle (1988a) showed that the Cd content of the calcitic shells of foraminifera records that of the water in which these animals have lived and can be used as a tracer of deep water paleocirculation.

Foraminiferal $\delta^{13}C$ and Cd data agree on the following features of the glacial ocean paleochemical distribution:

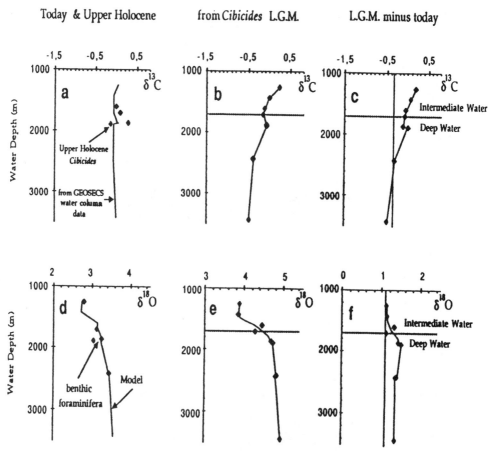

Figure 14.5. Reconstructed profiles of modern and LGM seawater $\delta^{13}C$ and benthic foraminifer $\delta^{18}O$ (expressed as deviation from the international PDB standard) and differences between LGM and modern conditions in (left) the Arabian Sea and (right) the Eastern Indian Ocean (Duplessy et al., 1989).

- Upper waters of the North Atlantic, tropical Atlantic, and northern Indian Oceans were significantly depleted in nutrients compared with the deepest waters in those basins.
- Deep waters of the North Atlantic were less ventilated (Figure 14.4b) and more enriched in nutrients than those of today (Duplessy et al., 1988).
- In the shallowest waters of the North and tropical Atlantic, and in intermediate waters of the northern Indian Ocean, nutrients were depleted compared with modern values (Boyle, 1992; Kallel et al., 1988).

However, $\delta^{13}C$ and Cd data strongly disagree in the Southern Ocean. Negative $\delta^{13}C$ values indicate poor ventilation and high deep water CO_2 content, whereas Cd data indicate minor changes in the nutrient content of glacial Antarctic deep waters compared with the modern ocean. They also disagree in the Pacific Ocean, where $\delta^{13}C$ displays a different vertical profile than that of the Holocene. When we account for the secular change of $\delta^{13}C$ during LGM, the data coincide with the modern curves deeper than 2 km in the North Pacific, but at shallower depth, $\delta^{13}C$ gradually increases

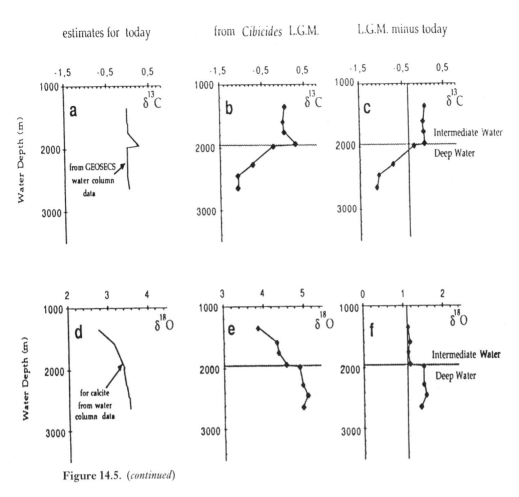

Figure 14.5. (*continued*)

by as much as 1‰ above the modern value (Keigwin, 1998). $\delta^{18}O$ values indicate the presence of a benthic front at $\sim$2 km in the Pacific Ocean. The isotope data suggest that during LGM, there was a better ventilated water mass at intermediate depth, and deep waters were as nutrient-rich as today. These data differ from Cd data, which indicate a source of nutrient-depleted deep water during glaciation (Boyle, 1992; Keigwin, 1998). A similar pattern is observed in the Indian Ocean (Kallel et al., 1988). $\delta^{18}O$ values indicate the presence of a benthic front at $\sim$2 km depth, with water more ventilated than today at intermediate depth and less ventilated below, noticeably in the eastern Indian Ocean (Figure 14.5).

14.2.5 Modeling the LGM Deep Water Circulation

Two-dimensional and three-dimension global ocean circulation models were forced by glacial SST and SSS estimates and wind stress derived from the ice-age response of an atmospheric general circulation model. Using CLIMAP SST and SSS estimated by Duplessy et al. (1991) in the North Atlantic and Nordic seas, and assuming a +1‰ anomaly elsewhere, models usually display deep convection in the North Atlantic around 50°N, reaching a maximum depth of 2,500 m. This gives rise to a thermohaline overturning of moderate strength that is essentially confined north of 40°S.

Below 2,500 m, most of the Atlantic basin is filled with Antarctic bottom water, so the relative contribution of water from the Southern Hemisphere as compared to that from the Northern Hemisphere is much greater than in the control case (Campin, 1997; Fichefet et al., 1994). However, sensitivity experiments with changes of $\pm 1\%_0$ in the glacial salinity boundary fields show circulation patterns ranging from one that is even stronger than the present day one to a near shutdown of the Atlantic deep sea circulation (Fichefet et al., 1994; Winguth et al., 1999).

Modeling experiments provide new insights on changes in rates of both thermohaline circulation and ocean ventilation. Two parameters are computed: the actual age of the water, which is the time elapsed since water has been in contact with atmosphere, and the ^{14}C age of the water, which is calculated from the difference of ^{14}C content between the deep water and the atmosphere. Both ages would be identical if the exchange of CO_2 and ^{14}C was instantaneous at the air-sea interface. There are significant differences between the modeled distribution of the ^{14}C age of the water and that of the actual age of the water that is only a function of the circulation intensity, because the ^{14}C content of water in deep water sources depends not only on the oceanic circulation but also on the rate of ^{14}C exchange at the air-sea interface (Figure 14.6). The differences between the ^{14}C and the actual ages of water increase significantly when glacial boundary conditions are used to drive the model (Figure 14.7). In particular, a larger age shift is simulated for the newly produced AABW, because the air-sea exchange of ^{14}C was drastically reduced near Antarctica as the permanent sea-ice cover extended northward beyond the Antarctic divergence in the glacial run (Campin et al., 1999). The thermo-haline circulation therefore runs faster than suggested by ^{14}C measurements performed in benthic foraminifera having lived in Pacific deep water or AABW. In fact, similar rates of modern and last-glacial ocean advection of North Atlantic deep or intermediate water into the Circumpolar deep water have been inferred from $^{231}Pa/^{230}Th$ ratios of deep-sea sediments (Yu et al., 1996). The reduced ventilation of Southern Ocean surface water during LGM may also explain part of the discrepancy between benthic foraminifera $\delta^{13}C$ and Cd content (Broecker, 1993).

14.3   Chemical Properties

14.3.1 pCO$_2$

Ice core studies have established that atmospheric carbon dioxide (CO_2) was about 80 parts per million by volume (ppmv) lower during cold glacial climates than it is during interglacial times (Barnola et al., 1987). The record of atmospheric CO_2 variation measured in the Vostok ice core indicates a strong similarity to other climate variables, such as global ice volume or south polar air temperature (Jouzel et al., 1993). On this long time scale, the atmospheric CO_2 variation is dependent on the surface ocean pCO_2, because there is 60 times more CO_2 in the ocean than in the atmosphere, and 20 times more CO_2 in the ocean than in the continental biosphere (including soil). The evolution of atmospheric pCO_2 during the last one million years may therefore be controlled mainly by the interaction of the cycle of biological production and decay of organic carbon with the vertical circulation of the ocean. This interaction depends on

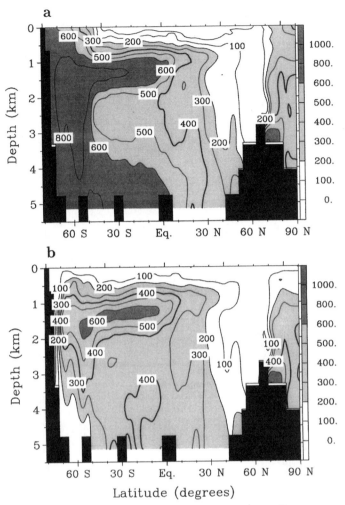

Figure 14.6. Latitude-depth distributions of zonal mean ^{14}C age (a) and actual age (b) of water in the Atlantic Ocean for modern conditions. The ^{14}C ages are derived from the simulated Δ^{14}C and are relative to the globally averaged surface value stemmed from the model (360 yr). Note that AABW and AIBW have an old ^{14}C age because of the sea-ice shielding of the air-sea ^{14}C exchange. This effect is not present in the actual age tracer simulation (Campin et al., 1999).

many factors, including net organic carbon production, nutrient utilization at high latitudes, global nutrient utilization, thermo-haline circulation, vertical mixing structure, export of particulate organic matter and carbonate particles from the surface, depth distribution of organic carbon oxidation, changes in denitrification areas, and so on (Archer and Maier-Reimer, 1994; Broecker, 1982; Francois et al., 1992; Ganeshram et al., 1995; Heinze et al., 1991; McElroy, 1983; Toggweiler and Sarmiento, 1985; Wenk and Siegenthaler, 1985).

As recognized by Broecker and Henderson (1998), the sequence of events recorded in Antarctic ice that occurred between the penultimate glaciation and the ensuing interglaciation provide some clues about the mechanisms that drive pCO$_2$ changes (Figure 14.8):

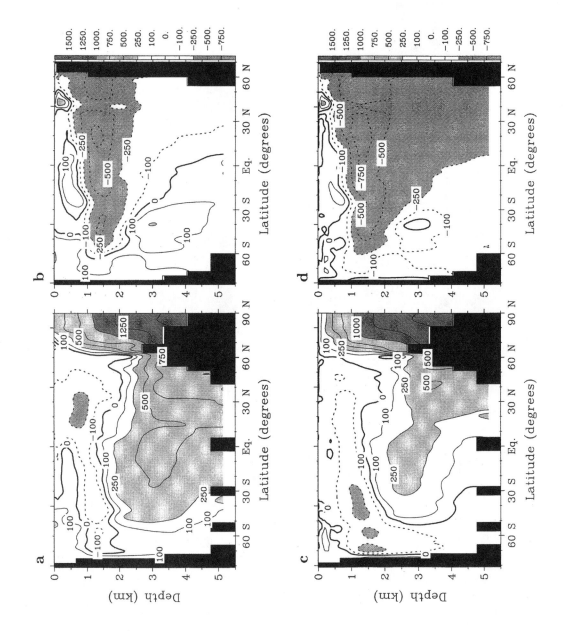

- The onset of the deglaciation is marked by the demise of the high dust flux that characterized the penultimate glaciation.
- An 8 kyr duration monotonic warming (as recorded by the rise in Deuterium content of the ice) and a parallel rise in the CO_2 content of the trapped air began at the time of this drop in dust input. A rise in global methane content also began at this time.
- The drop in the ^{18}O content of O_2 trapped in the ice ($\delta^{18}O_{atm}$) did not begin until the rises in Deuterium, CO_2, and methane were nearly complete. Assuming that the $\delta^{18}O_{atm}$ rise was responding primarily to changing $\delta^{18}O$ of ocean water in response to Northern Hemisphere ice-sheet melting, then ice sheets did not start to melt until the warming of Antarctica and the rise in CO_2 and CH_4 contents of the atmosphere were nearly completed.

These observations imply that the CO_2 rise preceded sea level change. It was gradual during 8 kyr, and it preceded the major nutrient reorganization in the North Atlantic associated with changing strength of the conveyor belt, which occurred during or after ice-sheet melting.

The first criterion, that of CO_2 rise occurring before sea level rise, rules out any mechanism that relies on sea level change to drive the CO_2, such as shelf nutrient hypothesis (Broecker, 1982) or shallow ocean $CaCO_3$ variation (Berger, 1982). Models involving nutrient redistribution associated with conveyor belt changes (Boyle, 1988b; Keir, 1993) are also ruled out by the last criterion, that of CO_2 rise preceding the major nutrient reorganization in the North Atlantic. In addition, all scenarios relying on better utilization of Southern Ocean surface water nutrient content (Knox and McElroy, 1984; Toggweiler and Sarmiento, 1985; Wenk and Siegenthaler, 1985) disagree with the low glacial productivity of the Southern Ocean recorded by deep-sea sediments (Francois et al., 1997; Mortlock et al., 1991). Additional constraints on the marine CO_2 cycle should therefore be taken into account to explain the low pCO_2 during the glaciation.

14.3.2 pH

The total dissolved CO_2 (ΣCO_2) is present in the ocean as dissolved molecular CO_2 only in very small amounts. Most of it is present as bicarbonate and carbonate ions, and the sum of all carbonate and bicarbonate charges must balance the excess cations so that the seawater maintains its electrical neutrality. Therefore, ΣCO_2, pCO_2, alkalinity (the sum of negative charges of carbonate and bicarbonate), and pH are linked by mathematical equations, which express the conservation of charges and carbon atoms. The chemical system is fully determined when two of these four quantities are known.

Records of past changes in the pH of the ocean should therefore provide insights into how the carbonate chemistry has changed over time. These changes may be recorded in

Figure 14.7. Latitude-depth distributions of the difference in zonal mean ^{14}C age of seawater between LGM and modern conditions in the Atlantic (a) and in the Pacific (b). For comparison are reported the latitude-depth distributions of the difference in actual age of seawater between LGM and modern conditions in the Atlantic (c) and in the Pacific (d) (Campin et al., 1999).

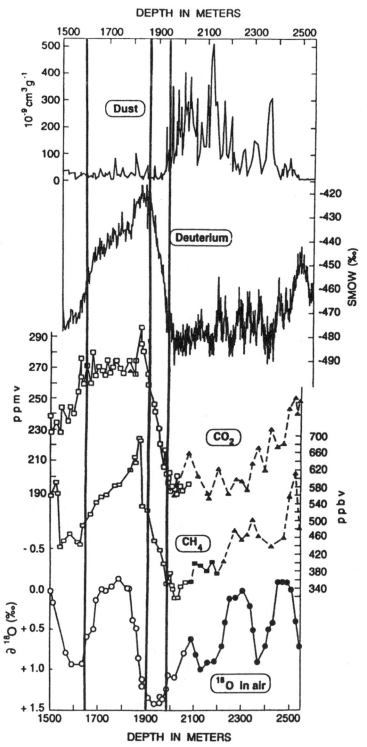

Figure 14.8. Events surrounding the end of the penultimate glaciation as recorded in the Vostok ice core (Jouzel et al., 1993). The right-hand vertical line is the beginning of the local temperature and global CO_2 increase and the drop in dust input to Antarctica. The middle line is the end of these increases and the beginning of the change in air $\delta^{18}O$ (proxy for ice volume). Melting of the Northern Hemisphere ice sheets occurred after the temperature and CO_2 change (from Broecker and Henderson, 1998).

deep-sea sediments, because the fractionation of Boron isotopes between seawater and precipitated carbonate is pH-dependent; the uncharged species $B(OH)_3$ is enriched in ^{11}B by $\sim 20\%$ over the charged borate species $B(OH)_4^-$. Because the fraction of the Boron present as these species changes with pH, so also must their respective isotopic compositions ($^{11}B/^{10}B$). Only the charged borate species is incorporated into carbonate minerals, and this incorporation occurs with a small isotopic fractionation, which is supposed to be constant. Thus, the $\delta^{11}B$ of the borate incorporated into foraminiferal shells should serve as an indicator of pH (Hemming and Hanson, 1992; Sanyal et al., 1995). Measurements performed on a few deep-sea cores in the equatorial Atlantic and Pacific Oceans indicates that the pH of deep waters was $\sim 0.3 \pm 0.1$ pH unit higher during the last glaciation than it is today. Surface water pH changes are more variable and depend on local nutrient utilization efficiency and calcite production. No significant pH change was recorded in the eastern tropical Pacific (Sanyal et al., 1997), whereas cores from the tropical Atlantic and western Pacific recorded glacial values 0.2 ± 0.1 pH unit higher than those measured during the Holocene (Sanyal et al., 1995).

The 0.3 pH unit increase of deep waters during the glaciation compared with the Holocene may be achieved by calcite dissolution of 135 μMole/kg of seawater, and the resulting carbonate ion concentration is about twice that of the modern ocean. A 0.15 pH unit increase in glacial ocean surface waters is required to achieve the 30% drop in pCO_2 that would correspond to a similar drop in glacial atmospheric pCO_2. Such an increase is indeed observed in the Atlantic and western Pacific surface waters, but not in the eastern equatorial Pacific. The cause may be either an enhanced calcite production in the euphotic zone or a reduced export of organic matter, resulting in a decrease in nutrient utilization efficiency (Sanyal et al., 1997), in agreement with low $^{15}N/^{14}N$ ratios measured in the total organic matter of sediment cores from the same basin (Farrell et al., 1995).

One mechanism to cause changes in atmospheric and surface water pCO_2 survives the timing constraint of Broecker and Henderson (1998): the hypothesis of Archer and Maier-Reimer involving changing sediment respiration (1994). These authors suggest that an oceanwide decrease in the $CaCO_3/Corg$ rain ratio to the ocean floor could bring about a significant change in the depth of the calcite saturation horizon without a significant deepening of the lysocline (the depth at which foraminiferal calcite dissolution becomes significant in deep-sea sediments). The mechanism responsible for such a decoupling of calcite saturation horizon from lysocline depth is respiration CO_2 driven $CaCO_3$ dissolution within the sediment. This is enhanced if a little more organic matter reaches the ocean floor. However, the amplitude of the depth decoupling between lysocline and carbonate saturation depth is so large that it disagrees with geological data and mass balance calculations (Broecker and Henderson, 1998).

The fact that falling dust supply is the first thing to change at the penultimate deglaciation (Figure 14.8) suggests that the supply of iron particles transported by wind to the sea surface may have increased phytoplankton productivity during glaciations. However, the slow response of the atmospheric CO_2 content implies an indirect link. It is not yet clear whether the availability of phosphate (Broecker, 1982) or that of nitrate (Eppley and Peterson, 1979; McElroy, 1983) limits phytoplankton growth, so Broecker and Henderson (1998) suggested that the nitrate concentration of the sea might have increased either because of elevated nitrogen fixation triggered by iron input to the ocean surface or reduced denitrification in the Pacific Ocean and the Arabian Sea (Altabet et al., 1995; Ganeshram et al., 1995). The relatively long residence time of oceanic nitrate offers an explanation as to how CO_2 can continue to increase after the dust flux has dropped to its interglacial rate, but many uncertainties about the nutrient cycles and paleoproductivity remain to be settled before we can establish a fully coherent theory of past atmospheric CO_2 changes.

14.4   Changes within the Coupled Ocean-Atmosphere-Ice System

We now turn to the temporal evolution of the oceanic circulation. The major components of the precessional and obliquity cycles are present in power spectra of all Quaternary paleoclimatic records derived from marine sediments. These records exhibit periodicities of 19, 23, and 41 kyr, which are statistically significant at the 95% confidence level and are invariant with respect to induced chronological errors in chronostratigraphy (Imbrie et al., 1993; Imbrie et al., 1992). However, superimposed on these long-term variations, abrupt climatic changes have recently been evidenced in most geological records. They suggest that the climate system consistently and frequently changes between near-glacial and near-interglacial conditions in periods of less than a few decades.

14.4.1 Response of the Climatic System to the Insolation Forcing

Most of the paleo proxies (foraminiferal $\delta^{18}O$, SST, atmospheric pCO_2, high-latitude air temperature, sea level) have a typical common first-order signature over the glacial-interglacial cycles of the last 800 kyr, modulated by the 100 kyr periodicity. Correlation coefficients between these signals are high, whatever their interrelationships, because they are all consequences of the evolution of the large continental ice sheets (Imbrie et al., 1993; Imbrie et al., 1992). High-latitude records usually exhibit a strong 41 kyr signal because of the major impact of obliquity changes at high latitudes. By contrast, the 20 kyr^{-1} frequency band of climate variability is associated with seasonal changes in the geographic distribution of insolation, and most paleo-proxy data present a near linear response in that band (Imbrie et al., 1993; Imbrie et al., 1992).

Labeyrie et al. (1998) made a detailed study of a sediment core from the northwest Atlantic (CH69-K09, 41°45′4 N, 47°21′ W, 4100 m water depth), which is located at the foot of the Newfoundland margin, between the northern limit of the North Atlantic current and the southern limit of the Labrador current (Figure 14.9, upper map).

--→

Figure 14.9. Upper map: location of core CH69-K09 and modern surface and deep circulation. EGC = East Greenland current, LC = Labrador current, NADW = North Atlantic deep water, AABW = Antarctic bottom water, NAC = North Atlantic current. Lower map: surface and deep circulation during glacial periods. The ice-sheets extension, maximum melting zone of the Heinrich events, and the 10 °C isotherm are represented. Large arrows indicate the location of potential iceberg discharges (from Labeyrie et al., 1998).

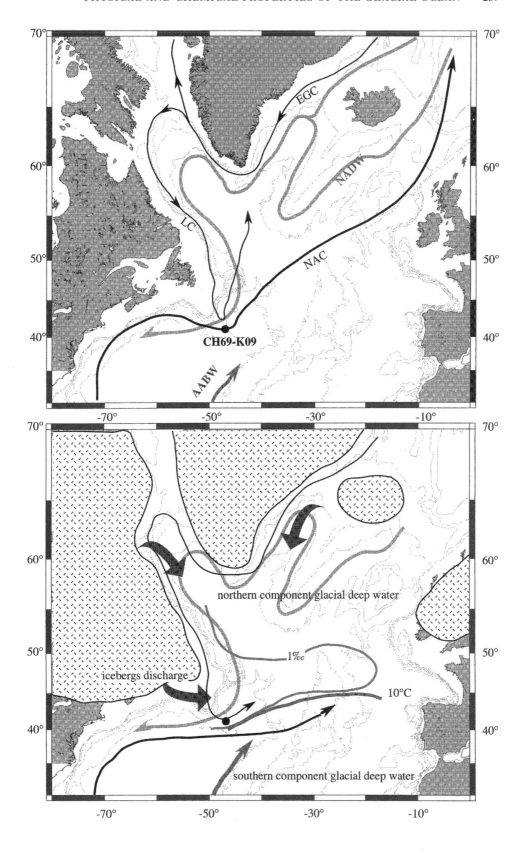

It is swept by the lower North Atlantic deep water (NADW) contour current. The Laurentide ice sheet was not far to the northwest during the last glacial period, with icebergs flowing from the north transported by the Labrador current (Figure 14.9, lower map). This core is thus at a good location for studying the temporal relationship between the Laurentide ice sheet, northward surface heat and salt transfers, and the evolution of the thermo-haline activity.

Following Imbrie et al. (Imbrie et al., 1993; Imbrie et al., 1992), Labeyrie et al. (1998) filtered the different paleoclimatic proxy records at 20 kyr and separated them into an early-response and a late-response group to the insolation forcing computed at 65°N following Berger (1978) (Figure 14.10). The early-response group is represented mostly by the $N.$ $pachyderma$ d^{18}O polar proxy and part of the benthic δ^{18}O signal. The precession band represents only a small part of their variability (the larger part is distributed over the 100 kyr^{-1} and 41 kyr^{-1} bands). But these proxies follow the changes in northern summer insolation with only a small lag. They represent the initial climatic reaction to changes in insolation: a change in high-latitude SST and salinity, snow and ice coverage, and ice volume. Open ocean thermo-haline convection and deep water convection were not affected, although the zones of deep water convection probably migrated south, out of the Norwegian-Greenland Sea, as low-salinity polar water expanded to lower latitudes. In the North Atlantic, significant lowering of the thermo-haline convection and deep water ventilation occurred only during glacial maxima and Heinrich events (Vidal et al., 1997).

The late-response group presents a strong variability within the precessional insolation band. It is represented by SST, the $G.$ $bulloides$ δ^{18}O, and the δ^{18}O difference between the planktonic species $G.$ $bulloides$ and the benthic foraminiferal record ($\Delta\delta^{18}$O). All are proxies for the North Atlantic current activity. The benthic foraminifera δ^{13}C record, which reflects ventilation changes of the deep water, also belongs to this group. Thus, these parameters, characterized by a late response to the insolation forcing, would represent the surface and deep water aspects of the thermo-haline conveyor belt. Their variations are in phase with winter Northern Hemisphere (and summer Southern Hemisphere) insolation. This suggests that the strength of the North Atlantic current is probably driven by low-latitude processes (Labeyrie et al., 1998). The 5 kyr lag of these parameters with ice volume (benthic δ^{18}O) corresponds to the derivative of the benthic δ^{18}O (growth and decay rate of the ice sheet) approximately in phase with the intensity of the thermo-haline conveyor belt, at least in the precession band: maximum growth occurs during maximum North Atlantic current activity, in agreement with the observation that surface waters maintained high temperatures during ice-sheet growth (Ruddiman and McIntyre, 1979, 1981). This feature has been simulated by the Louvain-la-Neuve climate model (Berger et al., 1998).

14.4.2 Rapid Climatic Changes

Superimposed on the long-term trends are short-term rapid climatic and ocean circulation variations, which may be either regional or global. For instance, the paleoclimatic record of core SU 81-18 (37°46′ N, 10°11′ W, 3135 m) raised off Portugal shows that at the beginning of the glacial-to-interglacial transition, about 14,500 radiocarbon

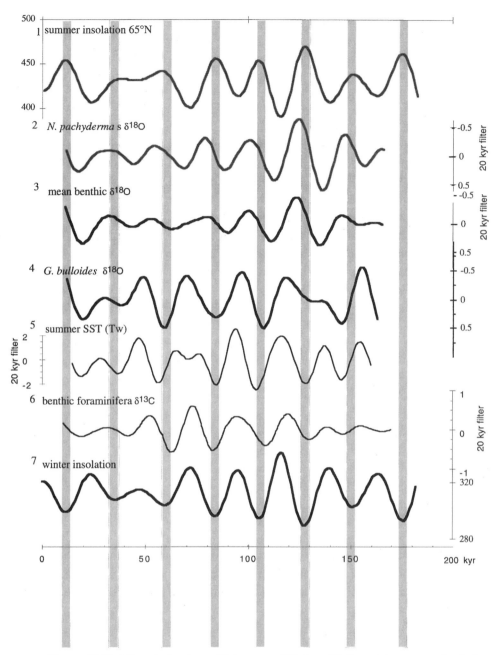

Figure 14.10. Filtered signals at 20 kyr over 200 kyr, (1) June–July summer insolation at 65°N (Berger, 1978), (2) $\delta^{18}O$ of *N. pachyderma* left coiling, (3) mean benthic stacked $\delta^{18}O$, (4) $\delta^{18}O$ of *G. bulloides*, (5) summer SST, (6) smoothed $\delta^{13}C$ of the benthic foraminifera, (7) December–January insolation (from Labeyrie et al., 1998).

years BP, winter SSTs off Portugal dropped from about 12 °C to 4 °C. Two thousand years later, the SST rose again to maximum values of 14 °C in winter at a minimum mean rate of 4 °C per century (Bard et al., 1987). This transition, which lasted less than four centuries, marks the beginning of the so-called Bölling interstadial and was also found in the Greenland ice record. Its duration was estimated to be only about one century. Oxygen isotope measurements performed on both the GRIP and GISP ice cores indicated that air temperature increased by at least 7 °C over Greenland during this rapid warming event (Dansgaard et al., 1993; Taylor et al., 1993a; Taylor et al., 1993b).

Oxygen isotope measurements in Greenland ice also demonstrate that a series of rapid warm-cold oscillations (called Dansgaard-Oeschger events) punctuated the last glaciation (Dansgaard et al., 1993). These oscillations were correlated with SST variations in several North Atlantic cores. A conspicuous feature common to both the ice and the ocean records is a bundling of the millennial-scale Dansgaard-Oeschger cycles into longer (7000 to 10,000 years) cooling cycles, each terminated by an abrupt shift from cold to warm temperatures. The most important finding of this correlation is the close relation between the ice core temperature cycles and one of the most prominent features of North Atlantic records, the Heinrich events (Figure 14.11). Every 7000 to 10,000 years, during times of brief sea surface cooling, exceptionally large discharges of icebergs from the Laurentide and European ice sheets took place, displaying conspicuous layers of detrital rocks in deep-sea sediments. Accompanying these events were large decreases in planktonic foraminiferal ^{18}O content, decreases that evidence lowered surface salinity probably caused largely by melting of drifting ice (Bond et al., 1993; Bond et al., 1992a). The bundles form a series of sawtooth-shaped cycles, each defined by a succession of progressively cooler interstadials (Figure 14.11), probably reflecting a progressive strengthening of the cold polar cell. The cooling trend ended with a very rapid, high-amplitude warming. The impact of Heinrich events on the climate system extends far beyond the northern North Atlantic area; at the time of major iceberg discharges, strong vegetation changes have been detected in Florida (Grimm et al., 1993), and changes in loess grain size, which were associated with atmospheric circulation changes, have been detected in China (Porter and Zhisheng, 1995).

A higher-resolution study of deep-sea cores shows the presence of ice-rafting cycles (referred to as Bond cycles) within the intervals between Heinrich events. They can be identified by peaks of lithic concentrations and magnetic susceptibility in North Atlantic sediments (Bond and Lotti, 1995; Kissel et al., 1997; Rasmussen et al., 1996). Their duration varies between 2000 and 3000 years, and they closely coincide with the Dansgaard-Oeschger events of the last glaciation. A sedimentological study of the

---

Figure 14.11. Variations of the $\delta^{18}O$ in the GRIP ice core interpreted in terms of air temperature variations (Dansgaard et al., 1993) compared with variations of SST and $\delta^{18}O$ variations of the planktonic foraminiferal species *N. pachyderma* in core ODP 609 from the North Atlantic Ocean (Bond et al., 1993; Bond et al., 1992b). Each millennial-scale $\delta^{18}O$ oscillation observed during the glaciation in the GRIP record is a Dansgaard-Oeschger (D-O) event. The D-O events are poorly represented in the marine record, which has a lower resolution than the ice record. Heinrich events, which have been defined in North Atlantic sediments as layers of ice-rafted detritus reflecting a massive iceberg discharge, coincide with the coldest events of the glaciation recorded in the GRIP ice core (Bond et al., 1993). During the glaciation, the warm event that follows an Heinrich event marks the beginning of a 7,000–10,000 year cooling cycle that ends by a new massive iceberg discharge. Such a cycle encompasses several D-O events.

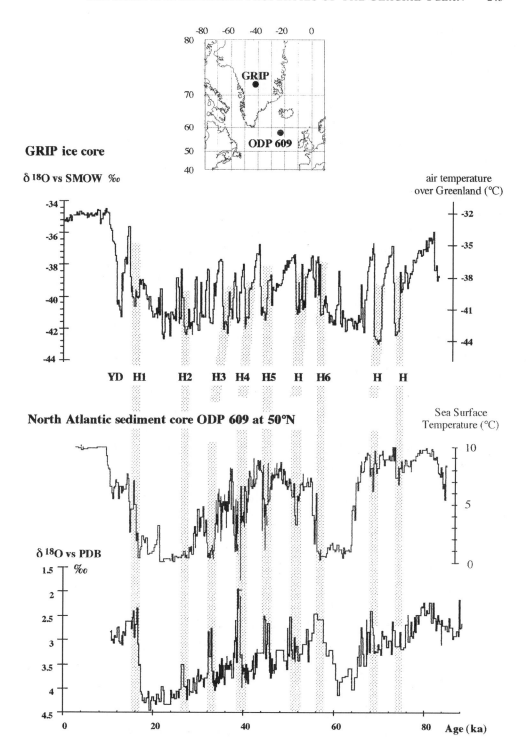

ice-rafted material deposited during Bond cycles shows that they contain basaltic glass derived from Iceland and grains with hematite coating (mostly quartz and feldspar) derived from any of the extensive red bed deposits that are found in Greenland, in Svalbard, in the Arctic Ocean islands, and near the Gulf of Saint Lawrence. In the Irminger Basin, the last glacial period was also characterized by numerous periods of increased iceberg discharges (Figure 14.12), originating partly from Iceland, as attested to by the presence of dark volcanic glass. A comparison with several mid-latitude sediment cores showed that ice-rafted material corresponding to the Heinrich events was deposited synchronously from 40 to 60°N within the error of AMS ^{14}C dates and bioturbation effects (Elliot et al., 1998). The more-frequent iceberg discharges correspond to instabilities at higher frequency of the coastal ice sheets and ice shelves in the Nordic area. By contrast with the mid-latitudes, the foraminifera content increases with the lithic content of the Irminger sediment (Figure 14.12). This is interpreted as reflecting retreats of the sea-ice edge synchronous with periods of iceberg discharges. The volume of injected meltwater was large enough to decrease glacial NADW formation (Kissel et al., 1997), and the associated reduced northward heat transport resulted in the Dansgaard-Oeschger temperature oscillations (Paillard and Cortijo, 1999).

A comparison between high- and mid-latitude sediment records of the North Atlantic Ocean suggests that the Heinrich events and the more frequent detrital events related to the Dansgaard-Oeschger cycles correspond to two oscillating systems: the massive iceberg armadas – which are released from large continental ice sheets every 5 to 10 kyr – and the more-frequent instabilities of the coastal ice sheets in the high-latitude Nordic regions, which occur every 1.2 to 3.8 kyr. At the time of the Heinrich events the synchroneity of the response from all the Northern Hemisphere ice sheets attests to the existence of strong interactions between the two systems (Bond and Lotti, 1995; Elliot et al., 1998). However, in the mid-latitudes of the North Atlantic Ocean, the volcanic debris originating from Iceland systematically precede the carbonate brought by icebergs released from the Laurentide ice sheet (Figure 14.13). This suggests that the relationships between the European and the Laurentide ice sheets were complex.

The changes in distribution of SST and SSS in the North Atlantic Ocean between 40°N and 60°N show that the meltwater input during deposition of Heinrich Layer 4 resulted in a 1–2 kyr temperature decrease of about 2 °C and a salinity decrease of 1.5‰ to 3.5‰ between 40°N and 50°N. Sites north of 50°N did not experience significant salinity variations. A much larger area was affected by the reduction in SST. The amplitude of the SST shift was, however, much smaller than the atmospheric temperature changes measured at GISP and GRIP sites (Cortijo et al., 1997).

High-resolution benthic δ^{18}O and δ^{13}C records from the same North Atlantic sediment cores were used to monitor the impact of Heinrich events on thermo-haline circulation and to estimate the sensitivity of the deep oceanic circulation to changes in freshwater input to the North Atlantic surface waters. Major rearrangements of deep water masses were directly associated with the massive iceberg discharges (Figure 14.14). Although deep water continued to form during the glaciation, the deep circulation was characterized by increased incursion of Southern bottom water, particularly in the eastern basin. Deep water production in the North Atlantic was reduced during the

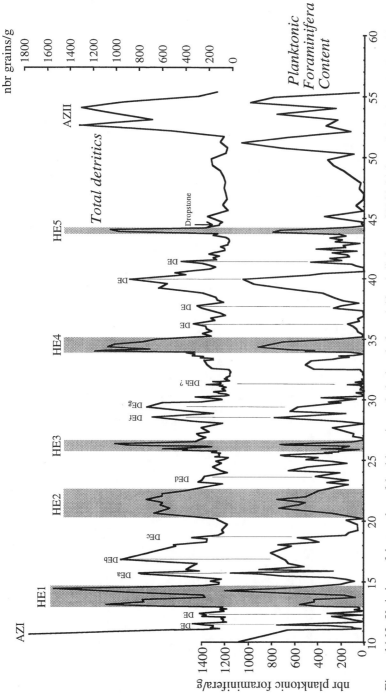

Figure 14.12. Variations of the number of detritic grains and planktonic foraminifer shells in core SU 90-24 from the Irminger Basin. This core received only ice-rafted detritus brought by icebergs originating from the Norwegian–Greenland Sea, characterized by large amounts of volcanic glass. HE indicates Heinrich events. DE indicates smaller detrital events in the sediment, coinciding with the cold phase of D-O events recognized in the GRIP ice core.

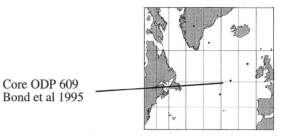

Core ODP 609
Bond et al 1995

- - - - . Detrital cabonate: Laurentite ice sheet

———— Volcanic ash: Iceland

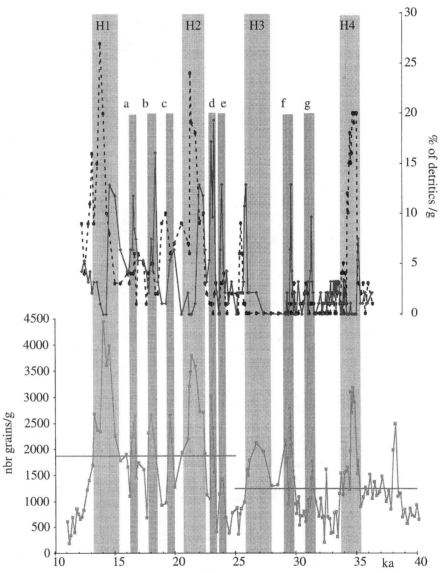

Figure 14.13. Variations of the number of total detritic grains, percent volcanic glass (Icelandic tracer), and detrital carbonate (Laurentide tracer) in core ODP 609 (Bond and Lotti, 1995). HE indicates Heinrich events. Letters a–g indicate smaller detrital events in the sediment, coinciding with the cold phase of D-O events recognized in the GRIP ice core.

iceberg discharge, but deep convection occurred in areas that were not affected by the salinity decrease. Soon after the Heinrich event, the oceanic pattern became similar to that before the event (Vidal et al., 1997). In addition, benthic records from deep-sea sediment cores raised from a depth shallower than 2 km exhibit light $\delta^{18}O$ peaks coinciding with Heinrich events. These peaks indicate a rapid transfer of $\delta^{18}O$-depleted meltwater to depth. However, meltwater has a low salinity and is unable to sink to great depth unless its salinity increases significantly. Paleo-oceanographers were therefore searching for a physical process allowing salinity increase for surface water without changing its $\delta^{18}O$ value. Evaporation and advection of highly evaporated tropical water are therefore ruled out, because evaporation results in large $\delta^{18}O$ variations. The most likely process responsible for a transfer of $\delta^{18}O$-depleted meltwater to depth is sea-ice formation during winter, because freezing occurs without significant isotope fractionation and, therefore, without variation of the seawater $\delta^{18}O$ value. Katabatic winds would maintain the process activity and would induce the formation of high-salinity surface water (brine), because sea ice is strongly depleted in salt. The more sea ice is formed, the saltier becomes surface water without $\delta^{18}O$ variation. Brine formation might have been particularly active along the Norwegian margin, where the $\delta^{18}O$ signals were also larger (Rasmussen et al., 1996; Vidal et al., 1998).

Rapid changes in ocean circulation are not limited to glacial conditions. During the last interglaciation, SST and SSS records from the North Atlantic Ocean and Nordic Seas show that summer SST and SSS decreased, sometimes rapidly, during the interval of minimum ice volume at high-latitude sites (>52°N), whereas they were stable or increased during the same period at low-latitude sites (31°N to 41°N). This increase in meridional gradients of SST and SSS may have been caused by changes in the latitudinal distribution of summer and annual-average solar radiation, and associated oceanic and atmospheric feedbacks (Cortijo et al., 1994; Cortijo et al., 1999; Fronval and Jansen, 1996, 1997). In addition, the end of the last interglaciation was marked by an abrupt event that begins the transition to a harsher climate within about 400 years at most (Adkins et al., 1997).

14.5   Concluding Remarks

The study of high-resolution deep-sea sediment cores, together with that of Greenland and Antarctic ice cores, has demonstrated that the Earth has switched repeatedly and abruptly between cold and warm climates over the course of ice-age cycles. These changes developed within a few decades. The paleo-oceanographic record of the North Atlantic Ocean revealed phenomena that may be related to changes in surface salinity much larger than the Great Salinity Anomaly in the 1970s (Dickson et al., 1988) during both glacial and interglacial conditions. These changes were closely correlated with sea surface temperature variations. Low-salinity events were associated with brine formation in the Norwegian Sea but with reduced convection in the North Atlantic, where surface waters remained stratified even during winter. These events were associated with low SST and low air temperatures over Greenland and the northern European continent. The atmosphere adjusted quickly to the perturbation, and most

The last glacial period:
Variability of air temperature
over Greenland and
North Atlantic hydrology

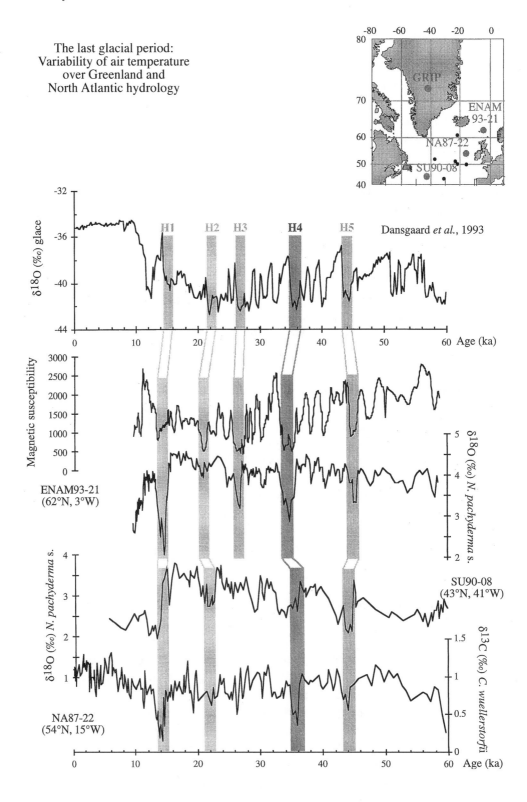

North Atlantic climatic changes were felt throughout the Northern Hemisphere either as a temperature or a hydrological cycle change. By contrast, warm episodes were associated with higher SST and convection in the Norwegian Greenland Sea.

It should be stressed that we still do not know whether the thermo-haline circulation has experienced significant changes during interglacial conditions, although some records suggest that this may be the case (Duplessy et al., 1992; Keigwin, 1996). It should also be stressed that interglacial climate variability is characterized primarily by major changes in the hydrological cycle (Gasse and Van Campo, 1994). As a consequence, the global warming induced by human greenhouse gas emissions might well result in precipitation changes over both continental and oceanic areas, possibly triggering lower North Atlantic surface water salinity. Such changes would reduce the mean flux of deep water formed every year during winter, the thermo-haline circulation, and the oceanic heat flux brought to the Norwegian Sea. This might deeply affect the European climate (Broecker, 1997; Manabe and Stouffer, 1995).

One of the most fruitful interactions between the modern and paleo scientific communities has been the study of multiple states of the thermo-haline circulation under various forcing factors. This characteristic feature of coupled ocean-atmosphere models would have been left aside without the numerous evidences brought by paleo-oceanographers and paleoclimatologists. Documenting the variability of the thermo-haline circulation on decadal-to-centennial time scales through the construction of high-resolution time series will lead to progress in our understanding of the thermo-haline circulation and will be useful for forecasting its future behavior.

However, we have discovered only a small part of the climatic puzzle. We are not yet able to explain the glacial-interglacial pCO_2 variations. We still don't know the complicated feedbacks that amplify North Atlantic circulation and SST changes to influence the Earth's climate on a large scale. How is it that millennial-scale variability appears nearly synchronously in California (Behl and Kennett, 1996) and in Indonesia (Linsley, 1996)? Or that the concentration of methane in the atmosphere (presumably a reflection of microbial activity on tropical land surfaces) varied along with other climatic variables (Stauffer et al., 1998)? Or that Antarctic air temperatures sometimes varied nearly inversely with those in the Arctic (Blunier et al., 1998)? These observations point to the involvement of the atmosphere and other oceanic processes, something that in turn raises the possibility that the North Atlantic oscillations were themselves the product of other, stronger sources of variability. In particular, the relationship between tropical and high-latitude oceans should be investigated.

Figure 14.14. Variations of the $\delta^{18}O$ in the GRIP ice core, a proxy for air temperature variations (Dansgaard et al., 1993), compared with magnetic susceptibility, a proxy for bottom water current activity, and $\delta^{18}O$ variations of the planktonic foraminiferal species *N. pachyderma* in core ENAM 9321 (Rasmussen et al., 1996) and core SU 90-08 (Cortijo et al., 1997). $\delta^{13}C$ in benthic foraminifera in core NA 87-22 show major circulation changes coinciding with Heinrich events and low salinity periods in the North Atlantic (Vidal et al., 1997). HE indicates Heinrich events recognized in deep-sea sediments. Their position in the GRIP record has been inferred from a detailed comparison with North Atlantic sediment records (Bond et al., 1993).

The paleoclimatic record has shown that the thermo-haline circulation and the Earth's climate are extremely sensitive to minor perturbations, such as freshwater discharge at high latitude into the North Atlantic Ocean. Any climate change similar to those that developed during the last climatic cycle and occurring in the near future would have profound repercussions for Earth's habitability. Identifying the processes responsible is necessary if we are to predict the evolution of the climate of the next centuries and to assess the likelihood of climate surprises associated with global warming.

REFERENCES

Adkins, J. F., Boyle, E. A., Keigwin, L., and Cortijo, E. (1997). Variability of the North Atlantic thermo-haline circulation during the last interglacial period, *Nature*, *390*, 154–156.

Altabet, M. A., Francois, R., Murray, D. W., and Prell, W. L. (1995). Climate-related variations in denitrification in the Arabian Sea from sediment $^{15}N/^{14}N$ ratios, *Nature*, *373*, 506–509.

Archer, D., and Maier-Reimer, E. (1994). Effect of deep-sea sedimentary calcite preservation on atmospheric CO_2 concentration, *Nature*, *367*, 260–263.

Bard, E., Arnold, M., Maurice, P., Duprat, J., Moyes, J., and Duplessy, J. C. (1987). Retreat velocity of the North Atlantic polar front during the last deglaciation determined by ^{14}C accelerator mass spectrometry, *Nature*, *328*, 791–794.

Bard, E., Rostek, F., and Sonzogni, C. (1997). Interhemispheric synchrony of the last deglaciation inferred from alkenone paleothermometry, *Nature*, *385*, 707–710.

Barnola, J. M., Raynaud, D., Korotkevitch, Y. S., and Lorius, C. (1987). Vostok ice core provides 160,000-year record of atmospheric CO_2, *Nature*, *329* (6138), 408–414.

Beck, J. W. E., Recy, J., Taylor, F. W., Edwards, R. L., and Cabioch, G. (1997). Abrupt changes in early Holocene tropical sea surface temperature derived from coral records, *Nature*, *385*, 705–707.

Behl, R. J., and Kennett, J. P. (1996). Brief interstadial events in the Santa Barbara basin, NE Pacific, during the past 60 kyr, *Nature*, *379*, 243–246.

Berger, A., Loutre, M. F., and Gallée, H. (1998). Sensitivity of the LLN climate model to the astronomical and CO_2 forcings over the last 200 kyr., *Climate Dynamics*, *14*, 615–629.

Berger, A. L. (1978). Long-term variations of daily insolation and quaternary climatic changes, *Journal of the Atmospheric Sciences*, *35*, 2362–2367.

Berger, W. H. (1982). Increase of carbon dioxide in the atmosphere during deglaciation: The coral reef hypothesis, *Naturwissenschaften*, *69*, 87–88.

Blunier, T., Chappellaz, J., Schwander, J., Dällenbach, A., Stauffer, B., Stocker, T., Raynaud, D., Jouzel, J., Clausen, A. B., Hammer, C. U., and Johnson, S. J. (1998). Asynchrony of Antarctica and Greenland climate change during the last glacial period, *Nature*, *394*, 739–743.

Bond, G., Broecker, W., Johnsen, S., McManus, J., Labeyrie, L., Jouzel, J., and Bonani, G. (1993). Correlations between climate records from North Atlantic sediments and Greenland ice, *Nature*, *365*, 143–147.

Bond, G., Broecker, W., Lotti, R., and MacManus, J. (1992a). Abrupt color changes in isotope stage 5 in North Atlantic deep sea cores: implications for rapid change of climate-driven events, in *Start of a Glacial*, edited by G. J. Kukla and E. Went, pp. 185–205, NATO ASI Series. Berlin: Springer-Verlag.

Bond, G., Heinrich, H., Broecker, W. S., Labeyrie, L., MacManus, J., Andrews, J., Huon, S., Jantschik, R., Clasen, S., Simet, C., Tedesco, K., Klas, M., Bonani, G., and Ivy, S. (1992b). Evidence for massive discharges of icebergs into the North Atlantic ocean during the last glacial period, *Nature*, *360*, 245–251.

Bond, G. C., and Lotti, R. (1995). Iceberg discharges into the North Atlantic on millenial time scales during the last glaciation, *Science*, *267*, 1005–1010.

Bonnefille, R., Chalié, F., Guiot, J., and Vincens, A. (1992). Quantitative estimates of full glacial temperatures in equatorial Africa from palynological data, *Climate Dynamics*, 6, 251–257.

Bonnefille, R., Roeland, K., and Guiot, J. (1990). Temperature and rainfall estimates for the past 40,000 years in equatorial Africa, *Nature*, 346, 347–349.

Boyle, E. A. (1988a). Cadmium: chemical tracer deepwater paleoceanography, *Paleoceanography*, 3 (4), 471–489.

Boyle, E. A. (1988b). The role of vertical chemical fractionation in controlling late Quaternary atmospheric carbon dioxide, *Journal of Geophysical Research*, 93, 701–715.

Boyle, E. A. (1992). Cadmium and $\delta^{13}C$ paleochemical ocean distributions during the stage 2 glacial maximum, *Annual Reviews of Earth and Planetary Sciences*, 20, 245–287.

Boyle, E. A., and Keigwin, L. D. (1982). Deep circulation of the North Atlantic over the last 200,000 years: Geochemical evidence, *Science*, 218, 784–787.

Broecker, W. S. (1982). Glacial to interglacial changes in ocean chemistry, *Progress in Oceanography*, 11, 151–197.

Broecker, W. S. (1993). An oceanographic explanation for the apparent carbon isotope-cadmium discordancy in the glacial Antarctic? *Paleoceanography*, 8, 137–139.

Broecker, W. S. (1997). Thermohaline circulation, the Achilles heel of our climate system: will man-made CO_2 upset the current balance? *Science*, 278, 1582–1588.

Broecker, W. S., and Henderson, G. H. (1998) The sequence of events surrounding Termination II and their implications for the causes of glacial-interglacial CO_2 changes, *Paleoceanography*, 13 (4), 352–364.

Broecker, W. S., and Peng, T. H. (1982). *Tracers in the sea*, 689 pp., Eldigio Press, New York.

Campin, J. M. (1997). Modélisation tridimentionnelle de la circulation océanique lors du dernier maximum glaciaire, PhD thesis, Université catholique de Louvain, Louvain La Neuve.

Campin, M., Fichefet, T., and Duplessy, J. C. (1999). Problems with using radiocarbon to infer ocean ventilation rates for past and present climates, *Earth and Planetary Science Letters*, 165, 17–24.

CLIMAP (1981). Seasonal reconstructions of the Earth's surface at the last glacial maximum, Geol. Soc. Am. Map and Chart Ser., MC-36.

Cortijo, E., Duplessy, J. C., Labeyrie, L., Leclaire, H., Duprat, J., and van Weering, T. C. E. (1994). Eemian cooling in the Norwegian Sea and North Atlantic ocean preceding continental ice-sheet growth, *Nature*, 372, 446–449.

Cortijo, E., Labeyrie, L., Vidal, L., Vautravers, M., Chapman, M., Duplessy, J. C., Elliot, M., Arnold, M., Turon, J. L., and Auffret, G. (1997). Changes in sea surface hydrology associated with Heinrich event 4 in the North Atlantic Ocean between 40°N and 60°N, *Earth and Planetary Science Letters*, 146, 29–45.

Cortijo, E., Lehman, S., Keigwin, L. D., Chapman, M., Paillard, D., and Labeyrie, L. (1999). Changes in meridional temperature and salinity gradients in the North Atlantic Ocean (32° to 72°N) during the last interglacial period, *Paleoceanography*, 14, 23–33.

Cox, R. A., McCartney, M. J., and Culkin, F. (1970). The specific gravity/salinity/temperature relationship in natural sea water, *Deep-Sea Research*, 17, 679–689.

Craig, H., and Gordon, L. I. (1965). Deuterium and oxygen 18 variations in the ocean and the marine atmosphere, in *Stable Isotopes in Oceanographic Studies and Paleotemperatures*, edited by E. Tongiorgi, pp. 9–122, Consiglio nazionale delle ricerche Laboratorio di geologia nucleare, Spoleto.

Curry, W. B., Duplessy, J. C., Labeyrie, L. D., and Shackleton, N. J. (1988). Changes in the distribution of $\delta^{13}C$ of deep water ΣCO_2 between the Last Glaciation and the Holocene, *Paleoceanography*, 3, 317–341.

Dansgaard, W., Johnsen, S. J., Clausen, H. B., Dahl-Jensen, D., Gundestrup, N. S., Hammer, C. U., Hvidberg, C. S., Steffensen, J. P., Sveinbjörnsdottir, A. E., Jouzel, J., and Bond, G. (1993). Evidence for general instability of past climate from a 250-kyr ice-core record, *Nature*, 364, 218–220.

Dickson, R. R., Meincke, J., Malmberg, S. A., and Lee, J. J. (1988). The "Great Salinity Anomaly" in the Northern North Atlantic 1968–1982, *Progress in Oceanography*, *20*, 103–151.

Duplessy, J.-C., Labeyrie, L., Juillet-Leclerc, A., Maitre, F., Duprat, J., and Sarnthein, M. (1991). Surface salinity reconstruction of the North Atlantic Ocean during the last glacial maximum, *Oceanologica Acta*, *14* (4), 311–324.

Duplessy, J.-C., Shackleton, N. J., Fairbanks, R. G., Labeyrie, L., Oppo, D., and Kallel, N. (1988). Deepwater source variations during the last climatic cycle and their impact on the global deepwater circulation, *Paleoceanography*, *3*, 343–360.

Duplessy, J.-C., Shackleton, N. J., Matthews, R. K., Prell, W., Ruddiman, W. F., Caralp, M., and Hendy, C. H. (1984). 13C record benthic foraminifera in the last interglacial ocean: implications for the carbon cycle and the global deep water circulation, *Quaternary Research*, *21*, 225–243.

Duplessy, J. C., Bard, E., Labeyrie, L., Duprat, J., and Moyes, J. (1993). Oxygen isotope records and salinity changes in the Northeastern Atlantic ocean during the last 18,000 years, *Paleoceanography*, *8* (3), 341–350.

Duplessy, J. C., Labeyrie, L., Arnold, M., Paterne, M., Duprat, J., and van Weering, T. C. E. (1992). Changes in surface salinity of the North Atlantic Ocean during the last deglaciation, *Nature*, *358*, 485–487.

Duplessy, J. C., Labeyrie, L., Kallel, N., and Juillet-Leclerc, A. (1989). Intermediate and deep water characteristics during the last glacial maximum, in *Climate and Geo-Sciences*, edited by A. Berger, pp. 105–120, Dordrecht, The Netherlands: Kluwer Academic Publishers.

Duplessy, J. C., and Shackleton, N. J. (1985). Response of global deep-water circulation to Earth's climatic change 135,000–107,000 years ago, *Nature*, *316* (6028), 500–507.

Elliot, M., Labeyrie, L., Bond, G., Cortijo, E., Turon, J. L, Tisnerat, N., and Duplessy, J. C. (1998). Millenial-scale iceberg discharges in the Irminger Basin during the last glacial period: Relationship with the Heinrich Events and environmental settings, *Paleoceanography*, *13*, 433–446.

Emiliani, C. (1954). Depth habitat of some species of planktonic foraminifera as indicated by oxygen isotope ratios, *American Journal of Sciences*, *252*, 149–158.

Emiliani, C. (1955). Pleistocene temperatures, *Journal of Geology*, *63*, 538–578.

Eppley, R. W., and Peterson, B. J. (1979). Particulate organic matter flux and planktonic new production in the deep ocean, *Nature*, *282*, 677–680.

Epstein, S., Buchsbaum, R., Lowenstam, H. A., and Urey, H. C. (1953). Revised carbonate-water isotopic temperature scale, *Geological Society of America Bulletin*, *64*, 1315–1325.

Farrell, J. W., Pedersen, T. F., Calvert, S. E., and Nielsen, B. (1995). Glacial-interglacial changes in nutrient utilization in the equatorial Pacific Ocean, *Nature*, *377*, 514–517.

Fichefet, T., Hovine, S., and Duplessy, J. C. (1994). A model study of the Atlantic thermo-haline circulation during the last glacial maximum, *Nature*, *372*, 252–255.

Francois, R., Altabet, M. A., and Burckle, L. H. (1992). Glacial to interglacial changes in surface nitrate utilization in the Indian sector of the Southern Ocean as recorded by sediment δ^{15}N, *Paleoceanography*, *7* (5), 589–606.

Francois, R., Altabet, M. A., Yu, E.-F., Sigman, D. M., Bacon, M. P., Franck, M., Bohrmann, G., Bareille, G., and Labeyrie, L. (1997). Contribution of Southern Ocean surface-water stratification to low atmospheric CO_2 concentrations during the last glacial period, *Nature*, *389*, 929–935.

Fronval, T., and Jansen, E. (1996). Rapid changes in ocean circulation and heat flux in the Nordic seas during the last interglacial period, *Nature*, *383*, 806–810.

Fronval, T., and Jansen, E. (1997). Eemian and early Weichselian (140–60 ka) paleoceanography and paleoclimate in the Nordic seas with comparisons to Holocene conditions, *Paleoceanography*, *12* (3), 443–462.

Ganeshram, R. S., Pedersen, T. F., Calvert, S. E., and Murray, J. W. (1995). Large changes in oceanic nutrient inventories from glacial to interglacial periods, *Nature*, *376*, 755–758.

Gasse, F., and Van Campo, E. (1994). Abrupt post-glacial climate events in West Asia and North Africa monsoon domains, *Earth and Planetetary Science Letters*, *126*, 435–456.

Grimm, E. C., Jacobson, G. L., Watts, J. W. A., Hansen, B. C. S., and Maasch, K. A. (1993). A 50,000-year record of climate oscillations from Florida and its temporal correlation with the Heinrich Events, *Science*, *261*, 198–200.

Guilderson, T. P., Fairbanks, K. A., and Rubenstone, J. L. (1994). Tropical temperature variations since 20,000 years ago: modulating interhemispheric climate change, *Science*, *263*, 663–665.

Hays, J. D., Imbrie, J., and Shackleton, N. J. (1976). Variations in the Eath's orbit: Pacemaker of the ice ages, *Science*, *194* (4270), 1121–1132.

Heinze, C., Maier-Reimer, E., and Winn, K. (1991). Glacial pCO_2 reduction by the world ocean: Experiment with the Hamburg carbon cycle model, *Paleoceanography*, *6*, 395–430.

Hemming, N. G., and Hanson, N. G. (1992). Boron isotopic composition and concentration in modern marine carbonates, *Geochimica and Cosmochimica Acta*, *56*, 537–543.

Hutson, W. H. (1977). Transfer functions under no-analog conditions: Experiments with Indian Ocean planktonic foraminifera, *Quaternary Research*, *8*, 355–367.

Imbrie, J., Berger, A., Boyle, E. A., Clemens, S. C., Duffy, A., Howard, W. R., Kukla, G., Kutzbach, J., Martinson, D. G., MacIntyre, A., Mix, A. C., Molfino, B., Morley, J. J., Peterson, L. C., Pisias, N. G., Prell, W. L., Raymo, M. E., Shackleton, N. J., and Toggweiler, J. R. (1993). On the structure and origin of major glaciation cycles. 2. The 100,000-year cycle, *Paleoceanography*, *8* (6), 699–735.

Imbrie, J., Berger, A., Boyle, E. A., Clemens, S. C, Duffy, A., Howard, W. R., Kukla, G., Kutzbach, J., Martinson, D. G., MacIntyre, A., Mix, A. C., Molfino, B., Morley, J. J., Peterson, L. C., Pisias, N. G., Prell, W. L., Raymo, M. E., Shackleton, N. J., and Toggweiler, J. R. (1992). On the structure and origin of major glaciation cycles. 1. Linear responses to Milankovitch forcing, *Paleoceanography*, *7* (6), 701–738.

Imbrie, J., and Kipp, N. G. (1971). A new micropaleontological method for quantitative paleo-climatology: application to a late Pleistocene Caribbean core, in *The late Cenozoic glacial ages*, edited by K. K. Turekian, pp. 71–181, New Haven, CT: Yale University, Press.

Jolly, D., and Haxeltine, A. (1997). Effect of low glacial atmospheric CO_2 on tropical African montane vegetation, *Science*, *276*, 786–788.

Jouzel, J., Barkov, N. I., Barnola, J. M., Bender, M., Chappellaz, J., Genthon, C., Kotlyakov, V. M., Lipenkov, V., Lorius, C., Petit, J. R., Raynaud, D., Raisbeck, G., Ritz, C., Sowers, T., Stievenard, M., Yiou, F., and Yiou, P. (1993). Extending the Vostok ice-core record of paleoclimate to the penultimate glacial period, *Nature*, *364*, 407–412.

Kallel, N., Labeyrie, L.D., Juillet-Leclerc, A., and Duplessy, J. C. (1988). A deep hydrological front between intermediate and deep-water masses in the glacial Indian Ocean, *Nature*, *333* (6174), 651–655.

Kallel, N., Paterne, M., Labeyrie, L., Duplessy, J. C., and Arnold, M. (1997). Temperature and salinity records of the Tyrrhenian Sea during the last 18,000 years, *Palaeogeography, Palaeoclimatology, Palaeoecology*, *135*, 97–108.

Keigwin, L. D. (1996). The Little Ice Age and Medieval warm period in the Sargasso Sea, *Science*, *274*, 1504–1508.

Keigwin, L. D. (1998). Glacial-age hydrography of the far Northwest Pacific Ocean, *Paleoceanography*, *13* (4), 323–339.

Keir, R. S. (1993). Cold surface ocean ventilation and its effect on atmospheric CO_2, *Journal of Geophysical Research*, *98*, 849–856.

Kissel, C., Laj, C., Lehman, B., Labeyrie, L., and Bout-Roumazeilles, V. (1997). Changes in the strength of the Iceland-Scotland overflow water in the last 200,000 years: Evidence from magnetic anisotropy analysis of core SU90-33, *Earth and Planetary Science Letters*, *152*, 25–36.

Knox, F., and McElroy, M. (1984). Changes in atmospheric CO_2: Influence of marine biota at high latitudes, *Journal of Geophysical Research*, *89*, 4629–4637.

Kroopnick, P. (1984). Distribution of ^{13}C and ΣCO_2 in the world ocean, *Deep Sea Research*, *32*, 57–77.

Labeyrie, L., Duplessy, J. C., Duprat, J., Juillet-Leclerc, A., Moyes, J., Michel, E., Kallel, N., and Shackleton, N. J. (1992). Changes in vertical structure of the North Atlantic Ocean between glacial and modern times, *Quaternary Science Reviews*, *11*, 401–413.

Labeyrie, L., Leclaire, N. J., Waelbroeck, C., Cortijo, E., Duplessy, J. C., Vidal, L., Elliot, M., Le Coat, B., and Auffret, G. (1998). Insolation forcing and millenial scale variability of the North Western Atlantic Ocean: surface vs deep water changes, in *Chapman Conference on Mechanisms of Millenial-scale Global Climate Change*, AGU, Snowbird, Utah.

Labeyrie, L. D., Duplessy, J. C., and Blanc, P. L. (1987). Variations in mode of formation and temperature of oceanic deep waters over the past 125,000 years, *Nature, 327*, 477–482.

Linsley, B. K. (1996). Oxygen–isotope record of sea level and climate variations in the Sulu Sea over the past 150,000 years, *Nature, 380*, 234–237.

Manabe, S., and Stouffer, R. J. (1995). Simulation of abrupt climate change induced by freshwater input to the North Atlantic Ocean, *Nature, 378*, 165–167.

McElroy, M. B. (1983). Marine biological controls on atmospheric CO_2 and climate, *Nature, 302*, 328–329.

Mortlock, R. A., Charles, C. D., Froelich, P. N., Zibello, M. A., Saltzman, J., Hays, J. D., and Burckle, L. H. (1991). Evidence for lower productivity in the Antarctic Ocean during the last glaciation, *Nature, 351*, 220–222.

Paillard, D., and Cortijo, E. (1999). A simulation of the Atlantic meridional circulation during Heinrich event 4 using reconstructed sea surface temperatures and salinities, *Paleoceanography, 14*, 716–724.

Porter, S. C., and Zhisheng, A. (1995). Correlation between climate events in the North Atlantic and China during the last glaciation, *Nature, 375*, 305–308.

Prell, W. L. (1985). The stability of low-latitude sea-surface temperatures: an evolution of the Climap reconstruction with emphasis on the positive SST anomalies, United States Department of Energy, Washington, D.C.

Rasmussen, T. L., Thomsen, E., van Weering, T. C. E., and Labeyrie, L. (1996). Rapid changes in surface and deep water conditions at the Faeroe Margin during the last 58,000 years, *Paleoceanography, 11* (6), 757–771.

Rind, D., and Peteet, D. (1985). Terrestrial conditions at the Last Glacial Maximum and CLIMAP sea-surface temperature estimates: are they consistent? *Quaternary Research, 24*, 1–22.

Rostek, F., Ruhland, G., Bassinot, F. C., Müller, P. J., Labeyrie, L. D., Lancelot, Y., and Bard, E. (1993). Reconstructing sea surface temperature and salinity using $\partial^{18}O$ and alkenone records, *Nature, 364*, 319–321.

Ruddiman, W. F., and McIntyre, A. (1979). Warmth of the Subpolar North Atlantic Ocean during northern hemisphere ice-sheet growth, *Science, 204*, 173–175.

Ruddiman, W. F., and McIntyre, A. (1981). Oceanic mechanisms for amplification of the 23,000-year ice volume cycle, *Science, 212* (4495), 617–627.

Sanyal, A., Hemming, N. G., Broecker, W. S., and Hanson, G. N. (1997). Changes in pH in the eastern equatorial Pacific across the 5-6 boundary based on boron isotopes in foraminifera, *Global Biogeochemical Cycles, 11* (1), 125–133.

Sanyal, A., Hemming, N. G., Hanson, G. N., and Broecker, W. S. (1995). Evidence for a higher pH in the glacial ocean from boron isotopes in foraminifera, *Nature, 373*, 234–236.

Shackleton, N. J. (1974). Attainment of isotopic equilibrium between ocean water and benthonic foraminifera genus Uvigerina: isotopic changes in the ocean during the last glacial, in *Les méthodes quantitatives d'étude des variations du climat au cours du Pleistocène*, pp. 203–209. Gif sur Yvette, France: CNRS.

Shackleton, N. J. (1977). Carbon-13 in Uvigerina: Tropical rainforest history and the equatorial Pacific dissolution cycles, in *The Fate of Fossil Fuel CO_2 in the Oceans*, edited by N. R. Anderson and N. A. Malahoff, pp. 401–428, Plenum, New York.

Shackleton, N. J. (1987). Oxygen isotopes, ice volume and sea level, *Quaternary Science Reviews, 6*, 183–190.

Stauffer, B., Blunier, T., Dällenbach, A., Indermühle, A., Schwander, J., Stocker, T. F., Tschumi, J., Chappellaz, J., Raynaud, D., Hammer, C. U., and Clausen, H. B. (1998). Atmospheric CO_2 concentration and millenial-scale climate change during the last glacial period, *Nature, 392*, 59–62.

Taylor, K. C., Hammer, C. U., Alley, R. B., Clausen, H. B., Dahl-Jensen, D., Gow, A. J., Gundestrup, N. S., Kipfstuhl, J., Moore, J. C., and Waddington, E. D. (1993a). Electrical conductivity measurements from the GISP2 and GRIP Greenland ice cores, *Nature, 366,* 549–552.

Taylor, K. C., Lamorey, G. W., Doyle, G. A., Alley, R. B., Grootes, P. M., Mayewski, P. A., White, J. W. C., and Barlow, L. K. (1993b). The "flickering switch" of late Pleistocene climate change, *Nature, 361,* 432–436.

Toggweiler, J. R., and Sarmiento, J. L. (1985). Glacial to interglacial changes in atmospheric carbon dioxide: The critical role of ocean surface water in high latitudes, in *The Carbon Cycle and Atmospheric CO₂: Natural Variations Archean to Present,* edited by E. T. Sundquist and W. S. Broecker, pp. 163–184, AGU, Washington, D.C.

Van Campo, E., Duplessy, J. C., Prell, W. L., Barratt, N., and Sabatier, R. (1990). Comparison of terrestrial and marine temperature estimates for the past 135 kyr off southeast Africa: a test for GCM simulations of paleoclimate, *Nature, 348,* 209–212.

Veum, T., Jansen, E., Arnold, M., Beyer, I., and Duplessy, J. C. (1992). Water mass exchange between the North Atlantic and the Norwegian Sea during the past 28,000 years, *Nature, 356,* 783–785.

Vidal, L., Labeyrie, L., Cortijo, E., Arnold, M., Duplessy, J. C., Michel, E., Becqué, S., and van Weering, T. C. E. (1997). Evidence for changes in the North Atlantic Deep Water linked to meltwater surges during the Heinrich events, *Earth and Planetary Science Letters, 146,* 13–26.

Vidal, L., Labeyrie, L., and van Weering, T. C. E. (1998). Benthic δ^{18}O records in the North Atlantic over the last glacial period (60–10 kyr): Evidence for brine formation, *Paleoceanography, 13* (3), 245–251.

Webb, R. S., Rind, D., Lehman, S. J., Healy, R. J., and Sigman, D. (1997). Influence of ocean heat transport on the climate of the Last Glacial Maximum, *Nature, 385,* 695–699.

Weinelt, M., Sarnthein, M., Pflaumann, U., Schulz, H., Jung, S., and Erlenkeuser, H. (1996). Ice-free nordic seas during the last glacial maximum? Potential sites of deepwater formation, *Paleoclimates, 1,* 283–309.

Wenk, T., and Siegenthaler, U. (1985). The high latitude ocean as a control of atmospheric CO₂, in *The Carbon Cycle and Atmospheric CO₂: Natural Variations Archean to Present,* edited by E. T. Sundquist, and W. S. Broecker, pp. 185–194, AGU, Washington, D.C.

Winguth, A. M. E., Archer, D., Duplessy, J. C., Maier-Reimer, E., and Mikolajewicz, U. (1999). Sensitivity of paleonutrient tracer distributions and deep sea circulation to glacial boundary conditions, *Paleoceanography, 41,* 304–323.

Yu, E.-F., Francois, R., and Bacon, M. (1996). Similar rates of modern and last-glacial ocean thermohaline circulation inferred from radiochemical data, *Nature, 379,* 689–694.

15 Ice Core Records and Relevance for Future Climate Variations

JEAN JOUZEL

15.1 Introduction

Any human-induced changes in climate will be superimposed on a background of natural climatic variations. Hence, to understand future climatic changes, it is necessary to document the spatiotemporal patterns and causes of natural climate variability. As useful as instrumental studies may be, they are limited by the available data record, which generally extends back less than 150 years. This duration is too short to extract the full range of possible climate system variability and to understand the mechanisms involved in climate changes operating at various time scales. For example, the paleoclimatic record has already provided multiple examples of how the climate system is capable of unprecedented abrupt changes that did not occur during the period of instrumental observations and thus can be studied only via the paleoclimatic record.

The aim of this chapter is to examine how our increasing knowledge of past climate changes contributes to gaining information relevant to the future of our climate. Our current understanding of past climate changes is based on a combination of two things: paleo data provided by ice core, marine, and continental records; and a modeling approach. It is this combination that is most useful for narrowing uncertainties associated with detecting and predicting climate change (see Overpeck, 1995, for a recent review).

In this context, ice core data have a key contribution because (1) they are unique in providing access to past atmospheric composition, in particular to changes in concentration of the main greenhouse gases, (2) they give access to other potential climate forcings (solar, volcanic, etc.), and (3) they allow a study of climate variability at very high resolution (annual or subannual) over long periods with very different climates (the longest ice core now extends back to 420 ka). We first examine how ice core data have contributed to the discovery of rapid climatic changes (typical of glacial periods and of the last climatic transition), drawing attention to the possibility of "climatic surprises." We next discuss how ice core studies can lead to a better knowledge of climate variability and climate mechanisms during warm periods, focusing on the last millenium, which must be much better documented for detection and attribution of climate change. Then we show how long-term glacial-interglacial records provide a way to assess the climatic role of greenhouse gases.

15.2   Rapid Climatic Changes

It was only during the 1980s that the possibility of rapid climatic changes, occurring at the time scale of a human life or less, was fully recognized, thanks largely to the oxygen 18 record (a proxy for local temperature change) measured along the Greenland ice core drilled at Dye 3 in southern Greenland (Dansgaard et al., 1982, 1984, 1989). A possible link between such events and the mode of operation of the ocean was then suggested (Oeschger et al., 1984; Broecker et al., 1985; see also Broecker, 1997, for a recent review). The occurrence of rapid changes was fully confirmed by the central Greenland ice record (GRIP and GISP2), which allows a resolution approaching annual over the entire period of the last climatic transition (Figure 15.1). The return to the cold conditions of the Younger Dryas from the incipient interglacial warming

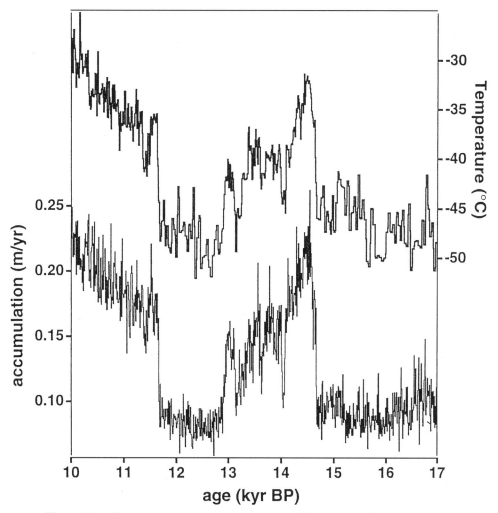

Figure 15.1. The environmental record during the Bölling-Allerod/Younger Dryas period, as revealed by the GISP 2 ice core data (adapted from Lorius and Oeschger, 1994).

13,000 years ago took place within decades or less (Alley et al., 1993). Finally, the warming phase at the end of the Younger Dryas that took place about 11,500 years ago was also very abrupt, and central Greenland temperatures increased in a few decades.

The classical interpretation – consisting of using the present-day spatial isotope/temperature relationship for interpreting past temperature changes – provides an estimate of $\sim 7\,°C$ for the associated temperature change (Dansgaard et al., 1989; Johnsen et al., 1992). This interpretation has recently been challenged (at least for central Greenland) by borehole paleothermometry (Cuffey et al., 1995; Johnsen et al., 1995; Dahl-Jensen et al., 1998) and by information deduced in using isotopic anomalies due to thermal diffusion in the gas phase (Severinghaus et al., 1998). The rapid temperature change occurring at the end of the Younger Dryas was probably higher than 10 °C in central Greenland, whereas temperatures were more than 20 °C colder than today during the Last Glacial Maximum. This warming coincided with a prominent rise in atmospheric methane concentration, indicating that the climate change was synchronous (within a few decades) over a region of at least hemispheric extent (Severinghaus et al., 1998). As indicated by these authors, this synchroneity provides constraints on proposed mechanisms of climate change at this time, supporting, in particular, an atmospheric transmission of the climate signal between low and high latitudes in the Northern Hemisphere.

Accumulation changed even more rapidly than local temperature, doubling at the end of the Younger Dryas in one to three years (Alley et al., 1993). Fluctuations in ice conductivity on the scales of <5–20 years reflect rapid oscillations in the dust content of the atmosphere. They reflect two preferred states of atmospheric dust load, a fact that would seem to require extremely rapid reorganizations in atmospheric circulation (Taylor et al., 1993).

Oxygen isotope measurements in Greenland ice also demonstrated that a series of rapid warm-cold oscillations (called Dansgaard-Oeschger events) punctuated the last glaciation (Dansgaard et al., 1993). These oscillations were correlated with SST variations in several North Atlantic cores (Bond et al., 1993). The most important finding of this correlation is the close relation between the ice core temperature cycles and one of the most prominent features of North Atlantic records: the Heinrich events (Figure 15.2). Heinrich events occur every 7000 to 10,000 years during times of sea surface cooling. They occur as brief, exceptionally large discharges of icebergs from the Laurentide and European ice sheets, and they leave conspicuous layers of detrital rocks in deep-sea sediments. These last glacial rapid climatic events, best documented in Greenland and the North Atlantic, have smoothed counterparts in Antarctica (Jouzel et al., 1994; Bender et al., 1994). The existence of a peak in the concentration of the beryllium 10 cosmogenic isotope (F. Yiou et al., 1997) and detailed records of atmospheric changes in the isotopic content of the oxygen (Bender et al., in press) and in the concentration of methane (Blunier et al., 1998) allow us to shed light on the complex link between Northern and Southern Hemisphere climates over this period. This latter study showed that Antarctic climate change leads that of Greenland by 1–2.5 kyr over the period 47–23 kyr BP. This makes a coupling between the hemispheres

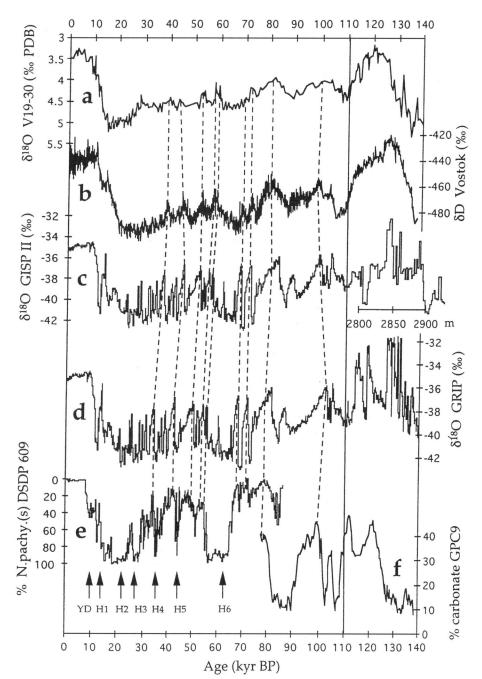

Figure 15.2. Comparison, over the last 140 ka, of various climatic records covering the last glacial-interglacial cycle (adapted from Jouzel, 1994). (a) The benthic $\delta^{18}O$ record from V19-30 showing continental ice volume variations. (b) The Vostok deuterium profile. (c) The GISP2 $\delta^{18}O$ record using a depth scale for the lower part of the record for which the GISP2 and GRIP records diverge. (d) The GRIP $\delta^{18}O$ record. (e) Records obtained from North Atlantic deep-sea cores with indication of the Heinrich layers. The dashed lines indicate correspondences between climatic events (Bender et al., 1994; Jouzel, 1994).

via atmosphere improbable. It points instead to a dominant role of the ocean controlling the past climate of both regions (Blunier et al., 1998).

One of the most fruitful interactions between the modern and paleo scientific communities is indeed the study of multiple states of the thermo-haline circulation that are quite probably associated with these rapid climatic changes, which occurred during the last glacial period and the following climatic transition (see Stocker, Chapter 9 in this volume). This characteristic feature of coupled ocean-atmosphere models would have been left aside without the numerous evidences gathered by glaciologists, paleo-oceanographers, and paleoclimatologists. General circulation models (GCMS) are now used to quantify the rapidity of swings between modes (Rahmstorf, 1994). Documenting the variability of the thermo-haline circulation on decadal-to-centennial time scales through the construction of high-resolution time series will lead to progress in our understanding of the thermo-haline circulation and will be useful for forecasting its future behavior (Stocker and Schmitter, 1997).

15.3 Climate Variability During Warm Periods

The study of climate variability during periods as warm or warmer than today focuses on the Holocene (the current interglacial) and on the Eemian (the last interglacial). Relevant to future climate change is the Eemian, which was probably slightly warmer than today with higher sea level (although this climate should not be taken as an analog of a future warm climate, inasmuch as climate forcings clearly differ). Study of the corresponding part of the GRIP and GISP2 records revealed rapid isotopic and chemical changes that have raised questions about the stability of this period (GRIP project members, 1993). Records of atmospheric composition in the air bubbles trapped in ice have later shown that such instabilities observed for the lowest part of both Greenland cores (GRIP project members, 1993; Grootes et al., 1993, Fig. 2) do not correspond to climatic instabilities during the last interglacial period (Bender et al., 1994; Fuchs and Leuenberger, 1996; Chappellaz et al., 1997). On the other hand, the Vostok record (Petit et al., 1999) provides a very detailed and undisturbed record of the Eemian (and of the two previous interglacial periods, stages 7.5 and 9.3).

This Antarctic record illustrates that the first part of the last interglacial was warmer than the Holocene. This first period of ~4 ka (Figure 15.2) is then followed by a relatively rapid cooling and then a slower temperature decrease somewhat parallel to pollen records from Western Europe (Cheddadi et al., 1998) and to some North Atlantic deep-sea core records (Cortijo et al., 1994; Jouzel, 1994). Paleo-oceanographers have now examined the Eemian in close detail. From a Bermuda Rise high-resolution sediment record, Adkins et al. (1997) showed that this period was relatively stable but began and ended with abrupt changes in deep water flow with transitions occurring in less than 400 years. Interestingly, during this "warm" period, the "conveyor belt" circulation, which today carries heat north from the Tropics and warms much of Europe, remained strong and relatively steady.

One of the striking features of the Vostok climatic record is that the Holocene is, by far, the longest (~11 ka) stable warm period recorded in Antarctica over the past

420 ka. The above comparison with North Atlantic and European records suggests that this may be true worldwide, with profound implications for the evolution and the development of civilization (Petit et al., 1999). It now appears that this relatively stable Holocene was indeed marked by a millenial-scale cycle observed both in ice core and in deep-sea core records (O'Brien et al., 1995; F. Yiou et al., 1995, 1997; Bond et al., 1997). As noted by Bond et al. (1997), such variability was demonstrated from measurements of soluble impurities in Greenland ice, showing that Holocene atmospheric circulation above the ice cap was punctuated by a series of millenial-scale shifts. Such Holocene events have now been documented in North Atlantic sediments, where they make up a series of climate shifts with a cyclicity close to 1470 ± 500 years (Bond et al., 1997).

The most prominent event seen in the Greenland record over this period was the cooling that occurred ~8200 years ago. This 200-year-long cooling is also seen in western Europe (von Grafenstein et al., 1998; see Figure 15.3). Widespread proxy records from the Tropics to the polar regions now show that this short-lived cooling event possibly had a worldwide extension (Gasse and van Campo, 1994; Alley et al., 1998; Stager and Mayewski, 1997). This abrupt cooling may have been triggered by a sudden freshwater pulse from the collapse of the Hudson Ice Dome, although there is no clear evidence for such an event at this time (von Grafenstein et al., 1998). Alternatively, Alley et al. (1998) assumed that the climatic oscillation was induced by a reduced conveyor belt strength, caused by weak changes in the freshwater supply to the North Atlantic. Such a weak forcing scenario would support models of the North Atlantic thermo-haline

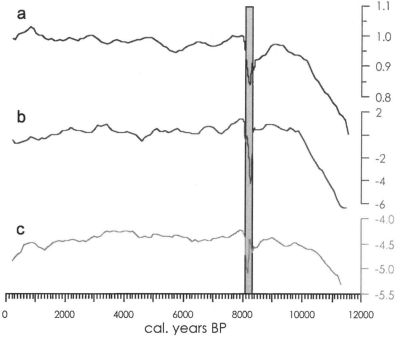

Figure 15.3. Variation, over the last 16 ka, of the Greenland ice oxygen 18 record (GRIP) and of the oxygen 18 content of shells of two different species of ostracods sampled in a core drilled in Ammersee Lake in Bavaria (adapted from von Grafenstein et al., 1998).

circulation to small changes in the freshwater influx, which could occur in response to global warming (Alley et al., 1998).

15.4   The Last Millenium

We now have a better, but still insufficient, knowledge of the climate variability over the past few centuries, knowledge that is important for climate change detection. It is also necessary in order to provide a comprehensive record of natural (non-anthropogenically forced) seasonal-to-interdecadal variability and to put the past 100 years, well documented by meteorological information, in the context of the last millenium (CLIVAR, 1997). Mann et al. (1998) recently inferred spatially resolved global reconstructions of annual surface temperature patterns over the past six centuries based on the multivariate calibration of widely distributed proxy climate indicators. Ice core data contribute to such global reconstructions, but their geographical coverage is obviously limited (Greenland and Arctic ice caps and tropical glaciers from the Andes, from the Himalayas, and, potentially, from Antarctica).

Despite such geographical limitations, ice core data should play a key role in our effort to better assess climate variability over the last millenium. First, they record changes of climate patterns over large regions and, second, they give unique access to relevant climate forcings.

There is a clear impact of events such as El Niño Southern Oscillation (ENSO) on the regime of precipitation in the Andes, and studies of ice cores such as that from the Quelccaya ice cap in the Southern Andes of Peru provide information on ENSO events over the last 1500 years (Thompson et al., 1984). More surprising are the striking similarities between the accumulation histories revealed by this Andean record and a record obtained in the Guliya ice cap in Tibet, inasmuch as the two sites are 20,000 km apart, lying on opposite sides of the Pacific Basin (Thompson, 1996). During the overlapping 1500 years the major periods of drought and wetness appear contemporaneous, suggesting teleconnections for ENSO events as well as for lower-frequency events.

It is also a record of accumulation change that shows the best potential for reconstructing a North Atlantic Oscillation (NAO) index from Greenland ice cores (Appenzeller et al., 1998) based on the connection between snow accumulation in western Greenland and mean pressure at sea level. This index extends back to A.D. 1650, indicating that the NAO is an intermittent climate oscillation with temporally active and passive phases. Greenland isotopic records provide another way to get information on conditions prevailing in the North Atlantic (Figure 15.4). One example is given by the deuterium-excess record at the GRIP site. The deuterium excess (a linear combination of deuterium and oxygen 18 ratios in water) is influenced by the temperature and humidity prevailing in the source regions for Greenland precipitation. The anomalous low deuterium-excess values in the 18th and 17th centuries (Figure 15.5) are interpreted as reflecting cooler conditions in the subtropical North Atlantic and thus correspond to a cool period in Europe known as the Little Ice Age (Hoffmann et al., unpublished).

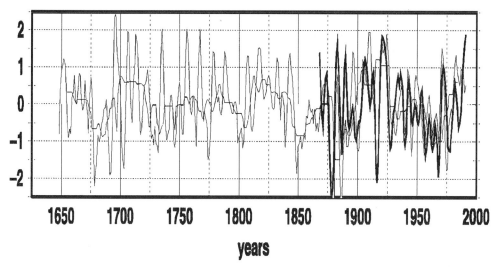

Figure 15.4. Normalized proxy NAO index based on western Greenland ice accumulation rates and normalized instrumental NAO index (thick line). The thin line corresponds to the 15-year running median of the proxy index (adapted from Appenzeller et al., 1998).

Polar ice core data are also key to an assessment of the origin of this variability because they allow us to track the increase in the concentration of the most important greenhouse gases since preindustrial times and give indirect access to changes in other climate forcings. One must realize that without the possibility of reconstructing changes in atmospheric composition from the analysis of air bubbles trapped in ice cores, we

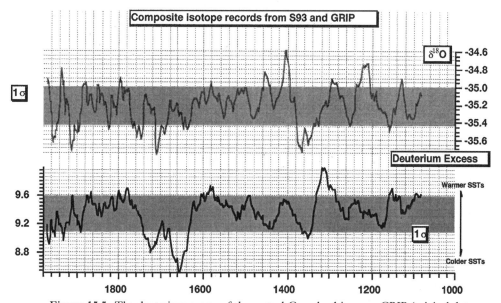

Figure 15.5. The deuterium excess of the central Greenland ice core GRIP (original data are smoothed by a running mean over one year). Adapted from Hoffmann et al. (unpublished).

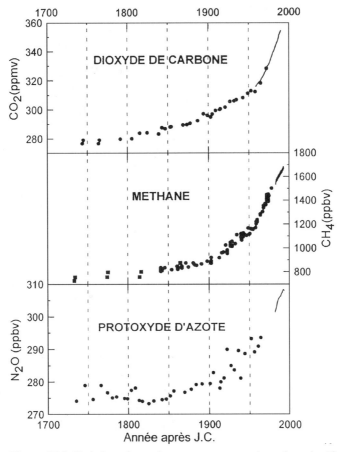

Figure 15.6. Variation of greenhouse gas concentrations since the 18th century (compiled by J. Chappellaz). These results are deduced from the analysis of air bubbles trapped in the ice (Blunier et al., 1993; Etheridge et al., 1996) and, for recent years, from direct measurements.

would have no clear information about the impact of anthropogenic activities on the atmospheric concentrations of CO_2, CH_4, and N_2O, inasmuch as direct measurements of these three important greenhouse gases are available only over the past few decades. Ice core data, obtained from high-accumulation sites, provide a unique and reliable record of these anthropogenically induced changes (Figure 15.6).

Ice cores also give indirect access to solar irradiance. This approach is based on the existence of a positive relationship between the 11-year cycle of solar magnetic activity and the output of the Sun and on the fact that magnetic fields of the solar wind deflect the primary flux of charged cosmic particles, something that leads to a reduction of cosmogenic nuclide (such as carbon 14 and beryllium 10). Using a detailed [10]Be record measured in an ice core drilled at South Pole Station, Bard et al. (in press) have extended the record of solar irradiance over the past 1200 years. Like others (Lean et al., 1995), these authors suggest that changes in solar forcing may have made a significant contribution to climate variability over the last millenium. Also, ice cores record the occurrence and magnitude of volcanic eruptions (Hammer et al., 1980, 1997; Legrand

and Delmas, 1987; Delmas et al., 1992), allowing us to estimate the associated strato-spheric loading and resulting changes in optical depth. Indeed, ice core data provide series of the three candidate external forcings for the past century (greenhouse gases, solar irradiance, and explosive volcanism) and are thus key to the attribution of climatic changes during the last millenium.

15.5 Glacial-Interglacial Changes

The fact that ice cores give unique and direct access to records of changes in atmospheric composition has played a key role in demonstrating the close association between CO_2 and CH_4, and climate variations at the glacial-interglacial time scale (Delmas et al., 1980; Neftel et al., 1982; Barnola et al., 1987; Stauffer et al., 1988; Chappellaz et al., 1990; Raynaud et al., 1993). This discovery comes, in particular, from the core drilled at Vostok in East Antarctica, which fully covered the last glacial-interglacial cycle in the 1980s. The later extension (Jouzel et al., 1993, 1996; Petit et al., 1997) of this ice core now permits an examination of this relationship between climate and greenhouse gases over four climatic cycles (Petit et al., 1997).

This extension of the greenhouse gas record (Figure 15.7) confirms that present-day levels of CO_2 and CH_4 are unprecedented during the last 420,000 years. The overall correlation between CO_2 or CH_4 and the Antarctic isotopic temperature is remarkable at the low resolution of this long record, which does not describe millenial-scale events ($r^2 = 0.71$ and 0.73, respectively, with only little change from one glacial-interglacial cycle to the next). This high correlation indicates that CO_2 and CH_4 may have con-tributed to the glacial-interglacial changes over this entire period in amplifying the orbital forcing, along with albedo and possibly other changes.

Interestingly, the comparison of climate changes and climate forcings at the glacial-interglacial time scale gives insight into the estimate of the climate sensitivity (Lorius et al., 1990), suggesting that amplification processes have, as predicted by climate mod-els, operated in the past. Petit et al. (1999) estimate that the direct radiative forcing corresponding to the CO_2, CH_4, and N_2O glacial-interglacial changes may reach up to 0.95 °C. This initial forcing is amplified by positive feedbacks associated with water va-por, sea ice, and possibly clouds. This total glacial-interglacial forcing is important (~ 3 Wm^{-2}), representing 80% of that corresponding to the difference between a $2 \times CO_2$ world and modern CO_2 climate. Results from various climate simulations (Berger et al., 1998; Weaver et al., 1998) make it reasonable to assume that greenhouse gases have, at a global scale, contributed significantly (possibly about half, i.e., 2 to 3 °C) to the globally averaged glacial-interglacial temperature change. Indeed, the fact that glacial-interglacial temperature changes are large (~ 5 °C) and that associated changes in greenhouse gas concentrations are well known offers an interesting opportunity to estimate global climate sensitivity. This approach, however, has limitations because cli-mate feedbacks may operate in a different way for a $2 \times CO_2$ than for a glacial climate (Ramstein et al., 1998).

Records of past climates also contain information about climate mechanisms such as those linked with changes in insolation, in oceanic and atmospheric circulation, and

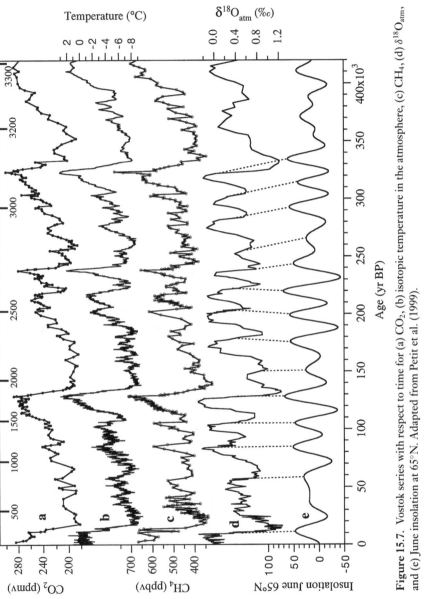

Figure 15.7. Vostok series with respect to time for (a) CO_2, (b) isotopic temperature in the atmosphere, (c) CH_4, (d) $\delta^{18}O_{atm}$, and (e) June insolation at 65°N. Adapted from Petit et al. (1999).

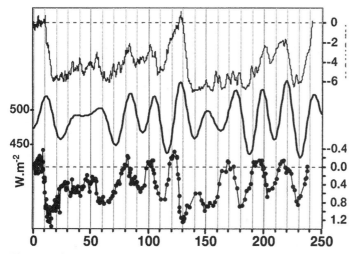

Figure 15.8. Variation during the last two climatic cycles of (a) the deuterium content at Vostok, (b) the 65°N mid-June insolation, and (c) the oxygen 18 content of atmospheric oxygen as measured in air trapped in the Vostok ice. This figure is adapted from Jouzel et al. (1996).

in atmospheric composition; they also shed light on the complex interactions between climate and biogeochemical cycles. The record of the oxygen 18 ratio in atmospheric oxygen (Figure 15.8) gives one example of such interaction between insolation changes and biogeochemical cycles. This parameter depends both on the isotopic composition of the ocean (and thus on change in ice volume) and on the rate of continental-to-oceanic biomass, which itself is largely governed by the hydrological cycle in mid-latitudes. The presence of a strong precessional periodicity in this record (23 kyr) is attributable, at least in part, to the influence of precession on monsoonal activity.

15.6   Conclusion

Over the past two decades, ice core data have made a significant contribution to the increased recognition of paleoclimatology by climatologists. In this review, we have focused on the relevance of these data to future climatic change and have pointed out how they are useful (1) to assess climate variability at various time scales, (2) to provide estimates of climate forcings and information on climate sensitivity, (3) to shed light on climate mechanisms, and (4) to draw attention to the possibility of "climatic surprises."

The need to further document past climates, to identify mechanisms involved, and to model them in a realistic way is now fully recognized by this scientific community involved in climate research. It is taken into account in international programs both within IGBP (International Geophysical Biological Programme) through the PAGES (Past Global Changes) project, and within WCRP (World Climate Research Programme) through the CLIVAR (Climate Variability and Predictability) project.

REFERENCES

Adkins, J. F., Boyle, E. W. , Keigwin, L., and Cortijo, E. (1997). Variability of the North Atlantic thermohaline circulation during the Last interglacial period, *Nature, 390*, 154–156.

Alley, R. B., Mayewski, P. A., Sowers, T., Stuiver, M., Taylor, K. C., and Clark, P. U. (1998). Holocene climatic instability: A large, widespread event 8200 years ago, *Geology*, *25*, 483–486.

Alley, R. B., and 13 others (1993). Abrupt increase in Greenland snow accumulation at the end of the Younger Dryas event, *Nature*, *362*, 527–529.

Appenzeller, C., Stocker, T. F., and Anklin, M. (1998). North Atlantic Oscillation Dynamics recorded in Greenland Ice cores, *Science*, *282*, 446–449.

Barnola, J. M., Raynaud, D., Korotkevich, Y. S., and Lorius, C. (1987). Vostok ice core provides 160,000-year record of atmospheric CO_2, *Nature*, *329*, 408–414.

Bard, E., Raisbeck, G., Yiou, F., and Jouzel, J. (in press). Solar irradiance during the last 1200 yr based on cosmogenic nuclides, *Geophys. Res. Lett.*

Bender, M., Sowers, T., Dickson, M. L., Orchado, J., Grootes, P., Mayewski, P. A., and D. A. Meese (1994). Climate connection between Greenland and Antarctica during the last 100,000 years, *Nature*, *372*, 663–666.

Bender, M., Malaize, B., Orchado, J., Sowers, T., and Jouzel, J. (in press). High precision correlations of Greenland and Antarctic ice core records over the last 100 kyr, *AGU Publication, Chapman Conference.*

Berger, A., Loutre, M. F., and Gallée, H. (1998). Sensitivity of the LLN climate model to the astronomical and CO_2 forcings over the last 200 ky, *Clim. Dyn.*, *14*, 615–629.

Blunier, T., Chappellaz, J., Scwhander, J., Barnola, J. M., Desperts, T., Stauffer B., and Raynaud, D. (1993). Atmospheric methane record from a Greenland ice core over the last 1000 years, *Geophys. Res. Lett.*, *20*, 2219–2222.

Blunier, T., and 13 others (1998). Asynchrony of Antarctic and Greenland climate change during the last glacial, *Nature*, *394*, 739–743.

Bond, G., Broecker, W. S., Johnsen, S. J., McManus, J. Labeyrie, L. D., Jouzel, J., and Bonani, G. (1993). Correlations between climate records from North Atlantic sediments and Greenland ice, *Nature*, *365*, 143–147.

Bond, G., and 13 others (1997). A pervasive Millennial-scale cycle in North Atlantic Holocene and glacial climates, *Science*, 278, 1257–1266.

Broecker, W. S. (1997). Thermohaline circulation, the Achilles heel of our climate system: Will man-made CO_2 upset the current balance? *Science*, *278*, 1582–1588.

Broecker, W. S., Peteet, D. M., and Rind, D. (1985). Does the ocean-atmosphere system have more than one mode of operation? *Nature*, *315*, 21–26.

Chappellaz, J., Barnola, J. M., Raynaud, D., Korotkevich, Y. S., and Lorius, C. (1990). Ice core record of atmospheric methane over the past 160,000 years, *Nature*, *345*, 127–131.

Chappellaz, J., Brook, E., Blunier, T., and Malaizé, B. (1997). CH_4 and $\delta^{18}O$ of O_2 records from Greenland ice: A clue for stratigraphic disturbance in the bottom part of the Greenland Ice Core Project and the Greeland Ice Sheet Project 2 ice-cores, *J. Geophys. Res.*, *102*, 26547–26557.

Cheddadi, R., Mamakowa, K., Guiot, J., de Beaulieu, J. L., Reille, M., Andrieu, V., Grasnoszewski, W., and Peyron, O. (1998). Was the climate of the Eemian stable? A quantitative climate reconstruction from seven European climate records, *Paleogeogr., Paleoclimatol., Paleoecol., 143*, 73–85.

CLIVAR (1997). CLIVAR initial implementation plan, *103*, World Climate Research Programme, 314 pp.

Cortijo, E., Duplessy, J., Laseyrie, L., Leclaire, H., Duprat, J., and van Weering, T. (1994). Eemian cooling in the Norwegian Sea and North Atlantic ocean preceding ice-sheet growth. *Nature*, *372*, 446–449.

Cortijo, E., Labeyrie, L. D., Vidal, L., Vautravers, M., Chapman, M., Duplessy, J. C., Elliot, M., Arnold, M., and Auffret G. (1997). Changes in the sea surface hydrology associated with Heinrich event 4 in the North Atlantic Ocean (40–60°N), *Earth Planet. Sci. Lett.*, 146, 29–45.

Cuffey, K. M., Clow, G. D., Alley, R. B., Stuiver, M., Waddington, E. D., and Saltus, R.W. (1995). Large Arctic temperature change at the Wisconsin-Holocene glacial transition, *Science*, *270*, 455–458.

Dahl-Jensen, D., Mosegaard, K., Gundestrup, N., Clow, G. D., Johnsen, S. J., Hansen, A. W., and Balling, N. (1998). Past temperatures directly from the Greenland ice sheet, *Science, 282*, 268–271.

Dansgaard, W., Clausen, H. B., Gundestrup, N., Hammer, C. U., Johnsen, S. J., Krinstindottir, P., and Reeh, N. (1982). A new Greenland deep ice core, *Science, 218*, 1273–1277.

Dansgaard, W., Johnsen, S., Clausen, H. B., Dahl-Jensen, D., Gundestrup, N., Hammer, C. U., and Oeschger, H. (1984). North Atlantic climatic oscillations revealed by deep Greenland ice cores, in *Climate processes and climate sensitivity*, edited by J. E. Hansen and T. Takahashi, Am. Geophys. Union, Washington.

Dansgaard, W., and 13 others (1993). Evidence for general instability of past climate from a 250-kyr ice-core record, *Nature, 364*, 218–220.

Dansgaard, W., White, J. W., and Johnsen, S. J. (1989). The abrupt termination of the Younger Dryas climate event, *Nature, 339*, 532–534.

Delmas, R. J., Ascencio, J. M., and Legrand, M. (1980). Polar ice evidence that atmospheric CO_2 20,000 BP was 50% of present, *Nature, 284*, 155–157.

Delmas, R. J., Kirchner, S., Palais, J. M., and Petit, J. R. (1992). 1000 years of explosive volcanism recorded at the South Pole, *Tellus, 44B*, 335–350.

Etheridge, D. M., Steele, L. P., Langenfelds, R. L., Francey, R. J., Barnola J. M., and Morgan, V. I. (1996). Natural and anthropogenic changes in atmospheric CO_2 over the last 1000 years from air in Antarctic ice and firn, *J. Geophys. Res., 101D*, 4115–4128.

Fuchs, A., and Leuenberger, M. (1996). $\delta^{18}O$ of atmospheric oxygen measured on GRIP ice core documents stratigraphic disturbances in the lowest 10% of the core, *Geophys. Res. Letters, 23*, 1049–1052.

Gasse, F., and van Campo, E. (1994). Abrupt post-glacials events in West Asia and north Africa, *Earth Planet Sci. Lett., 126*, 453–456.

GRIP project members (1993). Climatic instability during the last interglacial period revealed in the Greenland summit ice-core, *Nature, 364*, 203–207.

Grootes, P. M., Stuiver, M., White, J. W. C., Johnsen, S. J., and Jouzel, J. (1993). Comparison of the oxygen isotope records from the GISP2 and GRIP Greenland ice cores, *Nature, 366*, 552–554.

Hammer, C. U., Clausen, H. B., and Dansgaard, W. (1980). Greenland ice sheet evidence of post-glacial volcanism and its climatic impact, *Nature, 288*, 230–235.

Hammer, C. U., Clausen, H. B., and Langway, C. C. J. (1997). 50,000 years of recorded global volcanism, *Climatic Change, 35*, 1–15.

Hoffmann, G., Jouzel, J., and Johnson, S. J. (unpublished). The deuterium excess record from Central Greenland over the last millenium: Hints of a North Atlantic Signal during the little ice age.

Johnsen, S. J., and 13 others (1992). Irregular glacial interstadials recorded in a new Greenland ice core, *Nature, 359*, 311–313.

Johnsen, S. J., Dahl-Jensen, D., Dansgaard, W., and Gundestrup, N. (1995). Greenland paleotemperatures derived from GRIP bore hole temperature and ice core isotope profiles, *Tellus, 47B*, 624–629.

Jouzel, J. (1994). Ice cores north and south, *Nature, 372*, 612–613.

Jouzel, J., and 13 others (1993). Extending the Vostok ice-core record of paleoclimate to the penultimate glacial period, *Nature, 364*, 407–412.

Jouzel, J., Lorius, C., Johnsen, S. J., and Grootes, P. (1994). Climate instabilities: Greenland and Antarctic records, *C. R. Acad. Sci. Paris, t 319, série II*, 65–77.

Jouzel, J., and 13 others (1996). Climatic interpretation of the recently extended Vostok ice records, *Clim. Dyn., 12*, 513–521.

Lean, J., Beer, J., and Bradley, R. S. (1995). Reconstruction of solar irradiance since 1610: Implications for climate change, *Geophys. Res. Lett., 22*, 3195–3198.

Legrand, M., and Delmas, R. J. (1987). A 220-year continuous record of volcanic H_2SO_4 in the Antarctic ice sheet, *Nature, 327*, 671–676.

Lorius, C., and Oeschger, H. (1994). Paleo-perspectives: Reducing uncertainties in global change? *Ambio, 23*, 30–36.

Lorius, C., Jouzel, J., Raynaud, D., Hansen, J., and Le Treut, H. (1990). The ice-core record: climate sensitivity and future greenhouse warming, *Nature*, *347*, 139–145.

Mann, M. E., Bradley, R. S., Hughes, M. K., and Jones, P. D. (1998). Global temperature patterns, *Science*, *280*, 2029–2030.

Neftel, A., Oeschger, H., Schwander, J., Stauffer, B., and Zumbrunn, R. (1982). Ice core sample measurements give atmospheric CO_2 content during past 40,000 years, *Nature*, *295*, 220–223.

O'Brien, S., Mayewski, P. A., Meeker, L. D., Meese, D. A., Twickler, M. S., and Whitlow, S. I. (1995). Complexity of Holocene climate as reconstructed from a Greenland ice core, *Science*, *270*, 1962–1964.

Oeschger, H., Beer, J., Siegenthaler, U., Stauffer, B., Dansgaard, W., and Langway, C. C. (1984). Late glacial climate history from ice cores, in *Climate processes and climate sensitivity*, edited by J. E. Hansen and T. Takahashi, Am. Geophys. Union, Washington.

Overpeck, J. T. (1995). Paleoclimatology and climate system dynamics, *Review of Geophysics (supplement)*, 863–871.

Petit, J. R., and 13 others (1997). Four climatic cycles in Vostok ice core, *Nature*, *387*, 359–360.

Petit, J. R., and 13 others (1999). Climate and atmospheric history of the past 420,000 years from the Vostok ice core, Antarctica, *Nature*, *399*, 429–436.

Rahmstorf, S. (1994). Rapid climate transitions in a coupled ocean-atmosphere model, *Nature*, *372*, 82–85.

Ramstein, G., Serafini-Le Treut, Y., Le Treut, H., Forichon, M., and Joussaume, S. (1998). Cloud processes associated with past and future climate changes, *Clim. Dyn.*, *14*, 233–247.

Raynaud, D., Jouzel, J., Barnola, J. M., Chappellaz, J., Delmas, R. J., and Lorius, C. (1993). The ice record of greenhouse gases, *Science*, *259*, 926–934.

Severinghaus, J. P., Sowers, T., Brook, E., Alley, R. B., and Bender, M. L. (1998). Timing of abrupt climate change at the end of the Younger Dryas interval from thermally fractionated gases in polar ice, *Nature*, *391*, 141–146.

Stager, J. C., and Mayewski, P. A. (1997). Abrupt Early to Mid-Holocene climatic transition registered at the Equator and the Poles, *Science*, *276*, 1834–1836.

Stauffer, B., Lochbronner, E., Oeschger, H., and Schwander, J. (1988). Methane concentration in the glacial atmosphere was only half that of the preindustrial value, *Nature*, *332*, 812–814.

Stocker, T. F., and Schmitter, A. (1997). Influence of CO_2 emission rates on the stability of the thermohaline circulation, *Nature*, *388*, 862–865.

Taylor, K. C., Lamorey, G. W., Doyle, G. A., Alley, R. B., Grootes, P. M., Mayewski, P. A., White, J. W. C., and Barlow, L. K. (1993). The "flickering switch" of late Pleistocene climate change, *Nature*, *361*, 432–436.

Thompson, L. (1996). Climate changes for the last 2000 years inferred from ice core evidence in tropical ice cores, in *Climatic variations and forcing mechanisms of the last 2000 years*, NATO ASI Series, edited by R. S. Bradley, P. D. Jones, and J. Jouzel, Springer-Verlag, Berlin-Heidelberg.

Thompson, L. G., Mosley-Thompson, E., and Arnao, B. M. (1984). El-nino-Southern oscillation events recorded in the stratigraphy of the quelccaya ice cap Peru, *Science*, *234*, 261–264.

von Grafenstein, U., Erlenkeuser, H., Muller, J., Jouzel, J., and Johnsen, S. J. (1998). The short period 8,200 years ago documented in oxygen isotope records of precipitation in Europe and Greenland, *Clim. Dyn.*, *14*, 73–81.

Weaver, A. J., Eby, M., Fanning, A. F., and Wibe, E. C. (1998). Simulated influence of carbon dioxide, orbital forcing and ice sheets on the climate of the Last Glacial Maximum, *Nature*, *394*, 847–853.

Yiou, F., and 13 others (1997). Beryllium 10 in the Greenland Ice Core Project ice core at Summit Greenland, *J. Geophys. Res.*, *102*, 26783–26794.

Yiou, P., Jouzel, J., Johnsen, S. J., and Rögnvaldsson, Ö. E. (1995). Rapid oscillations in Vostok and GRIP ice cores, *Geophys. Res. Lett.*, *22*, 2179–2182.

Yiou, P., Fuhrer, K., Meeker, L. D., Jouzel, J., Johnsen, S. J., and Mayewski, P. A. (1997). Paleoclimatic variability inferred from the spectral analysis of Greenland and Antarctic ice core data, *J. Geophys. Res.*, *102*, 26441–26454.

PART FIVE

HOW TO MEET THE CHALLENGE

16 Toward a New Approach to Climate Impact Studies

WILL STEFFEN

16.1 Introduction

There is now a growing consensus that human activities are having a discernable effect on the Earth's climate (IPCC, 1996). An enormous amount of effort has gone into the scientific research that forms the basis for this judgment. There is also growing concern that such human-influenced changes to the Earth's climate could have negative effects on human societies and on the Earth's ecosystems and therefore that these changes should be avoided or slowed. This, in turn, has led to efforts to mitigate the human influence on climate, most notably through measures to slow the build-up of greenhouse gases in the atmosphere (e.g., the Kyoto Protocol). Critical to this line of argument, which calls for serious attempts to mitigate climatic change, is the view that the impacts of climatic change will be so serious and so difficult to deal with that mitigation is preferable to adaptation, that is, to accepting that climatic change is and will continue to occur and adapting societal activities to it. Thus, although more work is certainly needed for us to understand the nature of the climate itself, there is a rapidly growing demand to determine what the impacts of climatic change actually are and what they are likely to be: how serious they are, how much they will cost, what types of changes will be irreversible, and so on.

The objective of this brief commentary is to give a personal perspective on the evolution of climate impact research and perhaps to point the way toward some new approaches that may yield powerful and insightful results that will be of use to the policy and resource management sectors.

16.2 The Nature of Climate Change: "Smooth" vs. "Abrupt"

Until now, most climate impact studies of which I am aware have been based on GCM (general circulation model)–generated scenarios of change, rarely longer than a century. These studies – for example, the suite of IPCC climate change scenarios – show temperature changes in the range of +1.5 to +4.5 °C over that century and show changes in rainfall of about ±10% over the same period. However, we may have been somewhat trapped by the reliance on GCM-generated scenarios, and we may have missed some critical larger issues in terms of climate impacts.

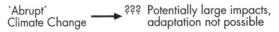

Figure 16.1. "Smooth" vs. "abrupt" climate change (as defined in the text) will have very different impacts on developed and developing countries. Adaptation will likely not be an option if human perturbations to the climate system trigger abrupt changes.

We now know from ice core records that there have been very large and sudden climatic changes in the past on at least a regional scale. For example, ice core records have shown that the mean annual temperature in the North Atlantic region has changed by as much as 10 or 12° over periods as short as a few decades, that is, within one human lifetime (e.g., Alley et al., 1993). In some periods, the climate undergoes such changes several times, as if it is oscillating between two very different states before settling on one for a longer period.

So the first step in approaching climate impacts studies is to determine how likely it is that the climate will undergo such sudden shifts, as a result of perturbations by human activities, before stabilizing. (I term such shifts "abrupt" change, as opposed to the nearly monotonic change in temperature projected by most GCMs for the next century at least, which I term "smooth" change.) To clarify this issue, we could make the very simplistic and rather provocative hypothesis, summarized in Figure 16.1, that if the Earth experiences smooth climatic change over the next century, then the developed world at least will be able to cope reasonably well with such change, as some technological optimists assert, and the emphasis should then be on adaptation instead of mitigation. It is a different situation for the developing world, however, where even smooth climate change will likely lead to severe impacts. The issue, then, is the transfer of technology and management expertise from the developed to the developing world.

On the one hand, if the climate system will undergo abrupt change, even the most technologically advanced nation will have extreme difficulty coping. In fact, it is hard to imagine any modern society surviving the kind of change recorded in the Greenland ice core records. And it is important to note that many of the Earth's most technologically advanced regions border the North Atlantic.

This line of argument suggests that the balance of resources for climate change research should perhaps be shifted somewhat toward understanding the fundamental nature of the climate system as a whole – its stability and the transition between states – instead of generating ever more detailed GCM scenarios for the next century only. We need to know what triggers these naturally occurring strong oscillations between climatic states and whether human activities are pushing the climate system toward or away from such rapid and severe transitions (see Schellnhuber, 1998). Two factors

"Linear, Pollution-pipe" Models

Figure 16.2. A schematic diagram of the "pollution pipe" approach to climate impact studies.

are worth noting in this context: (i) The climate system is now experiencing a mix of forcing functions that it has never experienced before and probably at a rate that it has not experienced for millions of years. (ii) Highly nonlinear systems, such as the Earth's climatic system, are probably more prone to sudden, sharp changes when the magnitude and rate of the forcing increase (see Schneider, Chapter 6, this volume). This is a more powerful argument in support of mitigation activities than are a whole bundle of economically oriented studies purporting to show that the costs of smooth climatic change outweigh the benefits.

16.3   From a "Pollution Pipe" to a "Systems" Approach

Now let's assume that for the next century at least, none of the Earth's regions will experience abrupt climate change. With respect to smooth climate change, the classic approach to studying impacts has been a rather linear paradigm consisting of a climate change scenario driving a biophysical impact, which, in turn, leads to a socioeconomic consequence; this approach is sometimes referred to as the "pollution pipe" approach (Larry Kohler, personal communication, see Figure 16.2). There are several serious flaws in such an approach.

1. Each of the modeling steps carries with it its own range of errors, and these can easily propagate through the modeling chain until the final result has such wide error bars that it could mean anything.
2. Because the focus is on climate change, the many other factors that affect the biophysical and socioeconomic systems are held constant, thus isolating the impacts of climate; this can often have the effect of magnifying the climatic impacts.
3. Because climate has at least some effect on many human activities, isolating and magnifying these effects and then listing them region by region or sector by sector can lead to a "sky is falling" syndrome, in which global change scientists appear to be the prophets of doom and gloom; this can easily lead to a loss of credibility of the global change scientific community.

Because of these problems, there has been a shift away from the pollution pipe approach toward a more sophisticated, systems-oriented approach. The emphasis in the latter is on the system being impacted rather than on the scenario of the forcing function (e.g., climate). The philosophy behind this approach is that ecological and socioeconomic systems are being impacted simultaneously by many different, interacting driving forces

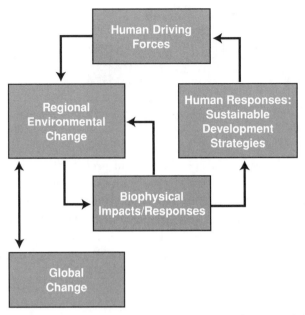

Figure 16.3. The conceptual framework for the Southeast Asian Integrated Study on global change and sustainable development (Lebel and Steffen, 1999). Note that here global change is viewed as a perturbation on an existing cyclical regional system involving human driving forces, impacts and environmental changes, and societal responses.

and not simply by climate in isolation. This approach is shown schematically in a simple way for the integrated Southeast Asian global change study (Figure 16.3; Lebel and Steffen, 1999) and in more detail for a subsistence rangelands research project in southern Africa (Figure 16.4; Odada et al., 1996). In both cases the focus of the approach is to understand a complex system involving interacting biophysical and socioeconomic components, one that is influenced by a number of factors from both the biophysical and socioeconomic realms. Climate is one of those factors, and climate change is most realistically viewed as a perturbation on that driver.

In adopting this systems-oriented approach, it is usually most effective to carry out a number of sensitivity studies as an initial step. These studies aim to examine ranges of driving variables – instead of a single predicted change in a variable – and interacting suites of variables. Thus, in this step, in terms of climate change, a series of plausible perturbations on present-day climate patterns (e.g., temperature, precipitation, storm frequencies) is used rather than a single scenario based on a GCM output. The focus is on understanding the dynamics of the system being impacted, to determine how it responds to such complex changes in driving variables. Such sensitivity studies are appropriate to tease out the existence of critical thresholds and nonlinearities in the impacted system, to determine when "state changes" can occur that may make adaptation very difficult. Information from sensitivity studies is often useful input to those who are developing the scenarios of the driving variables; it suggests which variables in which ranges are particularly important. Frequent exchanges of information between scenario developers

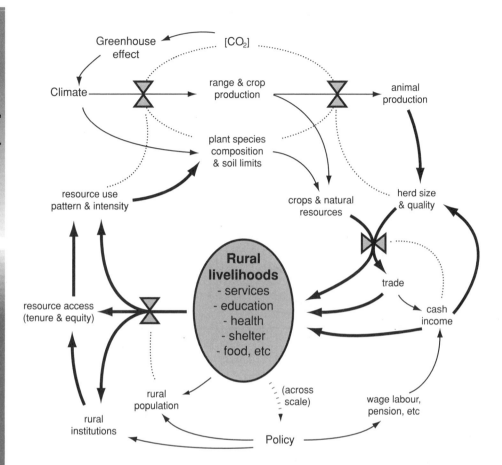

Figure 16.4. The modeling framework for the IGBP study of subsistence rangelands in southern Africa (Odada et al., 1996). Here, climate change is viewed as one of several interacting drivers of change, both biophysical and socioeconomic, impacting the ability of rural populations of the region to achieve sustainable livelihoods.

and experts on the affected systems is much more effective than a simple "handover" of information from scenario to impacts analyst, something that occurs in the pollution pipe approach. The best example of the integrated systems approach is the suite of studies on the impact of climate variability on agriculture, in which a crucial first step is to understand the agricultural system potentially impacted and the type of climatic information that is useful (Hammer, 1997).

16.4   Technological and Societal Adaptability

Adopting a systems-oriented approach and undertaking sensitivity studies lead naturally to a consideration of technological and societal adaptability. One of the great difficulties in projecting climate impacts to multidecadal timeframes is that human

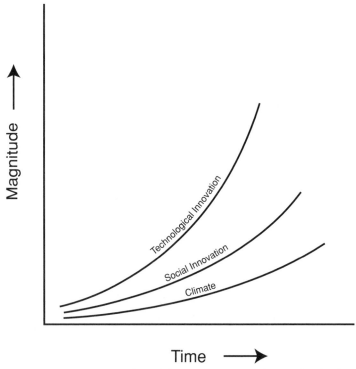

Figure 16.5. A schematic view of the relative rates of change of technology, social innovation, and climate (L. Kohler, personal communication).

societies are changing on even faster time scales (Figure 16.5). As the result of rapid changes in technologies and social systems, problems, identified in today's context and projected into the future, may simply disappear just when they are projected to be most severe. The classic example is the prediction that European cities would be buried in meters of horse manure as they grew and as transport demands increased. Another example is the patenting of more than 1000 "smoke-spark arresters" for wood-burning steam locomotives in the United States. None were effective enough to be adopted. This problem, which was projected to cause widespread, severe forest fires, was solved instead by a more fundamental shift in technology: the switch from steam to diesel and electric locomotives (Gruebler, 1998). The lessons from these and many other such examples are that (i) predicted environmental damage, projected from an understanding of present-day systems and drivers, often doesn't occur; (ii) ideally, impact studies should include a range of "technology trajectories" that attempt to project possible technology-driven changes to systems (e.g., food production systems); and (iii) more emphasis should be placed on sensitivity studies, as noted above, in an attempt to tease out those systems in which improvements in technology or management are unlikely to mitigate projected changes in the climate system. It may then be more fruitful (and believable) to focus climate impact research on those areas rather than take a blanket sector-by-sector or region-by-region approach.

16.5   Conclusion

This brief essay has highlighted the following as major issues in climate impact studies:

- Abrupt climate change has the potential to cause massive and irreparable damage to human societies. A first-order priority in climate system research should be to understand better the nature of these abrupt changes in the climate system and to discover whether human perturbations to the climate system are pushing it toward such changes.
- The earlier linear, pollution pipe approach to climate impact studies should be abandoned in favor of a systems approach.
- In a systems approach to climate impact studies, the emphasis is on the complex system being impacted and its response to multiple, interacting drivers. Climate change is probably best viewed as a perturbation to existing drivers of a system. The challenge is to determine when and for what systems global change may become a significant perturbation.
- The adaptability of human societies and their technologies to environmental challenges, although not always sufficiently effective, nevertheless should not be underestimated. Modern human societies have some potential, given an appropriate big picture of what global change is and what its potential impacts are, to use "adaptive management" or "fuzzy control" techniques to navigate toward a more sustainable future without an intervening environmental catastrophe (Schellnhuber, 1998).

REFERENCES

Alley, R. B., Meese, D. A., Shuman, C. A., Gow, A. J., Taylor, K. C., Grootes, P. M., White, J. W. C., Ram, M., Waddington, E. D., Mayewski, P. A., and Zielinski, G. A. (1993). Abrupt increase in Greenland snow accumulation at the end of the Younger Dryas event. *Nature*, 362: 527–529.

Gruebler, A. (1998). Technology and Global Change. Cambridge University Press, 452 pp.

Hammer, G. (1997). A general systems approach to applying seasonal climate forecasts. The Symposium of Applications of Seasonal Climate Forecasting in Agricultural and Natural Ecosystems: The Australian Experience. Volume of Abstracts, November, 1997.

IPCC (1996). Climate Change 1995: The Science of Climate Change. Contribution of Working Group I to the Second Assessment Report of the Intergovernmental Panel on Climate Change, ed. J. J. Houghton, L. G. Meiro Filho, B. A. Callendar, N. Harris, A. Kattenberg, and K. Maskell. Cambridge University Press, 584 pp.

Lebel, L., and Steffen, W. (1999). Global Environmental Change and Sustainable Development in Southeast Asia: Science Plan for a SARCS Integrated Study. Southeast Asian Regional Committee for START (SARCS), Bangkok, Thailand, 139 pp.

Odada, E., Totolo, O., Stafford Smith, M., and Ingram, J. S. I. (1996). Global change and subsistence rangelands in Southern Africa: the impacts of climatic variability and resource access on rural livelihoods. GCTE Working Document No. 20. Canberra, Australia: GCTE International Project Office, 99 pp.

Schellnhuber, H.-J. (1998). Earth system analysis – the scope of the challenge. In: H.-J. Schellnhuber and V. Wenzel (eds.), Earth System Analysis: Integrating Science for Sustainability. Berlin: Springer-Verlag. pp. 1–195.

17   Research Objectives of the World Climate Research Programme

HARTMUT GRASSL

17.1   Introduction

The World Climate Research Programme (WCRP) is the research component of the World Climate Programme (WCP). Soon after the launch of WCP at the First World Climate Conference in Geneva in 1979, the World Meteorological Organization (WMO) and the International Council for Scientific Unions (now called International Council for Science) (ICSU) agreed in 1980 on cosponsorship of WCRP. In 1993, the Intergovernmental Oceanographic Commission (IOC) of UNESCO joined WMO and ICSU as the third cosponsor of WCRP. This unique sponsoring structure has attracted both the scientific community and national meteorological, hydrological, and oceanographic services. Therefore, WCRP has developed into a global research program encompassing all those parts that need international climate research cooperation and coordination for a successful outcome.

17.2   The Overall Goal

WCRP has a clearly set goal: "To understand and predict – to the extent possible – climate variability and change, including human influences."

In reaching this overall goal, WCRP must

- Design and implement observational and diagnostic research activities that will lead to a quantitative understanding of significant climate processes;
- Develop global models capable of simulating the present and past climate and – to the extent possible – of predicting climate variations on a wide range of space and time scales, including the effects of human influences.

In practice, these tasks are executed by obeying a few principles for WCRP activities:

- Add value to climate research in general by international coordination of national research programs.
- Formulate science and implementation plans for projects that need global cooperation for success.

280

- Implement subprojects only if enough scientists are engaged and additional funding is secured.
- Assess progress regularly (once per year) by an independent scientific advisory body (Joint Scientific Committee) selected jointly by all sponsors.
- Liaise with the users from the beginning. That is, cooperate with numerical weather predictions centers, national meteorological and hydrological services, climate anomaly prediction centers, the Intergovernmental Panel on Climate Change (IPCC), environmental conventions of the United Nations, water resources managers, and so on.
- Integrate operational observation networks, new research observation networks, data management, process (field) studies, modeling and prediction, and applications in each project.

17.3   Present Structure

WCRP concentrates on the understanding of the physical parts of the climate system. However, chemical and biological processes are included if essential to an understanding of the climate processes on the time scales of major interest for a specific project. This is done in close cooperation with the respective projects of the International Geosphere Biosphere Programme (IGBP). As shown in Table 17.1, all climate system components and their interactions are at present part of WCRP projects, and it has been our strategy to limit the number and size of our projects. One earlier project has already been finalized, and two others are nearing completion. It is interesting to note that three projects were first discussed outside, but soon became WCRP projects: Tropical Ocean/Global Atmosphere (TOGA), World Ocean Circulation Experiment (WOCE), and Stratospheric Processes and Their Role in Climate (SPARC).

The WCRP modeling infrastructure is often not seen as being as prominent or important as the projects. However, it is through the Working Group on Numerical Experimentation (WGNE) that we liaise with weather prediction centers, which improve forecasting through new parameterizations of climate processes derived from field experiments. The same holds for climate variability predictions; the Working Group for Coupled Modeling (WGCM) helps improve these predictions, for example, through model intercomparisons.

17.4   Achievements

The successes of WCRP are crowned by the breakthrough to climate anomaly predictions on seasonal-to-interannual time scales, mainly for the areas strongly affected by the El Niño southern oscillation (ENSO) phenomenon. Table 17.2 highlights some of the achievements so far. In all projects, progress rests first on new observing systems and new combinations of existing systems and then on the use of models validated by the new data sets.

Table 17.1. WCRP Projects and Modeling Working Groups

	Projects		
Start of Implementation	**Project Name**	**Central Climate System Component(s)**	**Main Component Interactions**
1989	Global Energy and Water Cycle Experiment (GEWEX)	Atmosphere	Soil/Vegetation/ Atmosphere
1990	World Ocean (Ocean/Atmosphere) Circulation Experiment (WOCE)	Ocean	—
1992	Stratospheric Troposphere/Stratosphere Processes and Their Role in Climate (SPARC)	Stratosphere	(together with IGAC of IGBP)
1994	Arctic Climate System Study and Climate and Cryosphere (ACSYS/CLIC)	Cryosphere	Sea Ice/Ocean/Snow
1998	Climate Variability and Predictability (CLIVAR)	Ocean/ Atmosphere	Ocean/Atmosphere/Land
	Modeling Infrastructure		
Since GARP	Working Group on Surface Numerical Experimentation (WGNE)	Atmosphere	Atmosphere/Earth (together with GEWEX modeling group)
1997	Working Group on Surface/Coupled Modeling (WGCM)	Full Climate System	Atmosphere/Earth Ocean (together with modeling groups in CLIVAR, GEWEX, WOCE, SPARC, CLIC), relating WCRP to GAIM of IGBP and IPCC

17.5   Research Objectives for the Coming Decade

In 1995, one of the sponsors (ICSU) assessed WCRP and proposed to hold a conference on achievements, benefits, and challenges. At this conference the overall research priorities for the next 10 to 15 years were to be set. In August 1997, the World Climate Research Programme: Conference on Achievements, Benefits and Challenges took place in Geneva and asked WCRP to

- Assess the nature and predictability of seasonal-to-interdecadal variations of the climate system at global and regional scales, and to provide the scientific basis for

Table 17.2. Achievements of WCRP Projects

Project or Group	Duration	Main Achievement(s) So Far
TOGA	1985–1994	Breakthrough to seasonal predictions through new observing system and coupled modeling
GEWEX	1989–?	1. First quality-controlled global time series for precipitation, cloud parameters, water vapor column content, and surface radiation fluxes 2. Contribution of water storage in soils to medium-range weather and climate predictions
WOCE	1991–2002	1. First observation of global ocean structure (except polar seas) 2. Sea level change time series every 10 days
SPARC	1992–?	1. Stratospheric temperature trend analysis 2. Realistic stratospheric circulation modules as part of general circulation models (GCMs)
ACSYS	1994–2003	1. Sea-ice modeling for AOGCMs 2. Three-dimensional surveys of the Arctic Ocean
CLIVAR	1995–2010?	1. First initiatives on climate variability in the Americas and Africa 2. Asian monsoon prediction focus
WGNE	1980–ongoing	1. Extended weather forecasting 2. Realistic atmospheric circulation modules for coupled modeling
WGCM	1997	Coupled model intercomparison

operational predictions of the variations for use in climate services in support of sustainable development.

- Detect climate change, attribute causes, and project the magnitude and rate of human-induced climate change, its regional variations, and related sea level rise (as needed for input to the IPCC, UN FCCC, and other conventions).

How difficult it will be to reach these objectives can best be indicated by the formulation of scientific questions that must be answered before we can speak of success:

- Can we predict the onset, strength, and breaks of monsoons weeks and months ahead? Only joint action by the projects GEWEX and CLIVAR will be able to give an answer.
- Are the strength and frequency of ENSOs influenced by humankind?
- Would an increase in strength be a major consequence, and would there be new weather extremes? Only a reconstruction of climate history beyond the instrumental record, as planned in CLIVAR, can give an answer.
- Will the North Atlantic deep water (NADW) formation rate be slowed or stop if the greenhouse gas concentration increase continues unabated? The prerequisite

is better monitoring and improved coupled models. For WCRP CLIVAR, CLIC and WGCM must coordinate their plans.

- Can climate variability predictions and climate change projections be regionalized with confidence? Only better cloud feedback parameterizations and land surface modules in coupled models can give the answer. Again, at least GEWEX, CLIVAR, WGNE, and WGCM must coordinate their plans.

18   Panel Discussion: Future Research Objectives

MARTIN HEIMANN

At the opening of the discussion, Hartmut Grassl, director of the World Climate Research Programme (WCRP) of the World Meteorological Organization (WMO), presented an overview of the recently established Climate Variability and Predictability (CLIVAR) project. At its core this project includes the development, evaluation, and application of coupled numerical models of the atmosphere-ocean system. The goal of the project is the prediction and understanding of seasonal and interannual climate variability, such as the El Niño Southern Oscillation variations, as well as the assessment of climate change induced by forcing factors operating at centennial time scales, in particular the anthropogenic perturbation by the emissions of greenhouse gases and large-scale changes in land use. CLIVAR also supports the design and implementation of global observing systems for the physical state of the atmosphere, ocean, and land surfaces. Considerable progress has already been achieved in the atmospheric domain, spurred primarily by efforts to improve operational weather forecasts. Successes have also been noted in the oceanic domain with the implementation of the space-based TOPEX-POSEIDON system and the continuation of the in situ measurement arrays deployed as part of the previous Tropical Ocean Global Atmosphere (TOGA) project. However, deplorably, the Global Terrestrial Observing System (GTOS) is much less advanced. The main reasons for this failure are linked to a multitude of political difficulties in many parts of the world in establishing local components of a global observational system. In addition, there exists a widespread deterioration of established surface-based observational services in many parts of the world. Hence, any solution for a more effective observational system must rely heavily on remote sensing platforms.

The ensuing workshop discussion focused primarily on observational strategies for the better understanding of the global climate system, global biogeochemistry, their mutual interactions, and their perturbation by anthropogenic activities.

18.1   Methodological and Scientific Problems

Any efficient monitoring system must be based on a clear definition of the variables that are to be observed, which in turn must be chosen in view of the underlying problem that is posed. For example, a strategy to monitor changes in the terrestrial

285

carbon cycle on decadal-to-centennial time scales to understand its interaction with the climate will be different from a strategy directed at the determination of national carbon balances in the context of the Kyoto Protocol, and it will also be different if the goal is the assessment of regional impacts induced by various forcing factors.

A further fundamental problem consists of possible nonlinear, rapid transitions ("surprises") in parts of the global system, for which, almost by definition, an adequate observing strategy is impossible or very difficult to conceive – as for example, witnessed, by the detection history of the Antarctic ozone hole. Needed is an environment of good science that can detect surprises. Furthermore, observational systems should be designed to be flexible enough so that they can rapidly be adapted to new demands.

Monitoring large- and global-scale changes in biogeochemistry constitutes a formidable scientific challenge. For example, recent studies to assess continental-scale net carbon balances by inverse atmospheric transport modeling from global atmospheric CO_2 concentration measurements have yielded very contradictory results. The reasons for these discrepancies are related to largely unknown imperfections in the employed atmospheric models, to inaccuracies in the observational data because of measurement, calibration, and sampling errors, and also to the highly underdetermined nature of the underlying mathematical problem. The latter means that many different source/sink configurations will result in almost the same atmospheric concentration signatures at the relatively few observing sites, and that prevents an unambiguous discrimination among various source/sink scenarios. Significant progress might come from an integrative approach in which a meso-scale atmospheric model, combined with a process model of surface biogeochemical trace gas flux exchanges, is developed. This model would be run in an assimilation mode with high-density atmospheric concentration sampling including, for example, regular measurements from aircraft, surface-based flux measurements, and information from remote sensing platforms. Evidently, such an approach is based on the largely successful paradigm of the global meteorological data assimilation systems established under the auspices of WMO for operational weather forecasts. Several plans for feasibility studies to develop such an observing system for biogeochemical species (in particular for CO_2) have recently been proposed – for example, the Carbon America Plan (Tans et al., Global Change Biology, 2, 309–318, 1996) – or are being developed as part of new regional field projects – for example, the Large-Scale Biosphere Project in Amazonia (LBA, http://yabae.cptec.inpe.br/lba/, or http://www.PIK-Potsdam.DE/mirror/bahc/lba/) or the Eurosiberian Carbonflux project (http://www.bgc-jena.mpg.de/projects/sibir/flux01.html). Similar initiatives are also being planned for Australia and South Africa. Whether these initiatives will be successful remains to be seen. In any case, there exists a need to assess and optimize these regional initiatives into a design for a global monitoring strategy. The Task Force on Global Analysis, Interpretation and Modeling of the IGBP has recently taken up the challenge to take the lead in this endeavour. However, it is clear that the implementation and coordination of such a global observing system for biogeochemical trace gases ultimately will have to be established within a UN body, such as WMO.

18.2   Educational and Structural Problems

The workshop participants perceived a fundamental problem beyond the scientific challenge posed by the general lack of scientists from developing nations in global change research. Effectively, the concept of global change and the political pressure to act appear to be largely dictated by the science of the First World. This constitutes foremost an educational problem because of the large knowledge gap existing between the economically more highly developed and the developing regions of the Earth. Clearly, a lack of involvement of Third World scientists represents a huge waste of human potential. It also means that the knowledge base of global change science in many parts of the world is highly underdeveloped, while at the same time international treaties, such as the Kyoto Protocol, require an active participation of all nations. An efficient participation, however, can be achieved only if well-trained and internationally established scientists from all countries and cultures are able to critically assess the existing scientific knowledge and thus advise their political representatives in the negotiations in an unbiased and scientifically well-informed way.

Although the IGBP core project START (Global Change System for Analysis, Research and Training) aims at alleviating this deficit, there is a clear need to further enhance and develop participation of Third World scientists in major international projects in leading roles. A concern was also raised that the high-technology observational programs envisaged for certain developed regions of the Earth potentially will widen the knowledge gap between the First and the Third World.

Associated with this fundamental educational problem, the workshop participants also noted in many parts of the world an increasing deterioration of political and administrative structures, which, however, are needed to conduct field research and to maintain long-term measurements. For example, the dense network of rain gauges established during colonial times in Africa and maintained well into the 1980s is today practically no longer existing in many countries. However, monitoring global change requires observational systems that can be stably maintained over decades to centuries. A possibility might be provided by alternative structures, such as large private companies, which might be enticed to share their logistical networks in remote areas (e.g., oil platforms, mining camps, supply aircraft). The workshop participants noted in particular the excellent opportunity provided by a close collaboration with the church, which maintains, independent of local administrations, a large global network of highly motivated and educated people that has been relatively stable for many centuries. For example, the Global Atmosphere Watch (GAW) of the WMO Global Climate Observing System (GCOS) includes a new monitoring station at Assekrem near Tamanrasset in the central Sahara desert that is diligently maintained by a local monastery. Clearly, there is considerable scope for expanding such collaborations in many parts of the globe. In addition, the church may also provide a very welcome alternative means to reach and support scientists in the Third World.

Index